Modern ENGINE Blueprinting Techniques

Mike Mavrigian

S-A DESIGN

CarTech®

CarTech®

CarTech®, Inc.
6118 Main Street, North Branch, MN 55056
Phone: 651-277-1200 or 800-551-4754 • Fax: 651-277-1203
www.cartechbooks.com

© 2013 by Mike Mavrigian

All rights reserved. No part of this publication may be reproduced or utilized in any form or by any means, electronic or mechanical, including photocopying, recording, or by any information storage and retrieval system, without prior permission from the Publisher. All text, photographs, and artwork are the property of the Author unless otherwise noted or credited.

No portion of this book may be reproduced, transferred, stored, or otherwise used in any manner for purposes of training any artificial intelligence technology or system to generate text, illustrations, diagrams, charts, designs or other works or materials.

The information in this work is true and complete to the best of our knowledge. However, all information is presented without any guarantee on the part of the Author or Publisher, who also disclaim any liability incurred in connection with the use of the information and any implied warranties of merchantability or fitness for a particular purpose. Readers are responsible for taking suitable and appropriate safety measures when performing any of the operations or activities described in this work.

All trademarks, trade names, model names and numbers, and other product designations referred to herein are the property of their respective owners and are used solely for identification purposes. This work is a publication of CarTech, Inc. and has not been licensed, approved, sponsored, or endorsed by any other person or entity. The Publisher is not associated with any product, service, or vendor mentioned in this book and does not endorse the products or services of any vendor mentioned in this book.

Edit by Paul Johnson
Layout by Monica Seiberlich
ISBN 978-1-61325-047-1
Item No. SA251

Library of Congress Cataloging-in-Publication Data

Mavrigian, Mike, author.
Modern engine blueprinting techniques : a practical guide to precision engine building / by Mike Mavrigian.
pages cm
Includes bibliographical references.
ISBN 978-1-61325-047-1
1. Automobiles–Motors–Maintenance and repair. 2. Automobiles–Motors–Modification. 3. Tolerance (Engineering) I. Title.

TL210.M345 2013
629.25'04–dc23

2013011409

Written, edited, and designed in the U.S.A.
Printed in China
10 9 8 7

Title Page: *Every piston/rod assembly must be measured and weighed so they are within a couple grams of one another. To achieve maximum performance, you must have components on the rotating assembly that are precisely balanced.*

Back Cover Photos

Top Left: *Be sure the wall thickness of the cylinders is thick enough for rebuilding. If it isn't, you need to junk the block. To sonically test an engine, slowly run the sensor probe along the cylinder wall (in a number of clock positions and from top to bottom) while monitoring the gauge readout.*

Top Right: *Measure journal diameters in different clock positions so you take all the necessary measurements. It isn't a bad idea to perform a backup measurement with a different micrometer.*

Middle Left: *A connecting rod vise allows you to remove a rod cap from a new rod, or when tightening rod bolts with the rod off of the crank. Using a common bench vise can cause serious damage to the rod, and burrs or gouges can be created, which can lead to stress cracks and failures. A dedicated purpose-designed rod vise allows securing the rod without damage.*

Middle Right: *Spiral locks can be tricky at first but once you install a few it becomes second-nature. After gently spreading the spiral apart, insert one end and "walk" the remainder in a counterclockwise direction into the groove until the final end snaps into place. Some installers use fingers along with a small, flat-blade screwdriver; specialty installer tools are also available.*

Bottom Left: *Check the initial cut (note the witness cut through the dye) for uniformity and concentricity.*

Bottom Right: *Some LS cylinder heads have integral rocker pedestals; others require separate pedestal rails. This 5.3L LS features rocker rails. Hand snug a couple of rocker bolts to center the rail before fully tightening. Also, be sure to apply high-pressure assembly lube to all valve tips, rocker pushrod cups, and rocker valve tips.*

DISTRIBUTION BY:

Europe
PGUK
63 Hatton Garden
London EC1N 8LE, England
Phone: 020 7061 1980 • Fax: 020 7242 3725
www.pguk.co.uk

Australia
Renniks Publications Ltd.
3/37-39 Green Street
Banksmeadow, NSW 2109, Australia
Phone: 2 9695 7055 • Fax: 2 9695 7355
www.renniks.com

Canada
Login Canada
300 Saulteaux Crescent
Winnipeg, MB, R3J 3T2 Canada
Phone: 800 665 1148 • Fax: 800 665 0103
www.lb.ca

CONTENTS

CONTENTS

DEDICATION

Precision engine building involves both art and an extremely high level of applied science. My appreciation for the complexities and skill involved resulted from my early days of competition road racing, when I relied on various engine shops to prepare my race engines. Some of these experiences were good, and some were downright horrible. As time passed, I became more aware of what was involved, and out of necessity and an increased interest, I learned how to choose a skilled and experienced builder. In the process, I began to handle my own builds and to realize the enormous satisfaction that results from continually gaining knowledge.

The skills and insights that I've learned over the years (and continue to learn) are the direct result of my relationships with industry experts who share a common trait: the passion for the pursuit of perfection.

This book is dedicated to all professional performance engine builders for whom "good enough" is never good enough.

ACKNOWLEDGMENTS

Thanks go to the following for their input and expertise during the research for this book:

Trick Flow Specialties; Bob Fall, Fall Automotive Machine, Toledo, Ohio; Bill McKnight, Mahle Clevite; Scott Gressman, Gressman Powersports, Fremont, Ohio; Dave Monyhan, Goodson Tools; Tony Lombardi, Ross Racing Engines, Niles, Ohio; Jody Holtrey, Medina Mountain Motors, Creston, Ohio; Sean Crawford, JE Pistons; Dennis Ventrello, Jesel; Chris Oullette, BHJ; Chase Knight, Crane Cams; Ted Keating and John Keyes, Dart Machinery; Chris Raschke, ARP; Todd Ryden, MSD; Paul Hauglie, Melling Select; Tom Lieb, Scat Crankshafts; Duane Boes, Callies Crankshaft; Jeff Kaufman, Kaufman Racing; Randy Becker, Jr., Harland-Sharp; Randy Neal, CWT; Bill Tichenor, Holley; Trent Goodwin, Comp Cams; Ron Robart, Fox Lake Performance, Clyde Norwood, Lock-In-Tool; Neal Manton, Manton Pushrods; Barb Miller, Moroso; Debbie Fragola, Fragola Performance Systems; Schean Barrett, PRW; John Schwarz, Aviaid; Jeff Lance, Allan-M&S Auto, Wadsworth, Ohio; Scott Koffel, Koffel's Place, Huron, Ohio.

A special thanks to those individuals who took me under their wings and have generously shared their expertise, notably Bill McKnight, Bob Fall, Scott Gressman, Jody Hiltrey, and Tony Lombardi. These individuals, and many more within the engine community, have gained my utmost respect.

INTRODUCTION

The term "engine blueprinting" means different things to different people, so it's important to clarify it because it is often misused.

Many seasoned engine builders have an accurate and complete understanding of the procedures involved in blueprinting. But in many cases, enthusiasts and customers don't have the same understanding. Some believe that balancing and blueprinting are interchangeable terms, and that both procedures are performed when the term blueprinting is discussed. Some assume that if a component has been balanced, it has also been blueprinted, and vice versa. As an astute engine builder, you cannot make that assumption. If you review Craigslist or other online ads, you're bound to see multiple ads listing engines that have been "balanced and blueprinted," when in most cases the crankshaft has simply been balanced.

In most cases, owners and/or sellers of these engines are not intentionally trying to deceive. Rather, they simply have a misunderstanding of the blueprinting process. Blueprinting an entire engine assembly involves a high degree of skilled labor that goes far beyond a routine rebuild or a bolt-together assembly or reassembly. As a result, a blueprinting procedure adds substantial cost to an engine build. Balancing (which involves weight-matching pistons and rods and correcting the balance of the crankshaft to accommodate the weight of the pistons, rods, piston pins, rod bearings, and rings) is simply one aspect of a blueprinting procedure. An engine build can feature a balanced crankshaft without a blueprinting procedure, or an engine assembly may be blueprinted, which also involves crank balancing. Simply because a crankshaft has been balanced, it does not mean that the engine has been blueprinted.

Are OEM production engines blueprinted? No, they are not. Although a production engine may be based on specific design parameters, the reality of mass production results in deviations due to core shifts in castings and allowable tolerance ranges for parts, dimensions, and machining. Although a production engine may be perfectly suitable for use, it may not have been produced precisely to the design specifications. This isn't because car manufacturers don't take this seriously or are trying to short-change anyone, and this certainly doesn't infer that production engines are faulty. It's simply the reality of producing engines on a large scale.

The processes of blueprinting are performed when the performance and/or racing application calls for optimizing the engine's performance and durability.

In a nutshell, the basic goal of blueprinting is to gain a high degree of precision in order to achieve, as closely as possible, a "zero-resistance" engine assembly. This means all parts are properly aligned and clearances are achieved for optimum efficiency. In essence, you're trying to create the perfect engine by eliminating any variances that affect power and engine durability.

Of course, blueprinting *can* be utilized to achieve the OEM design specification, but the process is not limited to this. In most cases, a blueprinting approach is used to carefully accurize the block in order to locate all bore centerlines, bore diameters, and bore angles to eliminate deviation, even though you may be altering certain bore diameters. With the main bore corrected to a perfectly straight alignment, cylinder bore centerlines are corrected, lifter bores are corrected for centerline and angle, and decks are machined to be perfectly parallel to and equidistant in height from the main bore centerline. Regardless of the cylinder-bore diameter (whether at OEM diameter or overbored for increased displacement), blueprinting involves accurizing the block as the basis of the foundation.

Beyond block accurizing, blueprinting involves optimizing all clearances in the pursuit of performance and longevity. OEM production may allow a certain tolerance range for areas such as piston-to-wall clearance, bearing clearance, deck height, etc. But when blueprinting, you decrease the tolerance range significantly. For example, if an OEM specification allows a clearance of .020 inch, +/– .005 inch, you try to achieve *exactly* the optimal .020-inch clearance when blueprinting. You're refining all clearances while greatly minimizing allowable ranges.

Let's look at a specific example. Say that an engine manufacturer lists piston ring end gap at .003 to .005 inch for a specific engine application. Rather than varying end gap while staying within that tolerance range, you can tailor the gap to the application at hand. An ideal end gap for a road-racing endurance application might be .0045 to .005 inch. For a drag racing application this might be better suited to .0035 to .0040 inch. This is just one example.

By initially considering the published clearances, you can then fine-tune the gaps in order to obtain the best dimension for an application. Much of this depends on the engine builder's personal experience with various ring end gaps for specific racing applications. In blueprinting you narrow down various clearances in order to achieve the ideal clearance, instead of simply falling into the wider range acceptable for the average street engine.

Areas to Consider during Blueprinting

- Flaw inspection of all components (checking for cracks/flaws)
- Main bore position and alignment, which includes line boring the main bore
- Position (relative to the main bore) and alignment of the cam tunnel
- Block deck accurizing, including height (distance from the main bore centerline and correction of deck angle)
- Cylinder bore indexing (establishing accurate centerline)
- Lifter bore accurizing (establishing lifter bore centerline and angle)
- Bellhousing dowel location accurizing
- Correction of the crankshaft's main journal alignment and spacing, endplay, rod journal alignment and

spacing, stroke throw length, and clock-position of each throw (angularity of the rod throws)
- Indexing cylinder heads (correcting intake port centerline; scribing cylinder bores on the deck surface of the heads for consistent chamber layout
- Static weight matching of all pistons, rods, rod pins, and rod bearings
- Crankshaft dynamic balancing using bobweights to simulate static reciprocating weight packages
- Inspecting and correcting (if needed) crankshaft main and rod journal diameters
- Connecting rod dimensions and checks (big-end bore diameter, pin-end bore diameter, center-to-center length, checking for bend and twist)
- Checking piston pin bore centerline-to-dome distance (and verifying that this is equal on all pistons)
- Measuring the camshaft for journal diameter, straightness, lobe spacing, lobe lift, and ramp angles
- Measuring all pushrods for length and straightness
- Inspecting all lifters for length and diameter
- Measuring cylinder head combustion chambers for volume and (if needed) machining to equalize all chamber volumes
- Measuring all intake valves for height, stem diameter, and head diameter
- Measuring and equalizing all valve seating depths
- Valveguide sizing for desired valve-stem oil clearance
- Measuring all valvesprings for open and closed seat pressure and height, and checking for coil bind at the fully closed position
- Checking rocker arms for proper geometry and contact at the valve tips through the rocker arm's arc
- Inspecting and verifying crankshaft counterweight to block clearance, connecting rod big end to block clearance, connecting rod big end to camshaft clearance, etc.
- Intake port matching (of intake manifold-to-cylinder head)

Part and parcel of blueprinting involves inspecting everything. Never assume that any new or used component is dimensionally correct. Don't just take it out of the box and bolt it on. Blueprinting involves close examination of absolutely everything. Where variances exist, depending on the specific component, the part must be machined for correction or replaced with another that matches the desired dimensions. The golden rule is to assume nothing and measure everything.

Parts Selection

A true blueprinting job is very time consuming and labor intensive. So it doesn't make sense to start the job with inferior or questionable parts. Buy the highest-quality components that you can afford, and don't try to skimp.

Regardless of your component selection, if you're going to invest in a blueprinting approach, it's highly advisable to perform a flaw detection on components such as the block, cylinder heads, crankshaft, and connecting rods and check for cracks and porosity. For the block, this also means checking for cylinder wall thickness at each bore location. If you start the job with questionable or unknown parts, you're just defeating the purpose.

The quality issue aside, also consider the type of material and construction (for instance, when choosing between a casting, forging, or a machined-from-billet piece). As an example, depending on the manufacturer, a cast-iron crankshaft might be dependable up to, say, 400 hp. If you plan to produce more power, moving up to a forged crank is recommended. Select the parts based on the application. By choosing a component that is stronger and more resistant to failure, you increase your chances of avoiding a failure when producing increased power and/or increasing engine speed.

Can't Blueprint?

Certain race sanctioning body rules may not allow blueprinting (old SCCA showroom stock road racing as an example). However, there are ways around such nonsense. Even a stock engine can be improved, and mildly blueprinted by mixing and matching engine components in order to improve the engine's dynamic operation. Instead of machining various parts in order to optimize, an alternative may be to spend the time to locate stock parts that provide the improvement.

For example, an engine's original set of connecting rods may be specified as having a 6.000-inch center-to-center length. But, as a result of mass production and a factory tolerance range, your engine may actually have rods that vary in length from 5.996 to 6.002 inches. Instead of "illegally" machining to correct, you can take the time to create a matching set of rods that actually measure 6.000 inches. This may be aggravating and tedious, but it's a way around a silly ruling. The same holds true for OEM camshafts; if yours doesn't measure at the published specs, you can hunt through several others of the same part number until you find one that's on the money.

Remember: Blueprinting involves much more than simply balancing the crankshaft. This is serious business that goes far beyond the simple re-assembly of an engine with new parts. It's an investment in time, skill, labor, and money.

Flaw Detection

To achieve the highest level of precision engine building, you must perform diligent detective work and pay close attention to detail. When using flaw detection equipment, you are searching for damage that's more and less apparent. And many kinds of damage are not easily recognizable to the naked eye. Flaw detection equipment and technology help you discover potential problem areas in your engine components. A variety of testing methods are available for inspecting individual components for cracks or porosity, as well as prejudging cylinder wall thickness prior to block cylinder boring.

Sonic Testers

This specialty gauge is used to measure material thickness, and is especially useful for measuring cylinder wall thickness in an engine block. This hand-held tester has a corded probe that emits a sonar signal that travels through the material. When the signal reaches the opposite side of the material, the signal bounces back to the tester, which displays the thickness measurement.

Probe measurement checks are made from the top of the cylinder wall to the bottom, in all radius directions. It's especially important to check areas where water jackets exist behind the cylinder walls. Being able to determine existing cylinder wall thickness tells you how much material can be safely removed during an overbore. Walls that are too thin at critical areas deflect more during engine operation and degrade piston ring sealing. In addition, these walls can crack and/or blow out.

Ideally, you want to maintain at least .125-inch wall thickness at the thinnest areas. This requirement can vary depending on the specific block, but this is a reasonable ballpark. Any cylinder wall thickness concern increases if the engine is to be force-inducted with supercharging, turbocharging, or nitrous injection. The higher the cylinder pressures, the more critical wall thickness becomes.

Before using a sonic checker, the unit must be calibrated. Commonly, sonic testers include a machined calibration sample of steel that is milled at specific thicknesses of .200, .300, .400, and .500

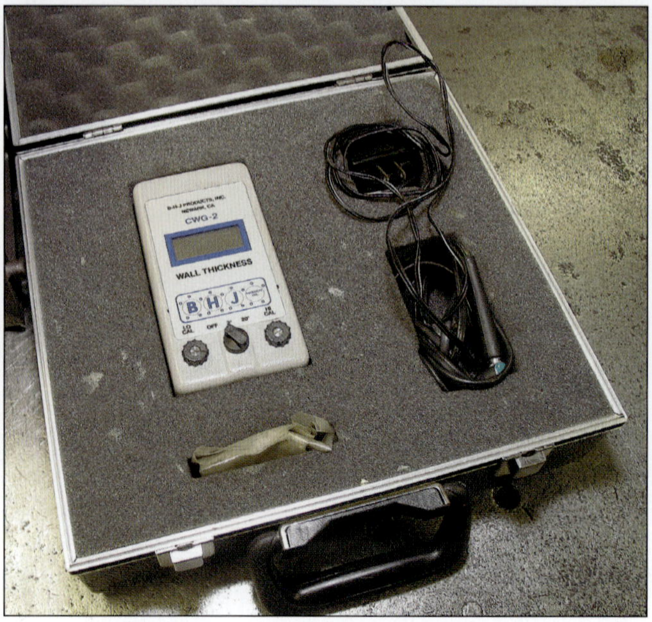

A sonic tester is a must-have in any engine shop, especially when considering a cylinder overbore job. Instead of guessing about existing cylinder wall thickness, this type of testing equipment allows you to accurately measure wall thickness anywhere in each cylinder. This is a sonic checker kit from BHJ.

With the cylinder surface clean, and after calibrating the tester with the thickness standards included in the kit, a bit of lithium grease is applied to the probe. While applying moderate pressure at the probe against the cylinder wall, the display reveals material thickness at that location. The goal is to check wall thickness at all clock positions and height locations to determine existing wall thickness, especially in areas adjacent to cooling passages.

inch. The calibration procedure is outlined with the specific brand of tester, but it's a simple process. Once the unit has been calibrated, apply a dab of lithium grease to the probe face. Contacting the probe against a clean cylinder wall (using moderate pressure on the probe) reveals material thickness.

Don't just measure one cylinder bore. It's very common for cast-iron blocks to have variations in wall thickness, so be sure to check each cylinder. If the wall thickness in any given cylinder is marginal, you have the choice of scrapping the block or (depending on the specific block) installing a sleeve/cylinder liner. Checking your block with a sonic tester helps you avoid investing in a block that doesn't hold up.

Hand-Held Magnetic Particle Testers

Applicable to cast-iron and steel materials only, this process involves placing magnetic poles at opposite sides of a suspected crack area and introducing a ferrous powder onto the inspection area.

The workpiece must be clean before testing. Only clean with solvents, and do not clean with abrasives or anything that might disturb the metal surface. In other words, don't use glass bead, steel blast, etc. because they may peen crack edges, which makes inspection more difficult.

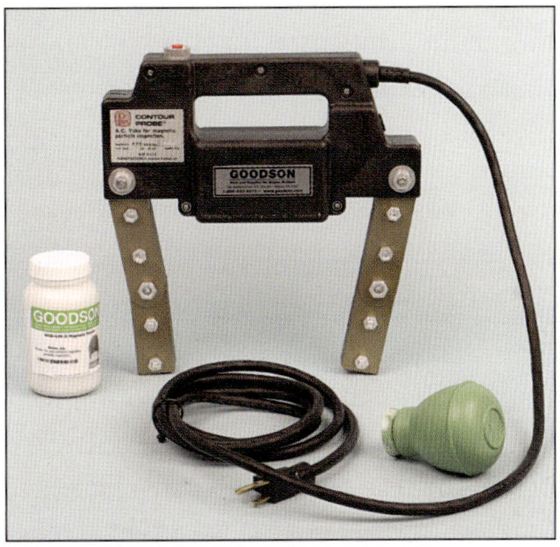

A hand-held magnetic checker has two magnetic poles. The poles are positioned on either side of the suspect area, and a special easy-to-view metal powder is sprayed onto the area. With the unit turned on, a field is created that draws the powder into the crack for easy identification. Shown here are Goodson's electric magnetic-particle tester MMP-210, a container of inspection powder, and a powder duster. Goodson supplies a wide range of tools and supplies dedicated to the needs of pro engine builders.

Using the squeeze duster, apply the powder to the surface.

A magnetic (or "mag") tester is a hand-held electromagnet used to check for cracks in ferrous materials, such as cast iron and steel. The surface to be checked must be clean and dry. Spray a light dusting of special iron powder onto the inspection area, and place the twin-poled magnet onto the surface. Press the momentary ON button to turn on the magnet. When using a twin-post magnetic tester, the best results are obtained if the bases are placed at a 45-degree angle to the suspected crack. Repeat the test 90 degrees from the first test (test in a criss-cross manner). This pulls the powder into both walls of a crack for easier viewing.

Place the magnetic particle tester onto the work surface and adjust the pivoting magnet posts to achieve contact.

With the magnet unit in place, press the ON-button to activate the magnets. The powder is pulled along and into any cracks. Note the small crack that runs between these two valve seats.

Wet Mag Bench Testers

Also referred to as a magnetic particle inspection station, a dedicated "wet mag" bench has a large-diameter magnetic ring and a black-light setup. Passing the tested part through the ring magnetizes it. A hand-held magnetic field tester verifies the integrity of the part. A special penetrating liquid (mineral spirits mixed with florescent particles) is rinsed over the part and then it's viewed under the black light. Any cracks show as a brightly colored line. Once the inspection is complete, the switch is flipped to de-magnetize mode to remove the magnetic field. If left magnetized, metal particles continue to cling to the part due to a potential residual field, even during engine assembly, which can obviously cause contamination and wear problems. The handheld field tester is used to verify that it has been demagnetized. The part is passed through the mag ring again. A wet mag testing unit is applicable to any ferrous (iron or steel) component, and is most commonly used for checking crankshafts and connecting rods.

A magnetic particle inspection station (a wet bench) has a large-diameter magnet that creates a magnetic field. The component being checked passes through this field.

Once the component has been lightly magnetized, a dye fluid is applied and inspection is done using a UV light (or black light). These machines usually have a curtain surrounding the testing area for better UV viewing. Here Bob Fall, of Fall Automotive Machine in Toledo, Ohio, prepares to inspect a crankshaft. The curtain is left open only for the sake of this photo.

Dye Penetrant

This photochemical inspection process is used with or without the need for ultraviolet light. It can be used to test ferrous (steel/iron) and non-ferrous (aluminum and titanium) materials. Dye penetrant testing uses special chemistry to reveal surface cracks. Because no magnetic field is involved, this process can be used on any steel, cast-iron, aluminum, or plastic part.

The process typically involves three chemicals: a cleaner, a dye, and a developer, but it also often uses a cleaning solvent to remove any grease. Spraying the special cleaner prepares the surface. A dye penetrant is then sprayed onto the part. This dye seeps into any cracks, pits, or other surface irregularities.

A dye penetrant kit. This is Goodson's Glo-Kit, complete with cleaner, penetrant, developer, a supply of rags, and an ultraviolet light.

Once the dye is given a few minutes to soak and dry, a special developer is then sprayed onto the part. The developer reacts with any concentrations of the dye, such as in a crack line. A readily visible, colored line reveals the crack. There are two types of these kits: One allows cracks to be seen under ambient room lighting; the other provides really bright-colored crack visibility under ultraviolet light. Once the crack is located and marked, a cleaning solvent is used to wash off the dye and developer.

Dye penetrant kits are available in affordable spray-can kit form from a number of sources, such as Goodson Tools. Additional spray cans are available individually when replacement is needed. Magnaflux Corporation also offers a similar Zyglo dye penetrant kit in a variety of sizes. It has a florescent dye that's viewed under UV light (some kits include the black light).

For example, the procedure for using Goodson's Glo Kit is simple. Clean the surface using a fast-drying solvent or hot water and soap. Rinse and dry. Spray the Goodson Cleaner onto the surface. This is a fast-drying solvent that prepares the surface. Allow this to completely dry. Then spray a liberal amount of penetrant. Allow this to soak for about 5 minutes. Using a clean rag, wipe off any

The three-chemical process involves a cleaner/solvent that prepares the surface, a penetrant that seeps into any cracks, and a developer that allows the black light to highlight any cracks.

Spray the cleaner onto the area and allow it to dry completely.

Spray a liberal amount of penetrant to the inspection area, and allow it to dry for about five minutes.

After wiping off any excess penetrant, spray several very light coats of developer until the area starts to look chalky-white.

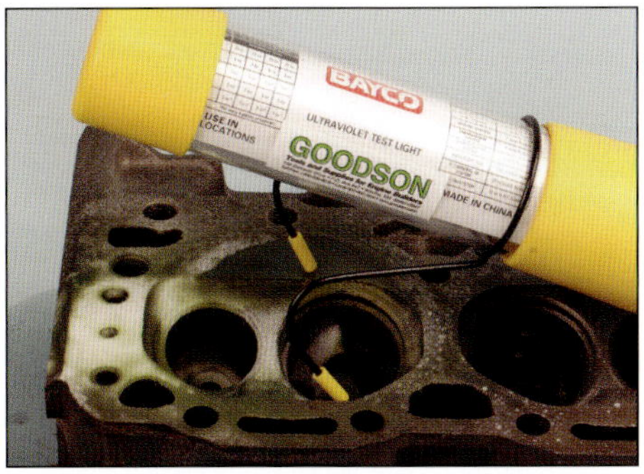

Using the black light, inspect for cracks.

excess. Next, spray several *light* coats of developer onto the area until you see a white chalky appearance. Place the UV light over the surface and inspect for cracks, which are revealed as contrasting lines.

A dye penetrant check found this crack leading from a spark plug hole.

Any of these kits are easy to use and come in handy for a variety of crack-checking jobs, including not only engine parts (blocks, cranks, rods, intake and exhaust manifolds, heads, etc.). They also work on socket wrenches, race car suspension parts, welded brackets, brake rotors, car frames, etc.

Pressure Testing

Pressure testing is indispensable for inspecting a cylinder head or other engine components for cracks, pinholes, porosity, or erosion. Two types of pressure testers are available. A positive pressure tester involves sealing all passages with special plugs. A gasketed sealing plate is mounted to the head deck. The head is then secured to a handling fixture and immersed in a water bath. Compressed air is introduced into the head revealing any leaks (cracks, faults, etc.) as bubbles escaping from the head.

A negative pressure tester (vacuum) can be used to check valve seating. With valves closed, special plugs are installed into the ports and vacuum is applied. Any drop in vacuum pressure confirms that a specific valve(s) is not sealed against its seat.

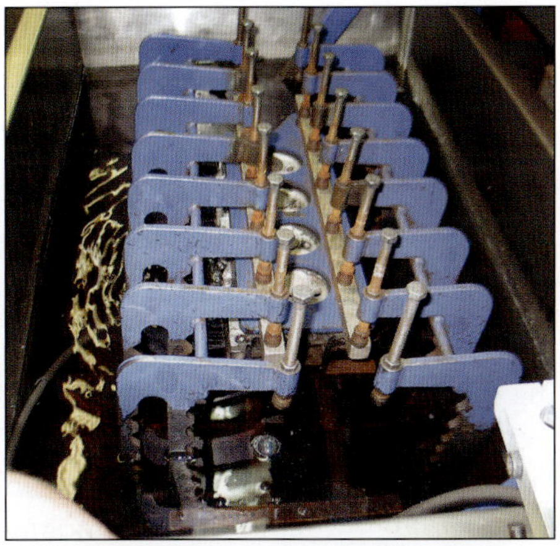

The head is immersed in a water tank, with compressed air pumped into an exterior water passage. A rubber fitting creates a seal for the passage that allows the compressed-air connection. Any leaks are evident as bubbles.

A bench vacuum tester allows testing for valve seating by applying vacuum to a spark plug port. Any drop in vacuum pressure indicates a poor seat seal.

CHAPTER 2

ENGINE BLOCKS

To blueprint an engine block, you must accurize it. You must achieve the proper bore dimensions, but also must consider the total geometric state of the block and correct any deviations from ideal geometry. This means that you inspect and machine in order to optimize the block geometry. This primarily focuses on the main bore, cylinder bores, lifter bores, and decks.

An OEM (original equipment manufacturer) mass-produced, cast block typically requires more corrective measures in a blueprinting approach, simply due to the nature of wide-tolerance mass-production practices. By contrast, today's high-quality aftermarket performance blocks, such as those produced by Dart, World Products, RHS, Brodix, and others have CNC-machined bore locations and angles from the very start. In essence, they are already "trued," but still need to be checked, typically requiring only deck height and bore diameter machining to the required sizes. OEM blocks, on the other hand, generally require more extensive machining in order to achieve the *design* geometry.

Generally speaking, leading-brand aftermarket high-performance blocks are made with superior materials and are made in lower numbers than OEM blocks, and therefore benefit from a higher level of attention to detail. An aftermarket block costs more, but you'll likely be faced with less corrective machining labor.

Aftermarket blocks from Dart, World Products, Brodix, and others are CNC machined to much tighter tolerances and quality standards than cast OEM blocks. (Photo Courtesy Dart Machinery)

The manufacturing process for OEM blocks have time and technology constraints. As a result, factory blocks have a range of tolerances that can and usually do deviate from the initial design. In general, the older the block, the wider the OEM tolerance ranges are. Late-model blocks generally have tighter machining tolerances thanks to advanced CNC technology. However, OEM blocks are produced in mass quantities so, considering machine calibration and tooling wear factors, you'd still be hard-pressed to find an OEM block that's precisely accurate to the design specifications. That's not to say that they're unusable, but they can always be corrected according to the design specifications

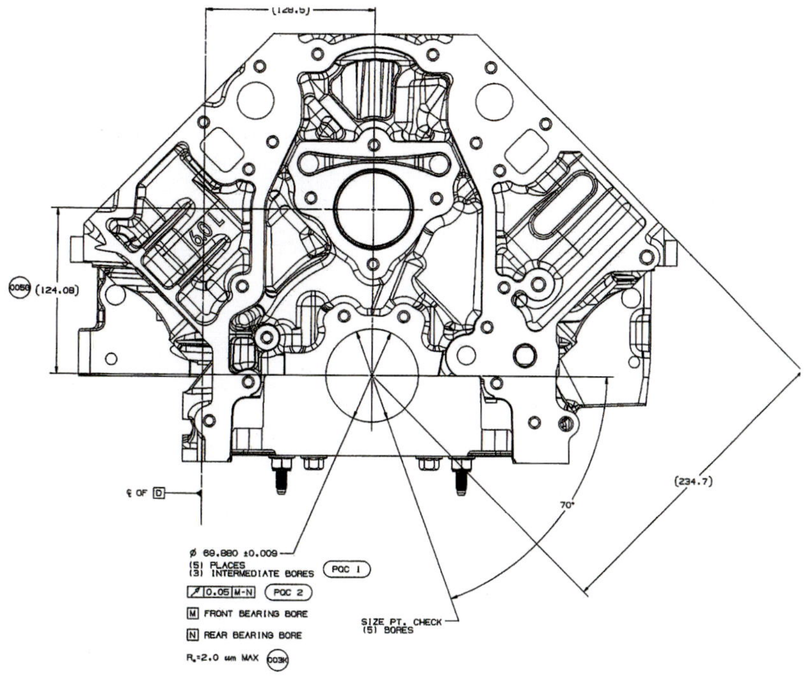

This drawing of an LS block clearly shows how the main bore centerline is used as the primary reference point. (Illustration Courtesy General Motors)

and performance and durability can be improved.

To accurize the block, make sure that the main bore is not only straight and round, but also exactly in plane front to rear, and corrects (where needed) the main bore height location. When accurizing the block, you are correcting all critical areas to eliminate the tolerance range that is otherwise deemed acceptable in mass-production runs.

When the average enthusiast claims that his or her engine has been "balanced and blueprinted," in reality it usually means that the crank has been balanced. True blueprinting involves much more attention to detail (and subsequently higher machining/prep costs) than merely honing the cylinders, performing a fresh cut of the decks, and balancing the rotating assembly. When someone claims that his engine was blueprinted, and, when asked how much he spent, he or she mentions something in the area of, say, $1,000 to $3,000, it's obvious that a blueprint has not been performed. Blueprinting requires skill and expertise, and is very time and labor intensive. That doesn't come cheap.

Flaw Checking the Block

Before any machining takes place, flaw check the block to make sure that no cracks are present. Any cracks or damage must be addressed by repairing the damage or finding another block. Also sonically test each cylinder wall to verify wall thickness and determine whether it can be overbored. This needs to be done before any machining and crack repair is done.

Minimum acceptable wall thickness may vary depending on the block, but in general, you should have at least .200-inch thickness after boring and honing. If the walls are too thin, they can distort enough to result in ring blow-by or even eventual cracking. Keep in mind that a too-thin cylinder wall may be corrected by overboring and installing a quality cylinder sleeve (again, this can vary depending on the type of block).

Once the block has been deemed serviceable, the first order of business involves checking and correcting the main bore if needed. In addition to making the main bore straight and round, it needs to be centered per its design. Specialty guide fixtures are available, but today, with the increasingly common use of CNC machining technology, the appropriate software can handle this duty. Once the main bore is verified as correct, all remaining machining basically uses the main bore centerline as the reference point.

When the block has been deemed serviceable, the first order of business

The sensor probe is slowly run along the cylinder wall (in a number of clock positions and from top to bottom) while monitoring the gauge readout.

Before measuring and/or align honing the mains, all caps must be fully installed (at spec'd torque value). If the block has additional side bolts (as in the GM LS block, or vintage-Ford FE side-oiler block, for instance), these must be installed as well. Fully tighten the primary main cap fasteners before tightening the cap side bolts. This is critical: The side bolts must be fully tightened to spec before measuring or honing, since they do affect main bore geometry.

If you plan to use main studs (instead of bolts), install the studs finger tight, then add a small bit of preload. Be sure to follow the stud maker's specs! Do not overtighten the studs into the block. There's no need to (since nut tightening achieves clamping load), and by overtightening studs, you can easily create a splayed stud problem. For instance, do not double-nut the studs and try to tighten to the main cap torque spec.

Utilizing specialty main bore and cam bore inserts (various diameters are available for specific blocks) and a dedicated micrometer, it's easy to accurately measure the distance between main bore and camshaft-bore centerline distance (measuring from the outside of each bore adapter and subtracting one-half of each guide's nose diameter). (Photo Courtesy BHJ Products)

TECH TIP

Block Inspection Check List

Here are the steps to perform a flaw detection to check for cracks:.

Use a sonic tester to measure and record all cylinder wall thicknesses. A minimum of three height locations and at four clock positions should be used. Record the data and use it for proper overboring.

Inspect and correct, if necessary, every threaded hole in the block for thread integrity and condition.

Align hone the block. This creates the centerline that everything else references.

Cut the decks to obtain a square block, where both decks are the same height from the main bore centerline and parallel to the main bore centerline.

Measure and correct cylinder bore spacing and angle. When correction is necessary, perform an overbore or overbore correction sleeve installation. The sleeves are resized to original or planned bore diameter.

Measure and correct camshaft bore centerline. If correction is required, overbore the cam tunnel and use the necessary oversize-OD cam bearings.

Check and correct all lifter bores for centerline location and angle. Correction may require overboring, installation of bronze liners, and resizing for proper lifter clearance. ■

involves checking and correcting the main bore if needed. Or, the block can be machined to correct any anomalies on a CNC-milling machine (with generic programs for specific blocks, bore sizes, and deck heights, or by custom-programming by the CNC operator). Accurizing fixtures are essentially precision templates, or guides, that establish proper geometric centerline locations and angles. If a shop isn't equipped with a CNC machine, these fixtures allow precision machining using traditional shop equipment boring machines, multi-axis milling machines, etc.

Decking the Block

Production engine block decks may be off specification by a few thousandths of an inch; in some, by as much as hundreds of thousandths. This means that both decks might be too high or too low or one deck may be higher than the opposite deck. Here's a case in point: A recent Pontiac 455 build started with an aged 1973 455 block. Although the OEM spec for deck height is listed as 10.210 inches, this block's left deck measured 10.2545 inches, which is .0445 inch taller than spec. The right deck measured 10.2456 inches, which is .0356 inch taller than spec. Luckily, the excess material allowed correction precisely to the desired 10.210-inch spec. In this case, both decks were too high, and had a different deck height at each bank. As is, the OEM excess in deck height decreases compression ratio. Also, the uneven heights from left to right and front to rear results in different compression ratios per cylinder and from bank to bank. Yes, the engine would run, but it wouldn't be running at *optimum* efficiency.

Keep this important point in mind: The OEM spec for block deck height should not be considered set in stone unless all other OEM engine dimensions and materials are used. If so, you must use the exact same crank stroke, the same connecting rod length and material, the same piston compression height and volume, the same thickness and material of head gasket, and the same volume and shape of cylinder head combustion chamber. Be sure to use the same intake and exhaust valve diameter and length and the same rocker arm ratio as specified in the OEM engine. You also mock the short block together (crank, rods, and pistons) to measure the pistons' top deck location at top dead center (TDC) relative to the block deck. If the deck height is too short, the pistons can hit the heads. If the deck height is too tall, you decrease compression.

When test fitting, be sure to consider the installed (crushed) thickness of the head gaskets. Be sure to use exactly the same brand and type of head gasket for test fitting that you plan to use during final assembly.

You want to be as close as possible to zero clearance between the piston dome and cylinder head during engine operation. However, obtain static clearances, because the clearances decrease as the engine operates because of heat expansion and metallurgical stress. The minimum clearances you should generally have in static state (with the engine not running) between the piston and combustion chamber at TDC are:

- Performance street application with steel connecting rods: about .040 inch
- Racing engine with steel rods: about .045 inch
- Racing engine with aluminum rods: about .060 inch

Remember that these approximated clearances are *minimum* clearances.

With this in mind, it's important to understand how block deck height relates to your crank, rod, and piston dimensions. In order to achieve a zero deck height (so the piston lies flush with the deck at TDC) consider crankshaft stroke, connecting rod center-to-center length, and piston compression distance.

A crankshaft's stroke refers to its total stroke from a rod journal at TDC all the way to that journal's bottom dead center (BDC). When determining where the top of the piston deck will be at TDC, consider one half of the crank stroke because you're only concerned here with how far the crank is pushing the piston upward. Connecting rod length refers to the distance from the big-end centerline to the small-end centerline, not the overall rod length. Piston compression distance (CD) refers to the distance from the centerline of the piston pin bore to the piston's flat deck surface.

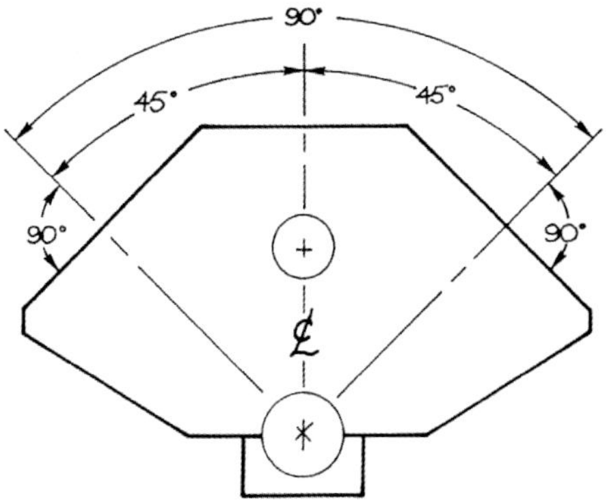

A CNC cutter spot faces lifter bore roofs in preparation of lifter bore centerline correction. With a programmed CNC machine, no indexing fixtures are needed for lifter bore correction.

Here's the formula for finding block deck height at zero deck:

1/2 crank stroke + rod length + piston CD

For example, the crank stroke is 3.500 inches, the rod length is 6.000 inches, and piston CD is 1.500 inches. When you plug the numbers into the formula, you get a 9.25-inch deck height.

$$(3.500 \div 2) + 6.000 + 1.500$$
$$1.750 + 6.000 + 1.500$$
$$9.25$$

If you want the piston top flat to be, say, .015 inch below deck, you add .015 inch. In the above example, the finished block deck height needs to be 9.265 inches. The decks may also be out of plane. You could have decks that are low at the front and high at the rear, high at the front and low at the rear, low inboard or low outboard, etc. In other words, decks may be flat, but they might be "crooked." For corrective machining, you need to index from the crankshaft main bore centerline to make both decks the same height (and the proper height) and 90 degrees to the main bore centerline in both front/rear and inboard/outboard planes.

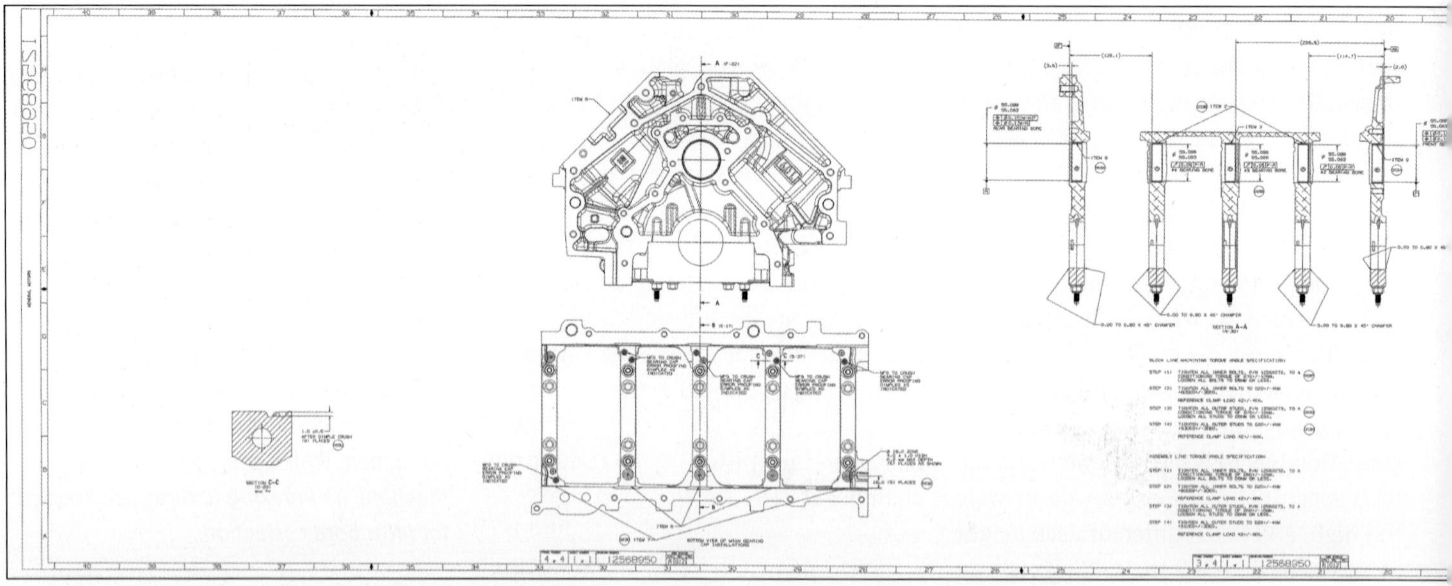

Correcting Cylinder Bores

Obviously, the cylinder bores must be round and the correct diameter for the intended pistons and rings. Also the centerline placement of each bore and the angle of each bore must be accurate. If not, they must be corrected in order to blueprint the block. When OEM blocks are made, mass production line tolerances may allow for the bore centerline to be placed slightly off-center, and the angle of the bores may be slightly offset from front/rear and inboard/outboard, which slightly deviates from the original engineering design.

Corrective machining, which is covered in Chapter 3, can relocate the cylinder bores to achieve exact centerline location and cylinder wall angle. As a result, removing material in order to accomplish this means you use oversize pistons and rings. But because most performance builds involve increasing displacement anyway, this is a moot point. The same holds true for lifter bores in overhead valve engines. The lifter bores may also be corrected, relocating their centerlines and angles. Oversizing the lifter bores in order to make these corrections simply means that bronze lifter bore liners are then installed and honed to the required diameter.

Machining decks to the proper height and angles equalizes and angle-corrects the base for cylinder volume. But you also equalize crank stroke, rod length, piston compression height, piston dome volume, and cylinder head chamber volume. By correcting cylinder bore centerline (and bore spacing), you place the bores and pistons in a no-compromise, as-designed, on-center travel path, which reduces operating friction and stresses. By correcting lifter bore centerline location and bore angles, you improve valvetrain efficiency. It all boils down to reducing friction and wasted energy, which translates into better performance and longer engine life.

Core shift commonly occurs during the casting process at the manufacturing level. It's not uncommon for cylinder bores to shift slightly beyond engineering specification. This is one of the primary reasons for thoroughly accurizing the block by determining where these shifts have occurred and machining to correct these out-of-spec centerlines. Out-of-spec tolerances may pass the manufacturer's acceptance range for common street applications, but for high performance and racing you want to correct these issues in order to achieve a high level of precision, both to accommodate increased horsepower and to aid in engine longevity.

Oil Restrictors and Screens

The purpose of oil restrictors is to reduce the amount of oil at the top end of the engine, so more oil is delivered to the rod and main bearings. These restrictors are threaded plugs that are installed in the oil passages, which feed oil to the lifters, valves, springs, and rockers. Windage drag, which is drainback oil accumulating on the crank counterweights, is also reduced. Opinions vary with regard to restrictors, and the need for them also varies depending on the type of engine.

In general, restrictors may be used with solid (mechanical) roller lifters, but should not be used with hydraulic lifters or with flat-tappet lifters, since they both need more lubrication. Also, while restricting oil to the upper end may be fine with full-roller rockers, stock OEM ball/pivot rockers need more oil delivery, so don't install restrictors if you're using ball-type rockers.

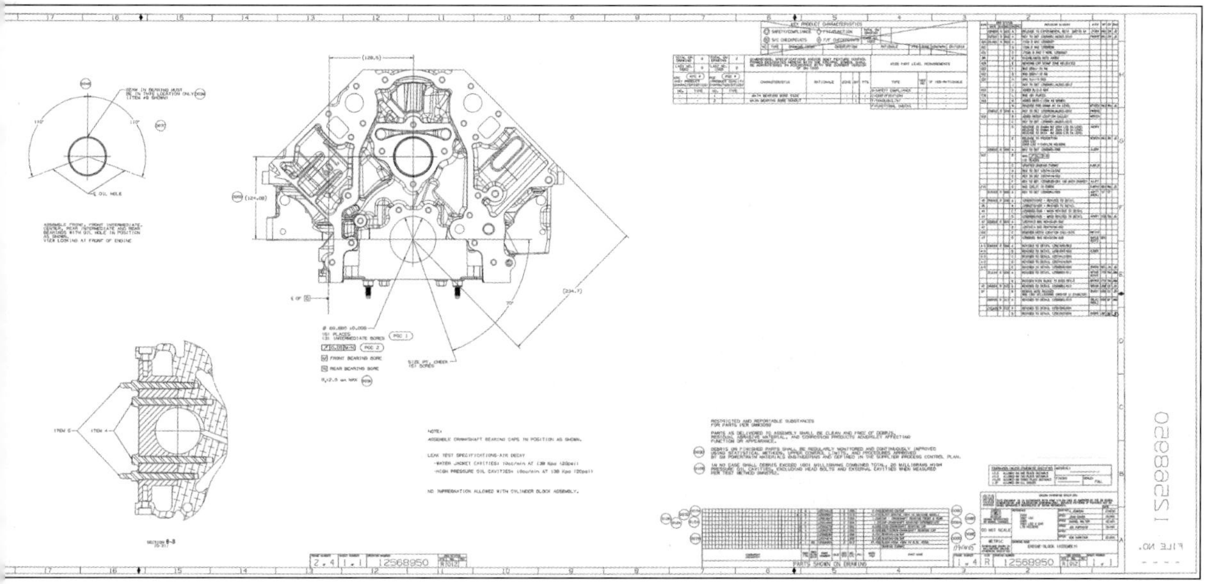

A blueprint illustration reveals all of the block dimensions according to the engineering design. Some points of reference may or may not be correctible, but a factory spec drawing allows you to double-check many critical dimensions to verify various areas of your OEM block. (Illustration Courtesy General Motors)

Depending on the type of engine, restrictors may be installed at the rear of the block. These are installed deep inside the lifter-valley oil gallery holes or in the lifter valley. In either case, thread tapping is required, so if restrictors are planned, the tapping and subsequent cleaning must be done prior to final wash and block assembly. The size of the oil holes in restrictors varies from about .040 to .065 inch. Again, this depends upon the application and is quite often based on the opinion and experience of the block machinist/engine builder.

The large oil drainback holes in the lifter valley of a V-type block obviously allow oil that was delivered to the upper end to drain back to the sump. However, if something goes awry at the top end, such as valvespring failure, valve failure, keepers falling loose, etc., the resulting fragments can travel across the lifter valley and drop into the oil sump. To avoid this, it's common for race engine builders to epoxy a piece of screen over these drainback holes. If you decide to do this, make sure that the block surface is clean and dry to provide good adhesion for the epoxy. If the screen pops loose, it can cause damage by interfering with lifters or by working its way down to the cam.

Many OEM blocks have ragged, unfinished drainback holes and/or slots that have jagged casting flash edges. It's always a good idea to grind these edges smooth, not for the sake of appearance, but to ensure that no flashing pieces break loose in the future.

Lifter Valley Surface

Residual oil that has done its job at the upper end needs to return to the sump in a timely manner. Because older cast-iron blocks commonly had fairly rough, raw casting surfaces in the valley, it was/is common practice to smooth out the surfaces or coat the valley with a high-build paint to fill the casting pores and promote faster drainback. Today's castings, especially quality aftermarket blocks, tend to have finely finished surfaces. In the case of the GM LS engine, there is no lifter valley.

If you decide to apply a coating to your lifter valley, the block must be absolutely clean and dry (serious hot-tank wash). The most commonly used coatings include Glyptal, an electrical armature winding coating, which is highly resistant to oils and acids. All machined surfaces, threaded holes, lifter bores, etc., must be plugged or masked to avoid paint contamination. In my and many others' opinion, the time and effort required as well as the risk of dried paint particles breaking loose just isn't worth it, considering the small increased rate of oil return. You're better off by simply grinding/smoothing any noticeable surface defects/lumps, sharp edges, etc., and ignore the paint idea.

Rod and Crank Clearance

Whenever you deviate from "stock," in crankshaft and rod selection, you *must* check for rotating and reciprocating clearance at the block. Specifically, if you choose a longer crankshaft stroke and/or beefier aftermarket connecting rods, clearance checking is an absolute must. The common clearance issues occur at the bottom of the cylinders and at the oil pan rail areas. Mock-install the crankshaft to the block with clean and lubricated main bearings, and using the same main caps that you plan to use in the final build. For interference checking, there's really no reason to fully torque the main caps at this time. Simply snug them to about 20 ft-lbs.

Slowly rotate the crank and, starting from the front, watch each crank counterweight as it moves close to the block (near the pan rails). You should have about .060 inch clearance between the counterweights and the block at the tightest locations. If you see component contact or too tight of a clearance, mark the block location using an ink or paint marker. After all counterweight clearances have been checked, remove the main caps, crank, and bearings; then remove block material as needed with a high-speed die grinder. Clean off the metal particles and re-install the crank and perform the clearance check again to verify.

When test fitting the crankshaft (with main bearings installed and lubed), check crank counterweight-to-block clearance, especially when using a stroker crank. This increased-stroke crank in a Pontiac 455 block had a fairly tight clearance between the forward area of the front counterweight and the block, just above the oil-pan rail. Once the crank was removed, hand grinding removed a small bit of material from the block to achieve a comfy .050-inch clearance at this tight spot.

During this test fit, you can see that the rod bolt is contacting the block. The area to be relieved was marked as a reference. The interference area of the block must be relieved to obtain a bare minimum of about .060-inch clearance.

Aftermarket performance blocks commonly provide a bit of additional rod clearance (more than an OEM block), since the block manufacturer is aware of the possibility that a longer stroke crank may be used. This Dart 351W block already has a convenient notch for big-end rod clearance. Depending on your specific crank and rods, some additional clearancing may still be needed, but it is minimal.

If using a stroker crank and/or a high-lift camshaft, don't forget to check for possible rod-to-cam lobe interference. This photo (taken without cam bearings or cam) helps to illustrate the potential for clearance issues. With cam bearings, camshaft, crank and rods, and timing gear test installed, carefully rotate the crank while observing cam lobe clearance. It may be difficult to see, but possible with a skinny inspection light. If clearance is a problem, you may be able to remove material from the rod big ends or move to a cam with lower lobe height. This is why aftermarket blocks are often available featuring a raised cam tunnel.

Once you're sure that the crank clears the block properly, install one rod/piston set, starting with the number-1 cylinder. There is no need to install rings at this time.

Slowly rotate the crank, check for the rod big-end clearance to the block during a full-stroke travel. Rotate the crank a full 360 degrees. Again, minimum clearance of the rod to the block at any given point should be about .060 inch. Typically, if you do find a rod clearance issue, it is likely in the rod bolt/rod cap shoulder locations. Don't be tempted to grind material from the rods; remove material (if needed) from the block.

Also, check the clearance between the crank counterweights and the bottom of the piston skirts. If clearancing is needed, the crank counterweights can be lathe-machined (prior to crank balancing). Mock-install the camshaft and timing assembly and check the rod big end

After carefully removing block material with a hand-held electric grinder (and after washing the block and reinstalling the crank and rod), clearance was again checked. It is a more-than-healthy clearance of approximately .080 inch. When checking rod clearance, you must be careful to rotate the crank slowly as you observe the rod approaching the block. In some cases (depending on the block), only a couple of spots may require clearancing. In some cases you can get lucky, with no additional clearancing needed, or you may need to provide clearance for the majority of rod-to-block areas. Just take your time and check each and every location.

clearance to the camshaft, especially if a high-lift cam is used (this is the reason that aftermarket performance/race block manufacturers often offer a "raised cam" block, which places the cam bore higher, farther away from the crank centerline). The cam bearings must be in place in order to properly center the cam in its bore.

Oil Hole Alignment

Never assume anything. Just because the block's main bearing saddles have oil feed holes, it doesn't necessarily mean that these holes properly align with the oil holes in the upper main bearings. Test fit the new upper main bearings to the block and check for oil hole alignment. If necessary, the oil holes in the upper bearings may be enlarged to achieve correct alignment. It's easier and quicker to do this than enlarging the saddle holes. Hole enlargement must be performed carefully to avoid damage to the bearing, and make sure that the new hole is carefully deburred to remove any high spots on either side of the bearing shell.

Whether you're modifying the upper bearings or not, once you obtain correct oil hole alignment with each upper bearing, mark each bearing for block location (number-1, -2, etc.) so that each bearing is installed in the same location during assembly. Mark each bearing backside with a felt-tip marker, or place each bearing in a clearly labeled plastic bag. In addition to checking for oil hole alignment, run a small-diameter rifle brush through each main saddle oil passage to make sure that there are no obstructions.

Threaded Holes

Check each threaded hole in the block for thread cleanliness and integrity, especially on a used block. A hole with contaminated or burred threads prevents you from achieving proper torque values and clamping loads during assembly. Prior to, during, and after block cleaning/washing, run the appropriate-size rifle brush through each threaded hole, along with a hot soapy water solution, followed by a rinse and a blow-out with clean compressed air.

When checking or correcting slightly burred threads, *do not use traditional cutting taps*; they are designed for removing metal. Instead, and especially on the most critical threaded holes, such as cylinder head holes in the block decks and main cap bolt holes, use a dedicated "chaser" tap. Chaser taps are designed to re-form threads instead of cutting threads. These taps are available through quality automotive machine shop supply companies such as Goodson Tools. Using a standard cutting tap, depending on the condition of the existing threads, can remove enough material to weaken the threads.

Threaded Inserts

If any less-critical threaded holes are stripped, cross-threaded, etc., use quality thread inserts to repair the hole. In basic terms, there are two common types of thread inserts: helically wound stainless steel (Heli Coil and other brands of the same style) and solid steel or stainless steel. In either case, first drill the damaged hole oversize following the spec provided with the intended repair insert. Then tap the hole, using the specific cutting tap supplied with the thread repair kit. Screw the thread repair insert into the newly tapped hole.

Install a helically wound insert using a specialty driver tool supplied in the kit. When fully installed with the top threads located immediately below the top surface, remove the driver tang at the bottom of the insert. The insert has a driver tang that's notched, which provides a stress point. A quick tap with a skinny, flat punch knocks the tang loose, which must then be removed from the hole. Install a solid-wall insert that has internal and external threads.

Solid inserts, depending on style, may or may not require application of a thread-locking compound to prevent it from backing out. Other styles may require staking at the top using a small, sharp punch.

There's nothing wrong with saving damaged threaded holes with inserts. Sensible applications include mountings for water pumps, oil pan, bellhousing, timing covers, etc. Thread-repair inserts also offer good methods of saving otherwise damaged holes for areas such as valve cover bolt holes, exhaust manifold/header flange bolt holes, intake manifold bolt holes, and spark plug holes.

High-tech threaded inserts are also available for high-stress areas, such as cylinder head bolt holes. Lock-n-Stitch makes reverse-angle-pitch threaded inserts that apply greater clamping force between the insert and bolt hole walls. This type of insert can actually result in stronger threaded connections than the original tapped holes.

The only real debate regarding thread inserts involves using dissimilar metals (for example, stainless steel inert in an aluminum cylinder head, especially in spark plug holes). Some builders claim that dissimilar metals can result in different rates of expansion and contraction during operating temperatures. My own experience with threaded inserts has always been successful. Others claim to have experienced problems with fastener loosening or spark plug performance. My opinion is that stainless steel inserts in aluminum components is a good idea, since the harder stainless steel material (as opposed to the parent aluminum) greatly helps to avoid galling or thread stripping issues.

CHAPTER 3

Block Machining

The engine block is the foundation of any build, and several critical machining processes are often required to bring the bores, deck, and dimensions of the block into precise and correct specification. When someone refers to an engine as balanced and blueprinted, this all too often means that the crankshaft has been balanced but the engine has not been blueprinted. Balancing the rotating and reciprocating components is simply one element in the process. Machining the block to achieve precise dimensions is the first and most important step.

Main Bore Align Honing

Once the block has been cleaned and flaw checked, the first order of business is to make sure that the main-bearing housing bores are properly aligned and sized to specification. Cast engine blocks, although seemingly stout chunks of cast iron or aluminum, have a tendency to "move" as the result of core-shifting because the block "seasons" through numerous heat cycles. In order to get the block to "settle" to a more stable condition, the block can be seasoned by use (and then machined and accurized), or the process can be hastened by treating the block to cryogenic or vibratory stress relief. This allows the molecules to settle and to become more uniform, reducing the potential for future stress-related movement.

The main bearing journals and caps are honed. After a few passes, the main bore diameters are regularly checked with a precision dial bore gauge.

High-quality aftermarket performance blocks may be stress relieved at the factory, which provides a more stable block over time. Heat and stress cycling over time improves the stability of the casting. For any used OEM block, the main bore must be reconditioned, even if no apparent damage is present. It should be obvious that the main bore must have perfectly round bores of the proper diameter, but beyond this, all the bores must be aligned to eliminate any potential crankshaft bind. Over time, an aged block may have tweaked enough to create a misalignment due to heat cycling or previous engine overheating (thermal-induced warp).

When aftermarket main caps are to be used, the radius is typically smaller than needed to provide the machinist with enough material to remove to accommodate a to-size main bore diameter.

Align honing is performed on a dedicated align-honing machine. With the honing mandrel centered to the block main saddles, the abrasive-stone-equipped mandrel is stroked fore and aft.

Aftermarket main thrust caps have a shallow thrust relief and must be finished-to-depth using a cutting bit on a mandrel to match the thrust area of the block's thrust saddle.

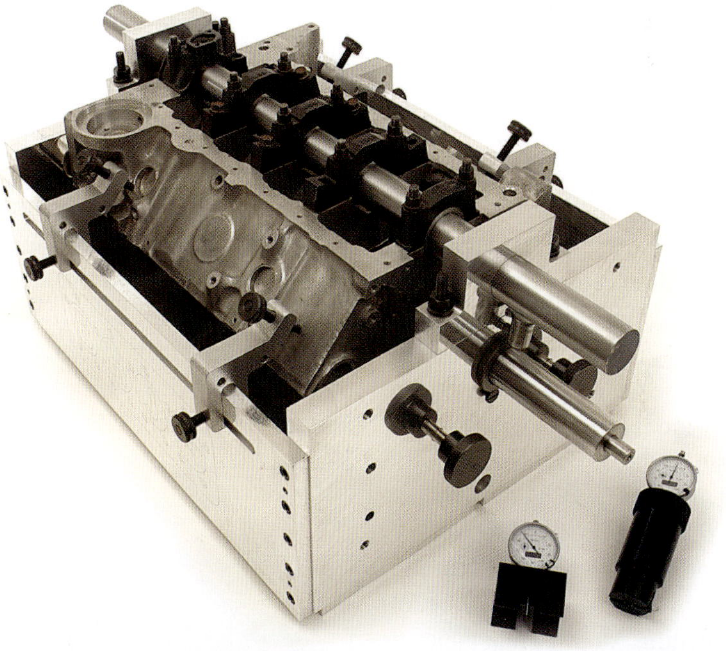

This is BHJ's LBF-1 (line boring fixture) for line boring main bearing or cam bearing bores, designed for roughing-in steel main caps or boring for installation of roller cam bearings. This allows obtaining correct cam tunnel location in all three planes, making the cam tunnel absolutely parallel to the mains at the correct center-to-center distance. (Photo Courtesy BHJ Products)

A typical procedure for any previously used block is to align hone the main bores. The block is rigidly mounted on an align-honing machine. A long honing mandrel that is fitted with abrasive stones passes back and forth through the main bores. In order to perform align honing, the bores must be reduced in diameter to allow material removal for achieving the original bore diameter. In order to accomplish this, use a cap grinder to grind the main bearing caps at the cap-to-block mating surfaces, removing about .003 inch from the caps. All housing bore diameters must be first established to exactly the same undersize before align honing.

The hone self-aligns to the housing bores and corrects any main bore distortion that may have been caused by block warpage and/or main cap stretch. Centering pins on the honing mandrel allow the mandrel to be positioned at the correct crank centerline. While rocking the mandrel on the centering pins, expand the honing stones until the unit doesn't rock. Remove the centering pins and install the main caps. Torque them to specification. Severe out-of-round condition can result if the mandrel is not properly centered.

With the honing machine's oil lubricating and cooling the stones, the mandrel spins and is stroked back and forth. After about five strokes, remove the mandrel and check the bores with a dial bore gauge. Continue the honing process until you achieve the final diameter (bores must be measured often to avoid overhoning).

The fore/aft movement of the crankshaft, which is the thrust clearance, deserves close attention. If the thrust clearance is excessive, the crankshaft eventually pounds out the bearings and the crank's journals. If thrust is too tight, the crank journals overheat and the bearings melt. A bad or incorrect torque converter on an automatic transmission often causes thrust bearing failure. If the converter body expands (pressure ballooning), moving by a mere .005 inch or so, it can apply excessive pressure to the rear of the crank, pushing the crank's thrust surface against the thrust bearing.

Once the main bores have been corrected, test fit the crank with bearings. Check thrust clearance by using the following procedure. Mount a dial indicator to the block and position the plunger to contact the crank snout or flat counterweight face. Move the crank rearward as far as it goes. Adjust the dial indicator with about .050-inch preload, and zero the indicator gauge. Move the crank fully forward and note the amount of movement. Compare this to the specification. Typical thrust clearance should be in the neighborhood of .004 to .006 inch. If thrust is too tight, the crankshaft's thrust surface may need to be reground to achieve the desired clearance.

If the block's thrust bearing surface is damaged, it can be refaced using an align-boring bar and a special tooling bit. If the block's thrust surface is resurfaced, a main thrust bearing with a thicker thrust face is required.

In order to bore (using cutters) or hone (using abrasive stones), mill the mating surfaces of the main caps by a few thousandths in order to create a slightly smaller, out-of-round condition (this allows material removal to achieve the final, proper bore diameter). Secure the main caps to the block, using the same main cap fasteners (bolts or studs) that will be used during final assembly. Torque the main caps to final assembly specification. This is critical.

Aftermarket performance steel main caps are generally made with a slightly undersized radius, allowing the machinist to establish a round hole of the proper diameter when the caps are fully fastened to the block. When aftermarket main caps are installed (or anytime main caps are replaced), first align bore the main bore to within about .005 inch of the desired final diameter. Then finish them by align honing to final size. When align honing, if the block and caps are cast iron, use a harder stone. For aluminum blocks fitted with steel main caps, honing requires a softer stone such as 150-grit J45 silicon carbide.

Deck Resurfacing

As the engine heats and thermally cycles, the head gasket must be allowed to move (slide) without grabbing/tearing. A surface finish of 60 Ra is generally okay for cast iron, but aluminum requires a smoother finish of about 12 Ra.

The block decks on a twin-bank block must not only be flat but must have the exact same deck height from the centerline of the main bore to the deck. Decks must also be parallel to the main centerline and must have the correct angle (90 degrees to the crank centerline). For blueprinting purposes, the block decks must be surfaced using specialty alignment fixtures such as those offered by BHJ, or the block must be surfaced on a programmed CNC machine.

Always keep in mind that removing material from the block deck changes deck height, which naturally affects piston deck clearance. Factoring in your

Resurfacing is best done with milling bits to achieve the desired surface finish.

crank stroke, connecting rod length, and piston compression distance, the block decks must be cut in order to accommodate the desired piston deck clearance and compression ratio. That's why most aftermarket performance blocks usually have taller decks to achieve exactly the block deck height required for a given setup. When dealing with an OEM/preused block, cleanup and accurizing the decks may results in less piston-to-deck clearance than desired, in which case you may need to order pistons with a shorter compression distance.

Although a surface grinder can be used, a milling operation is preferred for greater accuracy and because cutters don't create hazardous airborne dust. Typically, resurfacing cutter inserts are made of carbide, CBN (cubic boron nitride), or PCD (poly crystalline diamond). Carbide inserts are acceptable for surfacing cast iron or aluminum. CBN inserts are very durable and accept higher cutting speeds and feeds (for faster work time), but are best suited for cast iron. PCD represents more recent technology designed for surfacing all-aluminum blocks (alloy blocks with hardened cylinder wall treatment), but can't be used on alloy blocks that have steel or iron cylinder liners.

Inspect the decks for low spots that have not been cleaned up after deck resurfacing. The deck surfaces should be completely resurfaced, with no shadows (low spots). The head gaskets require 100-percent sealing contact.

Cylinder Honing

If the bores have taper wear and/or straightness/out-of-round issues are found, the bores must be enlarged to the next available oversize. Boring involves a dedicated boring machine (horizontal or vertical) and carbide cutters. Carefully measure your piston skirt diameter. Use the piston manufacturer's specified location on the piston. Factor in required piston-to-wall clearance. Once again, refer to the piston manufacturer's recommendations based on type of piston and type of application. Rough boring should be done to a smaller diameter than the finished size. In general, leave about .003 to .005 inch, which is then removed during final honing.

If a very slight oversizing is needed (a .005-inch oversize, for example), you can hone rather than machine the bores. To oversize hone the bores, start with 70-grit aluminum oxide or 100-grit diamond metal-bond stones. This leaves coarse scratches on the walls, which are then removed during final honing. If using 70-grit stones to rough hone, leave about .003 to .005 inch of material. If using 100-grit diamond stones, leave about .005 to .007 inch for final honing.

During either boring, rough honing, or final honing, stop to periodically check bore diameter with a high-quality dial bore gauge. Measure at a minimum of three locations in the bore (top, middle, and bottom) and in two directions 90-degrees apart. Taper should not exceed .001 inch, and out-of-round should not exceed .0005 inch.

Honing stone type (stone hardness) can affect bore geometry, so always check with the honing machine or honing stone manufacturer for recommendations regarding stone hardness for your block application.

If you're faced with slight out-of-round, using a softer stone can be beneficial. Thin-wall blocks may distort when using harder stones. Unsupported sections of the cylinder may tend to push out, resulting in less material removal, which results in a tight spot for the rings. Using the correct honing stones for the block material, and following the correct pressure and feed rates minimizes bore geometry problems.

Always final-hone cylinder bores by first installing deck plates to the block in order to simulate the stresses that the block sees when the heads are installed. The deck plates must be installed along with a precrushed (used) head gasket, and the fasteners must be torqued to the proper specification. Depending on the type of block and block material, as much as about .004 inch of bore distortion can occur when the cylinder heads are mounted and the head fasteners are fully torqued. Deck plates mimic this clamping load and bore distortion, so when the heads are installed, the bore geometry is established in a more uniform manner.

Cylinder Boring

Inspect the condition of the cylinder wall surfaces. If scratches or scoring are evident and more than .025 inch deep, overboring is necessary. If there isn't enough wall thickness for moving to the next size overbore, the likely option involves sleeving or replacing the block. During an overbore, the cylinders are bored to a diameter that is slightly less (or tighter) than the desired final diameter. This leaves enough material for honing, during which the final diameter or surface finish is achieved. An overboring operation typically results in an undersize of about .005 to .007 inch, leaving this amount to be removed during the honing process.

The main caps must be installed and fully tightened to spec, and then the block can be placed in the honing machine. This simulates the stresses introduced into the block, which affect cylinder bore geometry. Torque the main cap fasteners to specification and follow the torquing sequence used for final assembly. Also it's best to use the same fasteners that will be used during final assembly.

If you plan to use main cap studs, install them now, prior to honing. When installing studs, do not overtighten them into the block. The clamping load of caps to block is achieved when tightening the nuts. The studs should be installed finger tight, with an added nudge of no more than 10 ft-lbs. In every case, read the stud manufacturer's instructions regarding installation and any required preload. If the main caps have side bolts in addition to primary cap fasteners, be sure to install them as well, again following the recommended torque and sequence specs.

Always use a deck plate or torque plate for the honing process. It bolts to the block deck and is torqued to the same spec required for the cylinder head mounting. This plate simulates the installed cylinder head and places similar stresses inside the block, which affect cylinder bore shape. If you hone the cylinders without a deck plate, you may achieve a nice round hole, but when the heads are bolted on, some cylinder walls can distort into an out-of-round (or barrel) shape. By using a deck plate, you're making an effort to establish consistent round holes in the assembled state. When bolting a deck plate to a block, use a crushed head gasket of the same type that you plan to use during

Cylinder overboring is performed with high-speed cutters. Again, for purposes of block blueprinting, the best method is to take advantage of CNC programming to create bores that are precisely on-center per design specs.

Cylinder honing must always use deck plates. Deck plates simulate the installed cylinder heads, placing stresses in the block that it sees when assembled. This greatly improves cylinder wall geometry because the stresses imposed by clamping the cylinder head fastener slightly distort the bores slightly. When installing deck plates, precrushed head gaskets must be in place, and all cylinder head fasteners must be torqued to the same value (and in the same sequence) as the block experiences during final assembly.

Honing-stone pressure and feed rate is regularly monitored during hone-stroking steps. In between passes, a dial bore gauge is used to monitor current diameter and to check for out-of-round and taper. In order to maximize honing accuracy, cylinders are honed in a staggered sequence because of the heat generated by honing. For instance, after honing number-1 cylinder, honing commences at a cylinder location away from the first bore (the order might be number-1 cylinder, -3, -2, and -4). This gives each honed bore a chance to cool before honing the adjacent cylinder location.

Initial honing creates a microscopic series of peaks and valleys. The peaks, the high edges of the honing scratches, present a wear-in/break-in challenge for the rings.

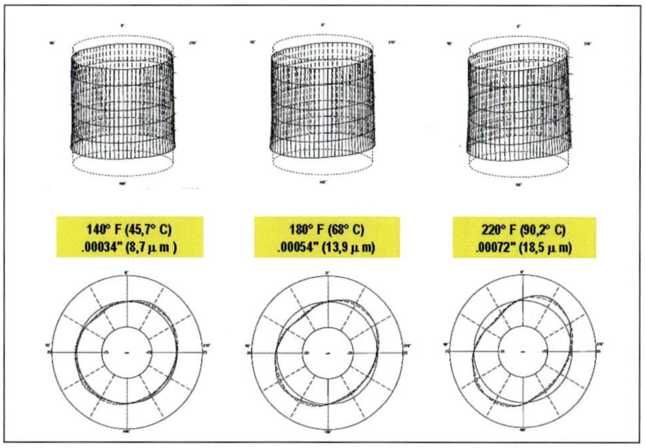

You may think that an engine block is a beefy, solid chunk of material, regardless of the block (OEM, aftermarket, cast iron, cast aluminum, or billet aluminum). However, while cylinder bores may be machined perfectly round from top to bottom, cylinder shape becomes "alive" under mechanical stress and thermal changes. Bores do change their shape during engine operation, however minimal. In the attempt to obtain maximum ring performance, it's important to understand this phenomenon and to do our best to address this and minimize these changes. (Photo Courtesy MAHLE Clevite)

assembly. Remember that you're trying to simulate the condition the block faces when fully assembled.

Depending on the design of the block, also consider other highly-stressed, component-mounting locations that can affect the shape of the cylinder bores. Here's a good example: A few seasons ago, my endurance race team ran a pair of Dodge Neons with the 2.0L Chrysler engines (4-bangers). As usual, we first honed the cylinders using a deck plate. After running the engines in track test sessions, we tore down the engines for inspection. We noticed that number-1 and -2 cylinder bores had retained their shape Number-3 was slightly out of round by about .0006 inch. Number-4 cylinder bore had been pulled out of round by about .015 inch, especially near the upper half of the bore. The reason was that the transmission bolted to the rear of the block with the upper two bolt locations very close to the rear cylinder.

After discovering this, subsequent blocks were honed with not only a deck plate, but with the addition of a thick plate bolted to the rear to simulate the installed transmission. After running a 24-hour race, teardown showed minimal out-of-round at number-3 cylinder at about .0001 inch and number-4 to the tune of about .0003 inch—a huge improvement.

Certainly, stressing the rear of the block during honing isn't needed for every engine block, but it's food for thought. Study the block and try to evaluate the need for stress simulation in various areas that could potentially affect cylinder bore geometry.

If you want to go a step further, consider mounting the engine mounts and the water pump as well, simulating all the mechanical stresses that a block experiences. Admittedly, for a street application, aside from a deck plate, stress simulating with other added parts is just going to be a waste of time. However, even for a race engine, where you're trying to squeeze every bit of power and durability out of that engine as possible, going to such lengths probably doesn't provide a real-world benefit, but in theory, it can't hurt.

Before honing, take a close look at the bottom of the cylinder bores. Depending on the specific block, it may be necessary to use a hand grinder to knock a bit of material from the webbing surfaces below the bores (for example, the center three cylinder areas where the

Most aluminum blocks have iron cylinder sleeves. OEM production blocks (a GM LS2 block is seen here) are notorious for featuring a rather sloppy tolerance range for cylinder-bore centerline locations. These bores were honed to size; notice the difference in liner thickness from one side to the other. This is due to small amounts of movement in the casting core. Because these liners are fairly thin to begin with, and due to the slight offset of the existing centerline, OEM blocks with integrally cast-in liners can only be slightly overbored or overhoned (usually about .005 to .010 inch oversize is all you can hope for).

Block accurizing on a CNC machine (providing the program is written for the block at hand) is relatively easy and requires no additional guide/indexing fixtures. Here a digital probe obtains existing cylinder-bore locations. With the theoretical centerline already programmed, the boring operation takes only a few minutes for each bank.

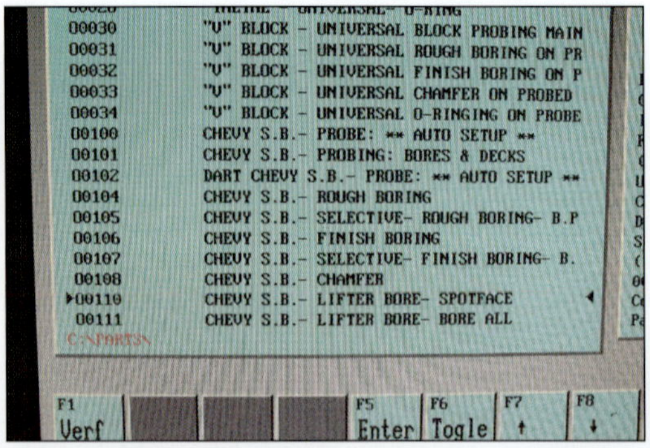

Several CNC equipment manufacturers provide already-written programs for popular blocks, including not only OEM blocks, but also several popular aftermarket performance blocks. An experienced machinist is also able to write a specific program for a given block, and is able to adjust machining operations from the prewritten programs.

Before honing, check the bottom of each bore. If a ledge or hump exists at the bottom of the bore (casting material of the main web area), the machinist should remove the obstruction with a hand grinder to prevent the honing stones from hitting this at bottom dead center.

With the block indexed at the crank centerline, an overhead cutter makes short work of deck surfacing that is parallel to the main bore and with each bank's deck equidistant from the main-bore centerline.

Each cylinder is constantly measured for roundness prior to and during the honing process. Diameter checks are measured near the top of the cylinder, at the middle height area, and at the bottom.

bores Siamese and have small humps in the casting), to prevent damage to the honing stones. This takes only about 10 seconds per spot, so it's no big deal.

Depending on the block material, the machinist must select the appropriate grade of honing stone. For a high-nickel-content block (Dart iron blocks, for example), you may need to begin with 500-grit diamond stones with a high load setting to hog out the bores to an initial diameter, followed by a final honing pass to remove the remaining material. Following honing-to-size, all cylinders should then be treated to four passes with silicon carbide brushes at 30-percent load for a plateau finish. A plateau finish essentially "evens out" the minute peaks and valleys created by the honing process. This provides a more uniform micro-finish and aids in faster ring seating as well as superior oil retention.

Cylinder honing must be performed on a dedicated honing machine (manufacturers include Sunnen, Peterson, Kwik-Way, and others.), where stone diameter adjustment, stone load pressures, and dwell time are easily adjustable and monitored. Do not attempt to hand-hone your cylinders. This isn't a backyard operation.

Never assume anything. Whether you plan to use OEM or aftermarket pistons, always measure each piston skirt diameter at the skirt height location specified by the piston manufacturer, for your final cylinder diameter, including desired piston-to-wall clearance. Final cylinder diameter is always based on the actual piston-skirt diameter.

The plateau brushes finalize the cylinder wall finish by averaging-out the miniscule peaks and valleys left by the honing stones. This provides faster and more consistent piston ring seating/sealing (the rings break-in quicker) and aids in cylinder wall oil retention.

Honing coolant is constantly supplied as the honing stones rotate and travel through the cylinder, keeping the stones clean and reducing heat buildup.

Plateau Finishing

Plateau finishing (or plateau honing) is a popular final step following finish-honing-to-size. Using dedicated plateau brushes on the honing machine shaves the tiny peaks left by honing scratches, to provide a better surface for the rings. Essentially, this honing step "breaks in" the cylinder wall finish (evens out the peaks and valleys), which provides a more uniform surface finish while maintaining proper ring bearing area for oil control and ring lubrication. This process also extends ring life (since the rings aren't forced to wear off these peaks). Today's piston rings are lapped at the factory for quicker break-in/seating, and don't require a rough cylinder-wall finish for break-in. Plateau honing immediately follows final honing-to-size and only requires a few short passes using 150- or 220-grit stones. This is followed by plateau finishing.

Cylinder Bore Distortion

An engine block's cylinder bores do not remain round and true during engine operation. Even though the cylinders may be machined perfectly round, during engine operation (and especially in the case of severe-duty operation such as racing), the cylinder profile can easily change due to heat and cylinder pressures. These changes also result from molecular changes in the block material as it ages or seasons due to thermal expansion and contraction. This is referred to as cylinder bore distortion. This is a naturally occurring phenomenon.

The goal of an engine blueprint job is to recognize this and attempt to minimize bore distortion. Cylinder bore distortion also results from engine assembly. As the cylinder heads are installed, stresses are placed on the block as the cylinder head fasteners are tightened. The pulling force that results from tightening the head bolts (or tightening the nuts on head studs) can cause slight shifts in the cylinder walls, which can lead to less-than-ideal piston ring seating. Much of this is dependent on the block material, placement of the cylinder head bolt holes, cylinder wall thickness, etc. Some blocks are more susceptible to geometric shifts in the cylinder bores than others. In addition to the stresses imposed by the cylinder head bolts, tightening bellhousing bolts and other attached components can affect the geometric shape of the block.

A good example of bellhousing-induced cylinder bore distortion relates

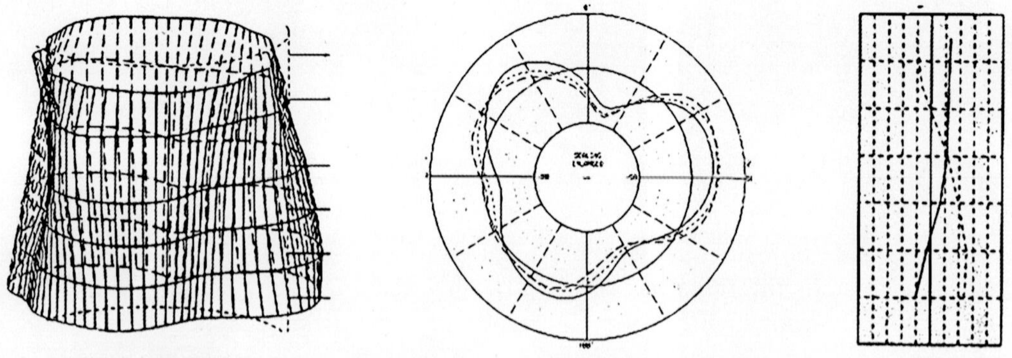

Three views of a typical dynamic bore distortion scenario. (Photo Courtesy MAHLE Clevite)

to the four-cylinder Chrysler Neon engine block. When my team ran a pair of Neons in 24-hour races, we found during post-race teardown that the rear cylinders (number-3 and -4) showed clear evidence of bore distortion, with unevenly worn piston rings and obvious pressure points in the cylinder walls. Even though the blocks had been honed with a deck plate torqued to the block to simulate the cylinder head, the stress imposed by the bellhousing bolts had pulled the two rear-most cylinders slightly out of round. Once we were aware of the problem, subsequent blocks were honed with a deck plate and a bellhousing plate, both torqued to assembly specification values. By simulating the final assembled stresses at the head deck and the rear of the block, our piston ring seating remained much more consistent, and blow-by and oil loss was nearly eliminated.

Even though you may think that an engine block is a massive and strong component, it does experience movement that can and does affect the geometry of the cylinder bores. Regardless of the type of engine, it's vital that you simulate, to the best of your ability, the assembled stress that the block experiences before the cylinders are final-honed to size.

Engine development researchers often take advantage of sophisticated laboratory equipment that allows them to "map" a cylinder bore in order to obtain a clear dimensional picture of how that bore is shaped, from top to bottom. APAT gauge system (inner contour meter) travels vertically through the cylinder bore centerline. It has a sensitive probe that monitors the cylinder bore, which provides a dimensional view of the entire bore. This provides an overhead radial view and an isometric view at various angles

Since most engine shops don't have access to this level of equipment, bore diameter checks are made with a calibrated bore gauge before, during, and after cylinder honing. Bore diameter measurements are taken from top to bottom of the bore in four height locations, and at four clock positions (12, 6, 3, and 9 o'clock as viewed from overhead). If you measure a cylinder with the block in a relaxed state (no deck plate), and then take measurements at the same locations with a deck plate installed and fully torqued, it is very common to find different readings. This is clear evidence that the stresses caused by the cylinder head can affect cylinder bore geometry. Once you realize this, it should become clear that you should always hone cylinder bores with a deck plate installed.

An additional factor that can affect cylinder bore shape relates to the cylinder head. Depending on the hardness level of the cylinder head material, different levels of stress can be placed on the block, although they are beyond your control. This becomes important when swapping heads (for example, changing from cast iron to aluminum) because it could result in a change, however slight, to cylinder bore geometry.

Cylinder bore shape can and does change when bolts are tightened and during engine operation. You can minimize the effect of cylinder bore shape changes by prestressing the block during cylinder honing. In addition, always follow the same routine with regard to component installation. For instance, the same torque level and the same tightening sequence/pattern should be followed every time a cylinder head is installed. This includes the installation of the deck plate for honing, and every time the cylinder head is installed to the block. Maintaining this consistency of head bolt tightening helps to minimize changes in stresses to the block.

A deck plate should be installed and torqued to the required spec for both cylinder banks before a hone is performed.

On a V engine, that means installing a deck plate on both banks. From a practical standpoint, however, most engine shops may only have one deck plate for a given type of engine. Using the same deck plate for both banks is common practice and certainly is acceptable. But if two identical deck plates are available, it's a good idea to install both at the same time.

Also, always install a head gasket along with the deck plate. This further simulates the stress that the block sees when assembled. Be sure to use the same type of head gasket that will be used during final assembly, whether composite or multi-layer steel (MLS). Preferably, use a head gasket that has already been crushed (a used one). It doesn't "settle" as much as a new gasket.

The type of head gasket is a consideration. MLS gaskets offer greater consistency of clamping load than a composite gasket with a wire ring "crush zone" around each cylinder bore. Although an MLS gasket typically relaxes by fewer than 10 percent after initial installation, a wire-ring type of gasket can relax by as much as 15 percent or so. In other words, MLS gaskets provide greater consistency in maintaining the clamping load between the head and block.

Two types of MLS gaskets are available: active and stopper. An active MLS cylinder-head gasket has a series of metal layers with gaps between the layers and an elastomer layer that allows the gasket to compress and seal with a slight spring effect. This helps to reduce block and head deck distortion. Stopper MLS gaskets include a "dead-stop" layer and rigid cylinder bore seals, which provide additional sealing under high cylinder pressure conditions. However, the stopper MLS gasket places more distortional stress against the cylinder head. When deciding between the two styles, it's best to follow the recommendation of the gasket manufacturer for your specific application.

Hot Honing

I've already discussed the importance of prestressing the engine block prior to honing the cylinders. The importance of mechanically stressing the block with a deck plate is paramount. Always hone the cylinders with a deck plate installed. With that said, you can also consider the stress imposed as a result of heat. Although the block experiences stress that affects cylinder bore shape when assembled (primarily as a result of cylinder head installation), cylinder bore geometry is also affected by engine operating heat.

In the pursuit of optimal shaped cylinder bores during operation, some builders use a process called "hot honing." In simple terms, the block is connected to a heater that circulates hot water at a predetermined temperature. Cylinder bores are then honed with a deck plate and with hot water that elevates the block to a temperature that approximates operating water temperature. In theory, this more closely mimics the stress and temperature that the block experiences when it runs. The practice of hot honing is embraced by some engine builders and viewed as unnecessary by others.

As engine temperature increases, a cylinder wall commonly tends to assume a slight barrel shape as a result of thermal expansion. This expanded area has inconsistent high and low spots. The result is a cylinder that is slightly out of round and no longer uniform. However, the engine's pressurized cooling system tends to partially counteract this as the water jackets that surround the cylinder try to push against the cylinder. If the builder plans to use a hot-honing process, the coolant must not only flow through the block, it must also be pressurized.

Lifter Bores

Damaged lifter bores can be overbored, fitted with bronze liners, and then honed for proper lifter oil clearance. In addition, lifter bores can and should be corrected to attain the proper centerline and angle by using specialty indexing fixtures or by CNC machining. OEM blocks typically require these procedures because sloppy factory machining tolerances are common in mass production. But high-quality aftermarket performance blocks already have correct geometry, and may require only final-honing for oil clearance.

If you're dealing with an old OEM block, pay attention to each lifter bore. Just because the lifter bores don't look scored doesn't mean that they're all

A CNC cutter spot faces lifter bore roofs in preparation of lifter bore centerline correction. With a programmed CNC machine, no indexing fixtures are needed for lifter bore correction.

the same size. When an OEM block is machined to correct lifter bores, all bores are likely machined at the same time using multiple tooling. On occasion, one or more lifter bores may have been machined slightly oversize (to correct tooling flaws), in which case the factory may have installed oversize lifters in certain locations.

Measure each lifter bore for diameter. If you have oddball sizing, your best solution is to have the lifter bores accurized. The lifter bores should be oversized using the theoretical (correct) bore centerline to correct any centerline off-center and/or lifter bore off-angles. The lifter bores *must* be in the correct plane to the camshaft. OEM lifter bores are usually held to within .010 inch or so of plane, but this can be corrected by realigning, using specialty fixtures or on a CNC machine. The bores are then fitted with bronze bushings and honed to size to achieve the oil clearance needed for the lifters.

Also pay attention to the lifter bore finish. If the block was shot peened, the lifter bores may have been peened-over. The resulting "dimples" reduce oil clearance. Honing the bores with a rigid hone allows you to control material removal. A brush-type hone skips over the dimples and doesn't solve the issue.

Lifter bore correction and/or reconditioning involves overboring the lifter bores using a precision reamer, followed by installing bronze sleeves that are then honed to size. For blueprinting purposes, the overboring is best accomplished by using a specialty lifter-bore accurizing fixture (such as one from BHJ) or by CNC machining. Either approach allows the machinist to correctly place lifter bores at the designed on-center axis and correct for any factory deviations in lifter bore angle.

Before installing bronze lifter bore inserts, oil feed holes must be drilled into the inserts (inserts are also available predrilled). Depending on the type of block involved, it is common to use a smaller oil feed hole to improve oil delivery to the main bearings. The example seen here is a Pontiac 455, where the excessively large lifter bore oil feed holes are reduced to about .040 inch.

Camshaft Bore

As with the main bore, the crankshaft bores must be accurately aligned to prevent camshaft bind and isolated camshaft bearing wear. Another aspect relates to crank-to-cam bore centerline distance and parallelism. Quality aftermarket performance blocks are CNC machined to accurately locate both centerlines, but OEM blocks may often have centerlines that are a bit off, or have a slight non-parallelism between the main bore tunnel and the cam bore tunnel. This aspect relates to the blueprinting/accurizing concern.

If a cam tunnel centerline is incorrect, or if the cam bore locations are not aligned, it can be corrected by centerline align-honing the cam tunnel in relation to the main bore. Since (in an overhead-valve engine design) there are no camshaft caps to shorten in order to recreate the original bore diameter, the cam bores can be bored or honed oversize, requiring oversize-OD cam bearings (thicker shells to accommodate the original cam-journal diameter).

CHAPTER 4

OILING SYSTEMS

Your engine must provide a strong and steady supply of oil to all critical components. In essence the engine must deliver the correct volume of oil under a certain pressure to reach all the critical components during operation. If this does not happen, the engine experiences oil starvation and this obviously degrades performance. The engine's vital components are also damaged, and this could lead to outright engine failure. In this chapter, I cover specific considerations for retaining and optimizing a stock-type wet sump oiling system and also the benefits of upgrading to a dry sump oiling system.

A "wet sump" system supplies pressurized oil to the engine's rotating and reciprocating assemblies. Engine oil is stored in the big reservoir section of the oil pan. A mechanically driven oil pump picks up the oil, which obtains the sump's oil via a submerged oil pickup. Depending on engine design, an intermediate shaft connecting the distributor shaft to the oil pump may drive the wet sump oil pump, or the crankshaft snout may drive a crank-mounted gerotor-style oil pump.

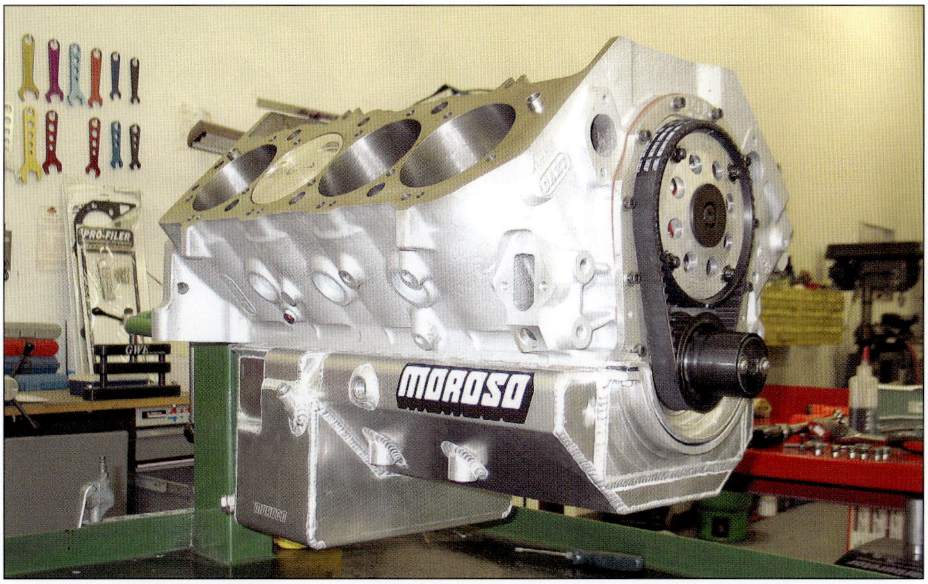

High-performance engines often benefit from a large-capacity custom oil pan. This oversize pan is installed on a Dart Big-M block fitted with a 4.750-inch stroker crank.

GM LS engines, as one example, have a crank-driven pump. Pressurized oil is distributed from this main source throughout the engine through the oil passages in the block, crank journals, main bearings, rod bearings, and eventually to the valvetrain. This system delivers oil to all required areas. In addition, the oil must be pushed through all of these passages to eventually route to the upper end of the engine. The oil is then free to drain back to the sump, with delivery and drain-back serving as an ongoing cycle during engine operation.

Wet Sump Systems

In a wet sump design, the pump's "pickup" is immersed in oil at all times and has a filtering screen. Depending on pump, engine, and oil pan design, the pickup has a short or long pickup tube that connects the pump to the pickup. If the oil pump is located directly over or very close to the oil pan's sump, the tube length is short. If the pump is located at one end of the block but the sump is

In-pan or wet sump oil pumps include (or require separately) a pickup assembly designed to be submerged in the sump reservoir. The distributor drives the pump, which is driven at half of crankshaft speed. In some cases, such as the old flathead Ford, the pump is driven by gears that engage with the camshaft gear.

Front-mounted crankshaft-driven oil pumps are commonly used on engines with distributorless ignition, and are driven at crank speed.

Billet aluminum pumps, designed for racing applications, have an integrated pickup and screen that require no external pickup tube. Because the pump bottom must be immersed in the oil supply, pump body heights are offered for specific oil-pan sump depths. This style eliminates the possibility of pickup tube leaks or vibrational damage.

located at the opposite end, a longer tube is needed to locate the pickup deep inside the sump.

When fitting the pickup to the pump, test fit to make sure that the pickup is located close to the floor of the pan sump. Clearance should be approximately 5/16 inch or so between the pickup and sump floor. If the oil pump design has a press-fit pickup tube, press the tube into the pump using a pickup tube installer tool. This has a crescent-shaped head that allows you to capture the tube's swedged end. With the pump mounted to the block, use this tool and a clean hammer to install the tube to the pump. Measure the distance from the oil pan rail to the sump floor. Then adjust the clock position of the tube at the pump to place the bottom of the pickup close to the pan's sump floor. It measures from the block's pan rail to the bottom of the pickup. With the pickup adjusted, use a marker to place matchmarks on the pump and tube. Remove the pump from the block and tack weld or full weld the tube to the pump. This ensures that the pickup tube doesn't move or loosen.

If the oil pump is crank driven, the pickup tube has a slip-in fit to the pump and is sealed with an O-ring. A mounting tab on the tube bolts to the pump body. If you're using a stock-type oil pan, no adjustment is needed because the pickup tube assembly is bolted down in a fixed position. The tube has another mounting tab or bracket that likely bolts to one of the main caps. Regardless of whether

Depending on design, the timing set must be installed prior to pump installation. Pictured here is an LS2 short block. Note that the pump is mounted ahead of the timing gear.

When interference-fitting a pickup tube to an in-pan pump, a special open-center tool is required for installation.

Once the pickup has been interference-fit to the pump and adjusted for pickup depth (to suit the pan), the tube should be welded to the pump body.

the pump is distributor driven or crank driven, if you're using an aftermarket oil pan with a deeper sump, you must use the specific pickup assembly recommended by the pan manufacturer.

Another type of gerotor pump is the billet style. It bolts to the block and has a built-in pickup with no external pickup tube. This style of pump is available for certain applications in which a specific oil pan is to be used. A big advantage of this type of pump is that there's no press-fit pickup tube that might loosen up during engine operation, so there's no concern about sucking air into the oil system due to a loosened or broken tube.

If you've decided to use an aftermarket oil pan with a deeper sump, pay attention to the oil level dipstick. Don't simply assume that a dipstick "made for that engine" has properly placed level

Crank-driven oil pumps have a driven gear that have a series of broad teeth.

The pump's driven gear slips over and engages the drive gear, which installs onto the crankshaft snout. This OEM-LS drive gear has an integrated timing chain gear.

Billet race pumps are easily disassembled for inspection and cleaning. As with traditional in-pan pumps, pressure adjustments can be made by changing pressure relief springs.

A crank-driven oil pump likely has a lone pickup tube that is necessitated by the location of a rear-pan sump. This type of pump has a pickup tube that bolts to the pump and is sealed with an O-ring, so no interference fitting or welding is needed.

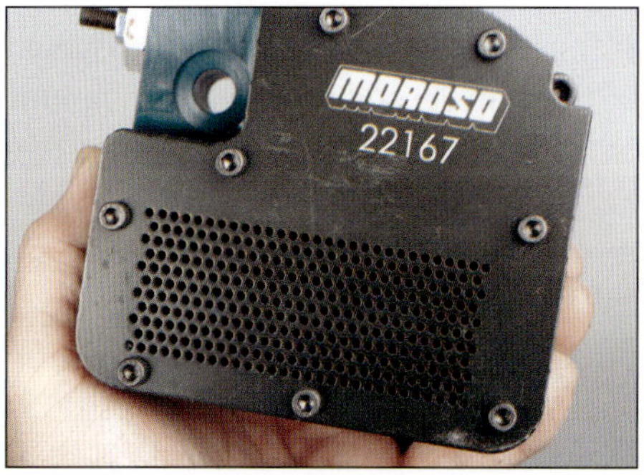

Billet oil pumps have a built-in pickup screen.

If the pump is located far from the sump, the tube will be long and require at least one additional support bracket to support the weight and length of the pickup tube. It is usually fastened to a main cap bolt or stud. Here is a stock type pickup on a GM LS engine. The sheet-metal tray seen here serves as a windage tray, which protects the crankshaft from oil splashback from the pan. This reduces parasitic oil drag on the crank counterweights and rod big ends.

The big advantage of the billet pumps is the elimination of a separate pickup, so pickup tube damage such as loosening or cracking due to engine vibration is eliminated.

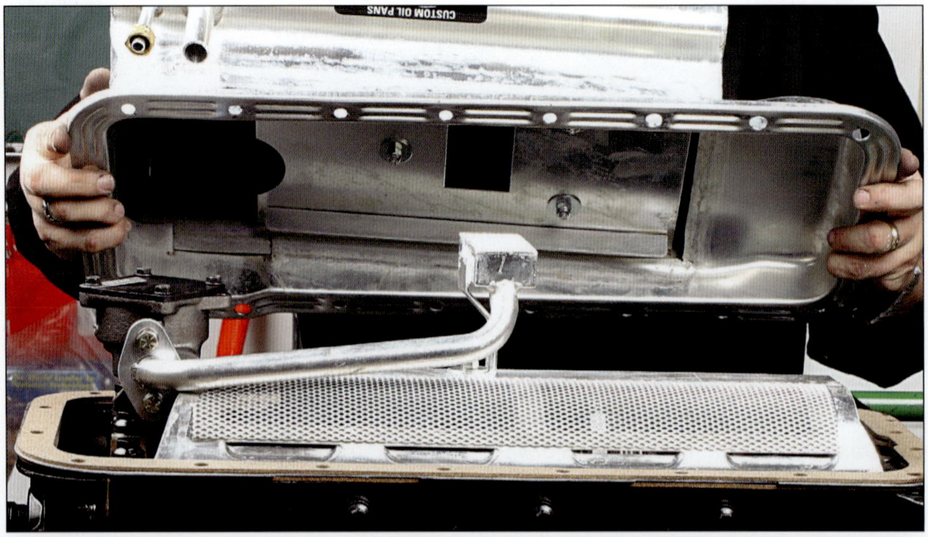

A screened windage tray reduces oil splashback while permitting faster oil drain to the sump. A built-in baffle/windage tray is also featured on this pan. Note the square passage hole for pickup entry. Vertical baffle wall(s) in the pan prevent sump/pickup starvation during severe acceleration, braking, and/or severe lateral turns, depending on design.

indicators. With the engine upright and at the level to which it would be installed in the vehicle, add the specified amount of oil that your pan calls for. Allow a few minutes for the oil to drain into the sump. Insert the dipstick and compare the fluid level to the marks on the stick. You may place different marks on the stick, or you may find that you need a longer stick. Aftermarket dipsticks are available that can easily be customized for fit. Flexible (woven stainless steel cable) sticks have a swedged-on tip that can then be cut to length at the top end. The upper end is secured in a billet-aluminum handle with a set screw.

Dry Sump Systems

A "dry sump" system uses a more direct manner to supply oil to the engine. A belt drives an externally mounted oil pump that uses external plumbing to flow oil to the engine. For a dry system, the sump is not placed in the oil pan; rather, it uses a remote reservoir. It's typically mounted in a variety of locations under the hood of a street vehicle or in the cockpit of a race car. In very basic terms, the external pump draws oil from the remote reservoir and forces the oil through the plumbing. In this manner, oil can be directly sent to the main bearings through a block port and delivered directly to the valvetrain with one or two plumbed hoses, then directly delivered to a turbocharger, for example.

Dry sump pumps are offered in several configurations, depending on how many direct-delivery and return routes are required. This provides direct oil delivery to specific areas, and elimination of oil starvation caused by angle and centrifugal forces placed on the car. Thus, the oil pump never sucks in air, and the potential for oil aeration is eliminated. The scavenge section(s) of the dry sump pump draws oil from the dry sump pan. Vacuum is created and that draws excess oil from the surfaces of the crank and rods, and this mitigates parasitic drag. As a result, the engine operates more efficiently.

When plumbing a dry sump system, size requirements may vary, but a general rule is to use –10, –12 and –16 hose, fitting, and hose end sizes. For some turbo applications, a –6 hose size may be recommended to feed the turbo. Generally, a larger size is required for the return hose that runs from the pump to the remote oil reservoir. For instance, if –12 hoses are used for all feeds, a –16 hose is used for oil return to the reservoir.

Dry sump oil systems use an externally mounted oil pump, a special pan design, and a remote oil reservoir from which the pump draws oil. External plumbing allows direct oil delivery to specific areas of the engine. Oil returning to the shallow pan is then pulled, or scavenged, back to the pump, which returns the oil to the remote reservoir. Dry sump systems allow you to deliver oil directly to specific areas of the engine instead of relying on one central pumping source to send oil throughout the engine. Since a dry sump system stores the oil in a remote reservoir, the pan is used simply as a seal to the bottom end, resulting in a shallower pan for improved ground clearance.

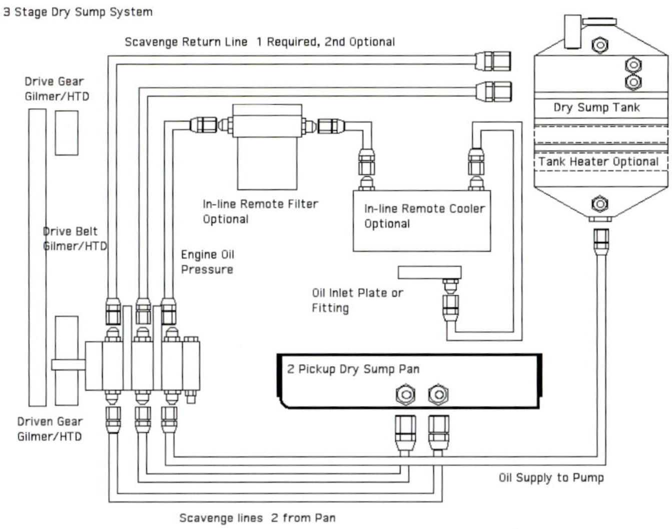

Typical belt-drive dry sump setup (three-stage shown here). Oil is drawn from the remote reservoir to the pump, which feeds oil to the engine. Scavenge plumbing pulls drained oil from the pan and sends it back to the remote reservoir. (Illustration Courtesy Aviaid Competition Oil Systems)

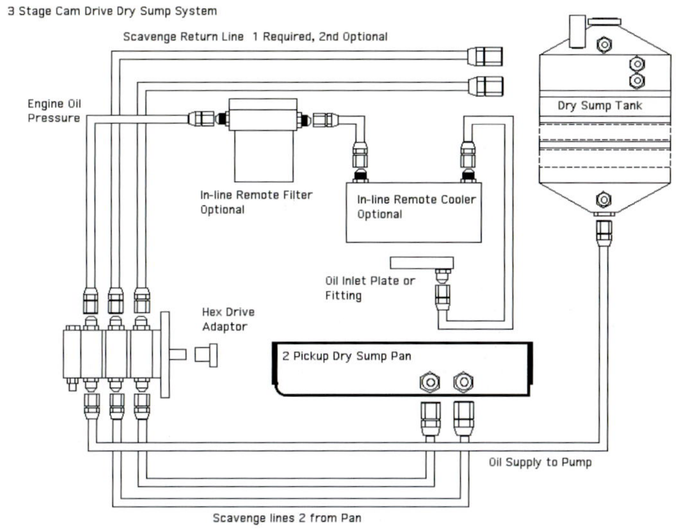

Another example of a three-stage system, but here the pump is driven by the camshaft instead of using a belt drive. (Illustration Courtesy Aviaid Competition Oil Systems)

Threaded bungs are installed on a dry sump oil pan for suction line fittings in which the pump draws drainback oil. Tapping and fitting special adapters may be required for plumbing to the block or other areas. The aftermarket typically offers a variety of adapters for all popular blocks to supply oil to the block's main galley at the stock oil filter location. By the way, if you're using the OEM filter location for oil feed, you remotely locate an oil filter.

A pulley on the crankshaft snout directly drives a notched belt that connects to an externally mounted dry sump oil pump with most setups. Dry sump pumps, via the proper pulley diameters, are generally driven at about half the crankshaft speed. Dry sump pumps are also available that are driven by the nose of the camshaft, with the pump mounted on the face of a special timing cover. A dry sump system has many advantages: It has consistent oil pressure; the oil pickup does not become uncovered and starve the engine for oil in severe turns, acceleration or braking; oil pressure is adjustable; capacity is increased; the short-depth pan allows lower mounting of the engine in the chassis; and positive oil delivery to vital engine components. The engine also enjoys a cooler oil supply because the oil is quickly returned to the remote reservoir instead of being stored in the hot oil pan.

TECH TIP: Dry Sump Stage and Section Designs

Type	*Pressure Sections*	*Scavenge Sections*
2-stage	1	1
3-stage	1	2
4-stage	1	3
5-stage	2	3
6-stage	2	4

TECH TIP: Dry Sump Oil System Components

- Dry sump pump
- Dry sump pump pulley
- Toothed pump drive belt
- Pump fittings (male –AN 37-degree flare to accept female –AN hose ends)
- Pump mounting bracket
- In-line screened oil filter(s)
- Remote in-line oil filter and filter mount
- Remote-mounted oil tank
- Breather (separate or on-tank)
- Oil filter block-off plate for engine block
- –AN hose assemblies as required

Sections and Stages

A dry sump pump has two sections. The pressure section delivers oil to the engine. The scavenge section pulls "leftover" oil from the dry sump pan and sends it back to the remote oil reservoir.

Dry sump pumps are built in stages, with one pressure section and one or more scavenge sections. The additional scavenge sections (as few as one, as many as six) allow oil to be scavenged more quickly and efficiently from specific areas of the engine, instead of waiting for the oil to be drawn into the pan for scavenge pickup.

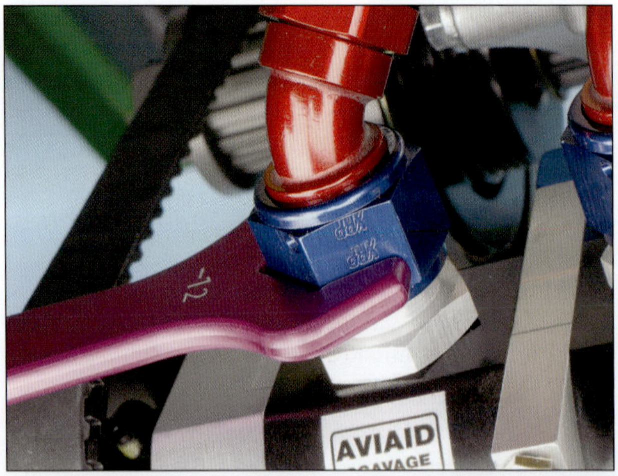

Whenever dealing with aluminum (-AN) fittings and hose ends, always use a dedicated aluminum wrench to avoid damaging the hose ends and fittings.

Dry sump pumps have a series of individual pumps stacked together. These individual pumps are referred to as stages. Pumps are available in two-, three-, four-stage, etc., depending on how many individual oil feed and scavenge lines are required for a particular engine.

Oil Pump Service

Any metal or foreign debris that enters the oiling system and travels through the engine leads to failure. You need to service a dry sump system during the rebuild process. This includes disassembling, cleaning, and inspecting the oil pump for damage.

You can plumb in-line filters in the oil return lines during system installation and the filters provide important protection. However, don't rely on the filter alone. Always inspect the pump.

A thorough cleaning of the system is necessary and many people often overlook this.

Flush all the hoses, especially if you're using hoses that you can't also visually inspect, such as a long hose or a hose fitted with angled hose ends. Often, the best approach is to simply replace the hoses because debris can stick to hoses' inner walls. If you're certain that the hose is clean, by all means feel free to reuse it. If you have any doubts, though, spend the money to replace it.

In order to relocate the oil filter to a remote location, a base adapter is installed to the block at the original filter boss. Using either -AN fittings or nipple fittings at this base, you can then run hoses to the remote filter adapter. Bases are usually marked for inlet and outlet locations. At the engine block base adapter, "IN" is for oil flowing from the filter to the block, while "OUT" is for oil flowing from the block to the filter.

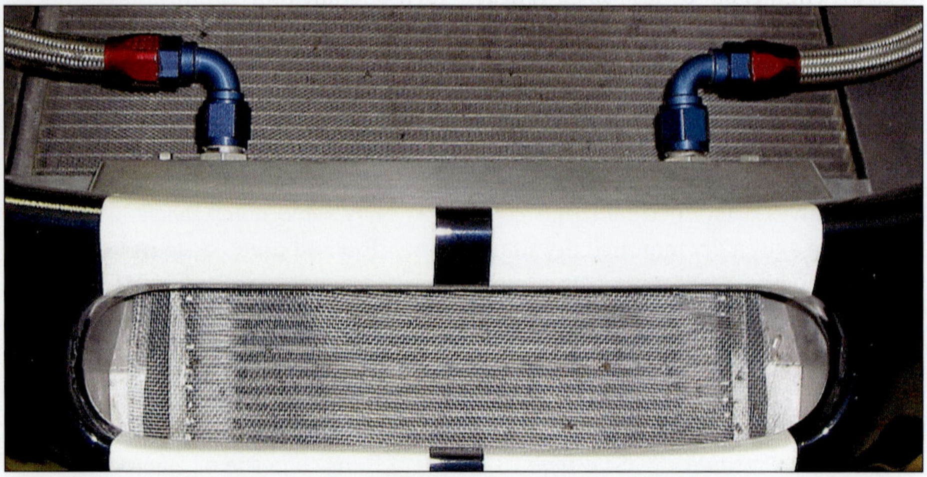

Remote engine oil coolers provide heat dissipation for engine oil. Similar in function to a cooling-system's radiator, an oil cooler serves as a heat exchanger, dropping oil temperature before oil returns to the block. In the event of an engine failure, where debris has flowed through the oil system, hoses and coolers may be very difficult to clean and difficult to verify as clean. If you scatter an engine, roach a few bearings, etc. do not place the existing cooler back into service. You're better off replacing the hoses and cooler to eliminate the possibility of circulating trapped metal particles back into the fresh engine.

When servicing –AN plumbing, disconnecting an oil hose can result in a mess. Quick-connect fittings, originally designed for aircraft and racecar applications, are available. This is a coupler from Jiffy-Tite. The coupler is similar to that found on a compressed-air hose. Pull the collar back to release, and pull the collar back when connecting. The nice thing about these fittings is that they self-seal when disconnected, which eliminates an oily mess during servicing. They also allow quick and easy plumbing removal or installation without wrenches.

External Oil Coolers

As you well know, engine oil lubricates the engine. But oil also absorbs heat from critical engine components and carries it away. The more available oil in the system, the more heat that can be absorbed. External oil coolers function as heat exchangers. The extra amount of oil in the cooler and cooler hoses adds more volume to the system. As oil circulates from the engine through the cooler and back to the engine more heat is released. This sends cooler and more viscous oil back into the engine. External oil coolers are generally plumbed with –10 AN hoses, which is equivalent to a 5/8-inch-inside-diameter hose.

The cooler must be clean. If metal debris has entered the cooler, you can't risk using it again because you can't be certain it's completely clean. It's not worth risking engine failure to use a contaminated cooler. Simply buy a new cooler.

Oil Restrictors

Adequate oil always needs to be delivered to the rod and main bearings. In some engines, restricting the oil delivered to non-critical areas improves oil delivery to vital components. For instance, in old Pontiac 455 engines, the factory oil holes in the lifter bores are larger than necessary so they steal some oil going to the main bearings.

When these lifters bores are bushed in order to resize the bores, a smaller oil hole is drilled into the lifter bore liners. This intersects with the original oil holes but reduces the passage to around .040 inch. This provides enough oil for the lifters, while sending more oil through the main gallery to the main bearings.

Certain aftermarket blocks have "priority main oiling" where oil feed passages are dedicated to the main bearings without bleed-off to other areas, so the mains get all of the oil they need. In some blocks, the lifter oil passages running from the rear are plugged by the factory. That means when these passages are initially drilled, they're drilled all the way through and then plugged at the ends.

Oil delivery to the distributor gear can be improved. You drill a small hole (about .020 inch or so) in the plug in-line with the distributor to provide a small amount of "dribble" lube.

Depending on application, oil may need to be restricted to reduce oil delivered to the top end to send more to the bottom end. Or an area that is normally blocked off may need a small bit of lubrication. Here a 3/8-inch NPT plug installed at the rear of the lifter galley in a Pontiac big-block is drilled with a .020-inch orifice to feed a bit of extra oil to the distributor gear.

Each block design has different possibilities and requirements.

Oil Drainback

The faster the oil returns to the sump, the better. Common methods of promoting oil drainback include smoothing all of the surfaces across which the oil travels. This includes slightly chamfering oil drain holes in the lifter valley and heads or taking advantage of coatings. Painting the lifter valley with Glyptol (a high-heat electrical armature paint) seals off the rough cast surface and provides a smooth oil return surface. If the surface isn't prepped properly, however, the Glyptol can flake off.

A distributor-driven, wet-sump oil pump requires an intermediate shaft. Be sure to install the shaft through the bottom of the block (in the oil pump shaft bore) before installing the pump. A built-in stopper, which can be a diameter change on the shaft, stopper bumps on the shaft, or a metal clip on the shaft, prevents the shaft from exiting through the top of the block. Make sure you lubricate the intermediate shaft before installation.

With the intermediate shaft and pump in place during test fitting, verify that the intermediate shaft has a bit of vertical endplay. If it's too long, this places stress on the distributor and pump.

If you're dead-set on smoothing the rough cast surface in the lifter valley, simply spend a few hours with a bunch of abrasives on a die grinder. That way, the Glyptol doesn't contaminate the engine over time.

Various specialty shops offer excellent oil-shedding coatings that provide a slippery surface to speed up oil travel. This type of coating can be applied to crankshaft counterweights, rods, the inside walls of the oil pan, and the inside wall of the timing cover. This application also helps to sling off oil from rotating parts to reduce parasitic drag.

With that said, for a street engine, don't mess with any of these mods. It's a waste of money, and chances are high that you're not going to see any benefit. For a race engine, though, anything you can do to speed up oil return and reduce oil-cling drag from the outer surface of rotating parts is a good thing.

Intermediate Shaft

If your engine uses the distributor to drive the oil pump, an intermediate shaft provides the connection between the distributor shaft and the oil-pump input shaft. For performance use, opt for a high-quality aftermarket shaft made of hardened steel or chrome-moly steel.

The shaft has either a male hex profile or a slot at one end and a drive tang on the opposite end. The slotted end engages to the distributor and the male tang end engages to the pump. In most designs, the intermediate shaft must be installed from the bottom of the block. The shaft is likely be designed to keep itself captive in the block, so that it can't be accidentally pulled out during a distributor removal. If the shaft has a hex body along the full length, it likely includes a stopper clip that prevents it from pulling up through the top of the block. Be sure to install the shaft before installing the pump.

During test fitting, with the block upside down, install the intermediate shaft and the oil pump. Carefully engage the pump-driven shaft to the intermediate shaft. Snug down the pump with its mounting bolts. You should be able to wiggle the shaft up and down in relation to the pump's driven shaft. A bit of vertical endplay ensures that the shaft isn't in a bind, placing undue pressure on the distributor and pump. Endplay clearance of around .060 to .080 inch is typically acceptable.

High-Pressure High-Volume Pumps

Bearing clearances create restriction to the flow of oil, which creates pressure. The oil pump produces oil flow and is regulated to promote pressure. Larger bearing clearances reduce pressure and flow. In turn, a pump with higher volume and higher pressure is required. Oil viscosity is also a factor. In very general terms, lower viscosity oil or lighter weight oil needs tighter bearing clearances. Larger clearances are better suited to higher viscosity oil.

About 10 psi for every 1,000 rpm is the desired goal. It's better to have higher oil volume and acceptable pressure than low volume and high pressure. If your oil pressure starts to drop at high RPM, it's an indication that a higher volume pump is needed. If you don't have enough volume, you can't generate enough pressure. Want more pressure? You need more volume. If you're running loose bearing clearances, say, in the .0035- to .004-inch range, you definitely need a higher volume pump in order to create more pressure.

A higher volume pump moves more oil under pressure. Pressure drops occur as the oil leaks past the bearings. You want enough flow to keep up with the pressure loss. Any additional pressure (after overcoming oil leaks) causes the pressure relief valve to open and dump excess oil straight back into the block.

In a gear-type oil pump, oil pump gear length affects volume. The larger (longer) the gears, the more volume. Pressure regulator springs (which can easily be changed) allow you to adjust when the pump's bypass valve opens (valve opens earlier, lower pressure; valve opens later, higher pressure). For example, Melling's small-block Chevy M55 standard volume/standard pressure pump has 1.200-inch-long gears and is regulated at 55 to 60 psi. The M55A standard volume/high-pressure pump has 1.200-inch-long gears and is regulated at 75-80 psi. The M55HB high-volume pump, which has a 1.500-inch-long gearset, is rated at 70 psi.

If pump pressure is too high and the engine has tight bearing clearances, the pressure rises until it finds an escape path. A pressure relief valve in the pump releases pressure. Otherwise, pressure can rise to the point of bursting an oil filter. Certain engines can benefit from a high-volume pump, such as old Ford big-blocks that have cam-bearing priority oiling, which tend to starve the mains.

In a nutshell, if the engine is modified from OEM stock form, a high-volume pump should probably be your choice. In

Engine pre-oiling requires sending oil throughout the entire flow circuit, including to main and rod bearings, cam bearings, cam, lifters, and rockers. If the engine has a distributor-driven oil pump, an oil primer adapter can be inserted into the distributor bore to engage and drive the oil pump using an electric drill. If the engine has a crank-driven oil pump, you need a pressurized oil source to deliver oil. A remote engine pre-oiler tank has an oil reservoir and an air bladder. Fill the tank with oil and use compressed air to pressurize the reservoir. Connect the hose to an oil port (or oil pressure port) on the block. When you open the tank's valve, pressurized oil is forced through the engine's entire oil circuit. It's best to remove the valve covers to view the rockers. When you see oil exiting from the rockers, you know that a full prime has been achieved.

a stock engine that doesn't need higher volume, you could do more harm than good, since you may be bypassing oil constantly, which increases oil temperature. In a dry sump system, it's easy to obtain more pressure, since you don't have a weak intermediate shaft to worry about (some dry sump systems run as much as 90 psi).

Pre-Lube Before Initial Startup

Always pre-oil any fresh engine prior to starting for the first time. *Never* fire a freshly built or freshly rebuilt engine dry! Oil *must* be delivered to all areas before starting any freshly built engine: main bearings, rod bearings, lifters, oil pump, rockers, etc.

Traditional Oil Pumps

If your engine has a traditional oil pump, the camshaft drives the distributor and the distributor drives the oil pump. First submerge the oil pump into a container of fresh engine oil, connect the intermediate shaft to the pump, and turn the shaft by hand. Oil should exit the outlet port of the pump. Finally, install the pump.

Before installing the distributor, use a pre-oiling adapter to engage the oil pump. With the valve covers off, turn the adapter driveshaft with an electric drill. Keep turning until you see oil at each rocker location. This may take a few minutes, so be patient. Install a temporary oil pressure gauge at the block's oil pressure gauge port to verify that you've built sufficient pressure. At this point you can install the valve covers and distributor, and you're ready to fire.

Gerotor Oil Pumps

If the engine has a gerotor-style, crankshaft-driven oil pump, you can't drive the pump in order to pre-oil the engine. You need a remote oil pressure canister, which is a pressure tank with an internal bladder. Melling's pressure canister prelube kit (MPL-101), for example, holds 4 quarts of engine oil.

First, add oil (depending on the model, this may take 4 quarts or more). Charge the tank with compressed air. Connect the outlet hose to a fitting that threads into a main oil galley port (located on the driver's side of the block, toward the front, just behind the timing cover). LS blocks have a 16-mm x 1.5 threaded hole.

Connect the canister hose to the appropriate fitting (this should be supplied with the canister) and open the canister valve. Pressurized oil is then pushed throughout the engine's oil passages.

CHAPTER 5

ENGINE MATH

This chapter provides information that will aid in understanding essential dimensions and volumes that are considered when planning and executing an engine build.

Bore and Stroke

In general, a long-stroke engine (an engine with a relatively long stroke in relation to bore diameter) revs slower but produces more torque at lower RPM. A short-stroke engine (short stroke in relation to a larger bore diameter) revs higher, and produces peak power at a higher RPM range.

A crankshaft's stroke dimension is the total stroke of the crankshaft. This is measured from the rod pin's BDC to TDC positions. When selecting a crankshaft, connecting rod, and piston combination, you use one half of the crankshaft's published stroke dimension in your decision making. The distance from the centerline of the crank rod pin at TDC, plus rod length, plus piston compression distance is the length that must fit within the block's available deck height dimension (the distance from the main bore centerline to the block's cylinder head deck surface).

As an example, if you install a 4.000-inch-stroke crankshaft, the crank and connecting rods have a 6.125-inch length and the pistons have a compression distance of 1.115 inch. The block has been square-decked to a deck height of 9.234 inches and achieves 403.13 ci with 4.005-inch cylinder bores. With that stroke/rod/piston combination, the pistons protrude above the decks by .006 inch, which is more than compensated for by using cylinder head gaskets with a crushed thickness of about .045 inch.

The stroke package must fit within the block, so you must always consider the block's deck height. LS factory blocks, as an example, are notorious for having unequal deck heights (high/low side-to-side and/or front-to-rear). So before choosing your stroker combination it's wise to first have the block decks surfaced in order to establish equal deck distance from the crank centerline. You can probably fudge this and assume that the decks are okay, but if you want absolute precision, correct (or at least carefully measure) the block deck height at all four corners (right-front, right-rear,

The stroke package must fit within the confines of the block, so you must always consider the block's deck height. Deck height is the distance from the main bore centerline to the block's cylinder head deck surface. Most OEM blocks (because of a wider range of manufacturing tolerances) have uneven block decks, especially deck height and deck taper. High-quality aftermarket blocks will commonly provide extra deck height to allow you to cut the decks to exactly suit your piston-to-deck clearance.

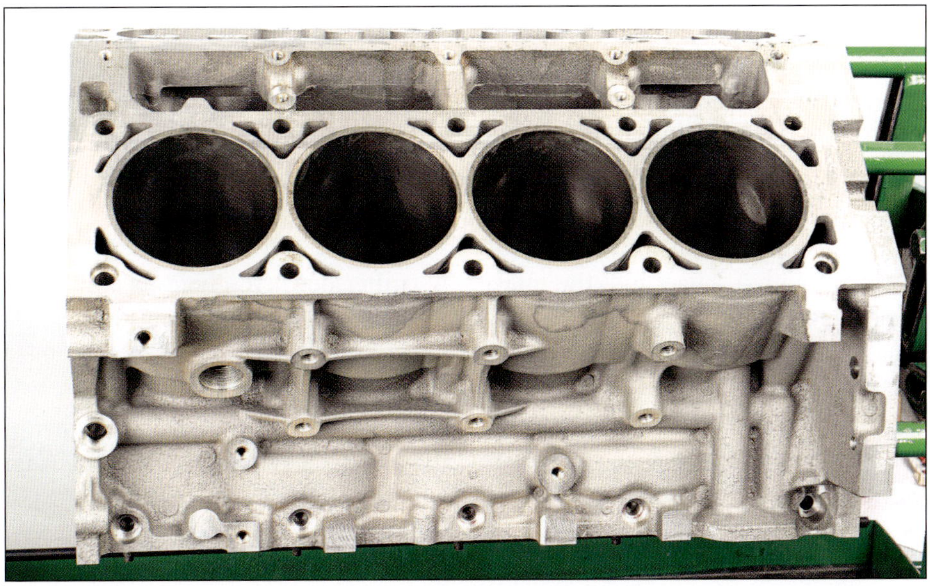

Even today's OEM LS factory blocks, to cite but one example, are notorious for having unequal deck heights (high/low, side to side, and/or front to rear), so it's wise to first have the block's decks surfaced in order to establish equal deck distance from the crank centerline before choosing your stroker combination. You can probably fudge this and assume that the decks are okay, but if you want absolute precision, correct (or at least carefully measure) block deck height at all four corners (right front, right rear, left front, left rear) before spending money on rods and pistons for a stroker combination.

Performance aftermarket cranks are usually stamped or etched on the face of the front counterweight indicating the crankshaft stroke. In this example, the crank stroke is 4.750 inches.

Pictured here is a forged 4.000-inch stroker crankshaft from Lunati. This crank was coupled with 6.125-inch-long connecting rods and pistons with a compression distance of 1.115 inches in a factory LS2 block. This block was square-decked to a deck height of 9.234 inches and resulted in 403.13 ci with 4.005-inch cylinder bores. With this stroke/rod/piston/block combination, the pistons protrude above the decks by 0.006 inches. Using cylinder head gaskets with a crushed thickness of approximately 0.045 inches more than compensates for this difference.

left-front, and left-rear) before spending money on rods and pistons for a stroker combination.

Connecting Rod Length

When discussing the "length" of a connecting rod, it is not simply the total length. Instead, I am referring to the distance from the centerline of the crankshaft pin bore (the big end) to the centerline of the wrist pin bore (the small end). Precision rod-length specialty tools are designed to precisely measure this. However, if you want to perform a rough check of an existing rod, you can use a long caliper with a dial or a digital caliper with a range greater than the length of the rod.

Using either method, you measure from the bottom of the wrist pin bore (6 o'clock) to the top of the big-end bore (12 o'clock). Record this distance. Then measure the diameter of the wrist pin bore and the diameter of the big-end bore. Record these diameters. And plug the figures into this formula:

$$\text{Rod Length} = A + 1/2B + 1/2C$$

Where:
A = distance from bottom of the wrist pin bore to top of the big-end bore
B = diameter of wrist pin bore
C = diameter of big-end bore

Although this isn't the most precise way to measure, it gives you a rough idea of your rod's center-to-center length, for rod identification purposes. Trying to measure rod length with a ruler by guessing at the two bore diameter centers is a waste of time.

Today's quality aftermarket connecting rod manufacturers (such as Scat, Lunati, Oliver, GRP, Crower, Callies, and more) produce rods at extremely tight tolerances for center-to-center length. It is very uncommon to find a set of rods that are not precisely at the specified length and all the rods in a set are not matched.

Although you can rely on performance rod dimensions in general, when blueprinting, you don't want to assume anything. Measure each rod for length, small-end diameter, and big-end diameter. Knowing exactly what you have is better than guessing or assuming. If you're using OEM rods, you *must* check all dimensions due to the greater potential for tolerance deviations.

Rod length has a direct relationship to engine performance characteristics. Granted, the rod length is part of the TDC dimension (from centerline of the crank rod journal at TDC to piston dome location relative to the block deck), but the rod length can be selected in combination with crank stroke and piston compression height in order to tailor the engine for certain performance characteristics. A shorter rod is slower at the BDC range, but faster at the TDC range. A longer rod is faster at the BDC range but slower at the TDC range.

Here's an explanation from Stahl Headers: "With a longer rod, the intake stroke draws harder on the cylinder head from 90-degrees after top dead center (ATDC) to BDC. On the compression stroke, the piston travels faster from BDC to 90-degrees before top dead center (BTDC) with a longer rod; but travels slower from 90-degrees BTDC to TDC, which may change the ignition timing requirement. It is possible that a longer rod could have more cylinder pressure at 30-degree ATDC but less crankpin force at 70-degrees ATDC."

On the power stroke, the piston is farther down the bore for any given rod/crank pin angle. At any crank angle from 20- to 75-degrees ATDC, less force is exerted on the crank pin with a longer rod. However, the piston is higher in the bore for any given crank angle from 90-degrees BTDC to 90-degrees ATDC, so cylinder pressure could be higher. A longer rod spends less time from 90-degrees BTDC to BDC, which allows less time for

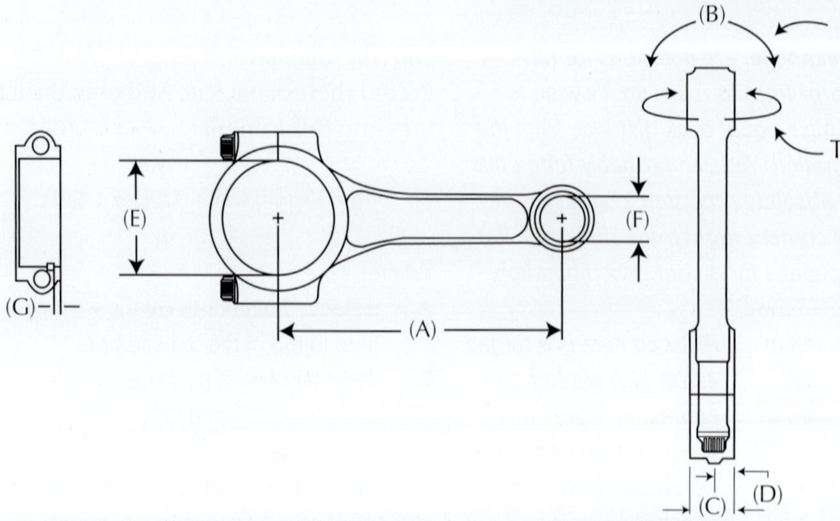

Rod dimensions: A) center-to-center length; B) bend and twist; C) rod width; D) rod offset; E) big end bore; F) small end bore; and G) bearing tang location.

Check each piston location at TDC, relative to the block deck (after the block has been decked). If there is any deviation from bore to bore, you may be able to swap rods (if there is any CTC length deviation in your rod set) in order to equalize all piston deck heights.

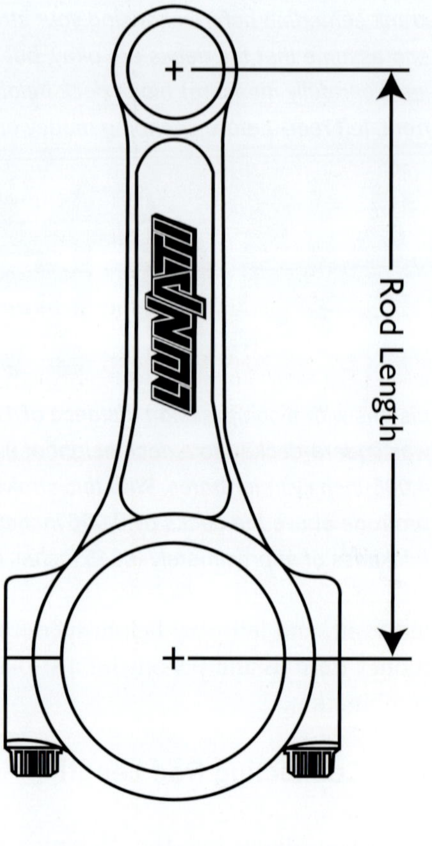

Connecting rod length always refers to the distance from the center of the wrist pin bore (small end) to the center of the crank journal bore (big end). (Illustration Courtesy Lunati)

exhaust to escape on the power stroke and forces out more exhaust from BDC to 90-degrees BTDC. If the exhaust port is not efficient, a longer rod helps produce peak power.

In order to place the piston at or near the block deck on TDC, the rod and crank stroke combinations can include a shorter stroke crank with a longer rod or a longer stroke crank with a shorter rod.

Long Rods

A longer connecting rod provides a longer dwell time at the TDC range. This helps to extend the compression state by keeping the combustion chamber volume small, which is good for mid- to upper-RPM torque. A longer rod reduces the rod angle, which helps to reduce friction. Also, with a longer rod, you can run a shorter piston compression height (that means a lighter piston), which helps to gain RPM.

However, longer rods are less efficient at promoting volumetric efficiency at low engine speeds. The piston moves from TDC (downward) at a reduced rate, gaining its maximum speed at a later point of crank rotation. Longer duration camshaft profiles tend to reduce cylinder pressure during the closing period of the intake cycle. Longer intake manifold runners with slightly smaller port volumes may be needed. Longer rods also pose more of a clearance issue (camshaft, bottom of cylinders, and pan rails).

Short Rods

Shorter rods provide higher intake and exhaust speeds at lower engine RPM, which improves low-end torque (and promotes higher vacuum). Shorter rods increase piston speed as it travels from TDC on the power stroke, which increases chamber volume. This delays the point of maximum cylinder pressure, which is a good match for forced induction (supercharger, turbo, and nitrous injection). Shorter rods also allow more radical camshaft timing. However, since a shorter rod increases piston travel from TDC, at high RPM the piston can run away from the flame front faster, which can decrease total cylinder pressure.

In a nutshell, run a shorter rod for the street or whenever low-end torque is the priority and run a longer rod where you want peak torque to occur higher in the engine RPM band.

Keep in mind that some rod, crank stroke, and piston CD combinations don't work or are impractical due to clearance constraints or unavailable piston compression heights.

Rod Ratio

Rod ratio refers to the relationship of the rod length to the crankshaft stroke. Theoretically, a higher rod ratio produces more torque at peak RPM, and a lower rod ratio produces more torque at lower RPM.

Depending on the type of engine being built, there is a target range for rod ratio. So, a higher rod ratio for racing and a lower rod ratio for street performance seem to make sense.

Here's the formula for calculating rod ratio.

Rod Ratio = rod length ÷ crank stroke

For example, you have a rod length of 5.700 inches with a 3.000-inch stroke. Using the formula:

Rod Ratio = 5.700 ÷ 3.000
1.90:1

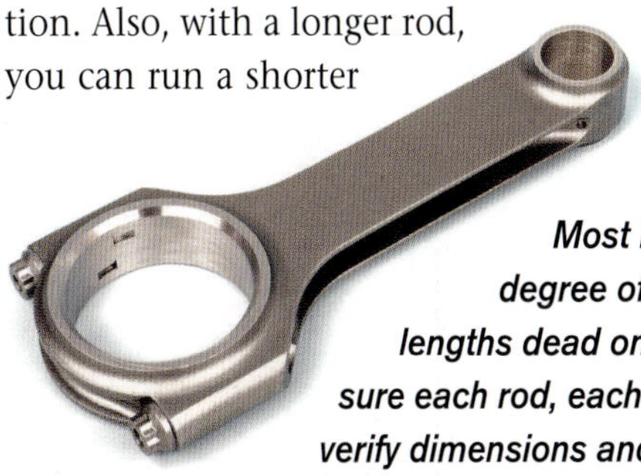

Most high-quality aftermarket rods have a high degree of precision machining, with center-to-center lengths dead on or within 0.0005 inch. Even so, always measure each rod, each crank rod journal stroke, and each piston to verify dimensions and to avoid a stack-up of tolerances. For example, if one rod is 0.0005 inch shorter than the other rods, dedicate that rod to the crank throw that is 0.0005 inch longer, or with a piston that has a compression height that's 0.0005 inch taller. If there are any deviations in tolerances, measuring each component will allow you to mix and match to optimize equalization of all cylinders.

TECH TIP

Rod Ratio Recommendations

Short Rods (1.45:1 to 1.75:1)

- Large intake port volume compared to engine size
- Single-plane or 360-degree intake manifolds
- Large carburetor size compared to engine size
- Moderate engine speed (street, RV, light truck, towing)
- Tall axle ratio (2.76, 3.23, 3.55:1 for example)

Long Rods (1.75:1 to 2.1:1)

- Small intake port volume compared to engine size
- Dual-plane 180-degree intake manifold
- Small carburetor size compared to engine size
- High engine speed (priority on peak power)
- Short axle ratio (3.91, 4.10, 4.55:1 for example)

Computer Software

Computer software doesn't build an engine. Machining and assembly skills are still required for the build itself. However, software related to engine building can serve as a tremendous aid in your quest to design and, in some cases, test a virtual engine build combination. These programs allow you to experiment with various engine component and dimensional combinations. That gives you theoretical performance insight into horsepower and torque via different bore and stroke combinations, cam profiles, cylinder head and intake flow, rocker arm ratios, and more.

Most programs include handy calculations that allow you to plug in various data and obtain quick answers, so you don't have to perform the math on your own. This type of computer program allows you to play "what-if" games by trying different combinations of components.

For pro engine shops that utilize CNC equipment, highly sophisticated programs are available for digitizing, designing, and machining individual components. This type of program (Mastercam, etc.) applies only to design and machining processes and are highly technical in nature. They are not discussed here.

The following manufacturer list includes programs suited to the needs of the enthusiast. It does not, however, include "games" designed to let you play drag racer, oval track racer, or road racer. I've limited the findings strictly to those that apply specifically to the process of engine component selection, engine math, modifications, and dyno simulation. They each have different capabilities. The information here should save you time and aid in your selection of various component and dimensional choices.

Comp Cams

The Desktop Dyno 5 software program is reportedly designed to apply to any four-cycle engine, ranging from four-cylinder to twelve-cylinder applications. It has an interface capability that provides a series of DirectClick menus. This permits the selection of specific components and the ability to enter parts with custom specifications.

A series of built-in calculators includes a CamMath Quick-Calculator, Induction-Flow calculator, and an Air Flow Pressure-Drop calculator. Test combinations are illustrated with detailed graphs that display projected horsepower, torque, volumetric efficiency, and engine pressures. The program's automated testing tool provides additional support as you attempt to determine optimum component combinations. Another useful feature is a combustion chamber modeling program. Windows 7 compatible.

ProRacing Sim

The DynoSim5 is a Windows-compatible program designed to try out your engine build in a virtual state. According to the manufacturer, this software package allows you to experiment with simulations of a variety of build platforms, including forced induction (turbocharging, supercharging, and nitrous injection).

Additional features include ignition curve modeling and rocker arm ratio calculation. The DynoSim5 is also designed for program updating online. A CamDisk8 supplement provides data applications for more than 6,000 camshafts. A specific cam profile can be selected and added to your virtual engine build for testing and simulated dyno results.

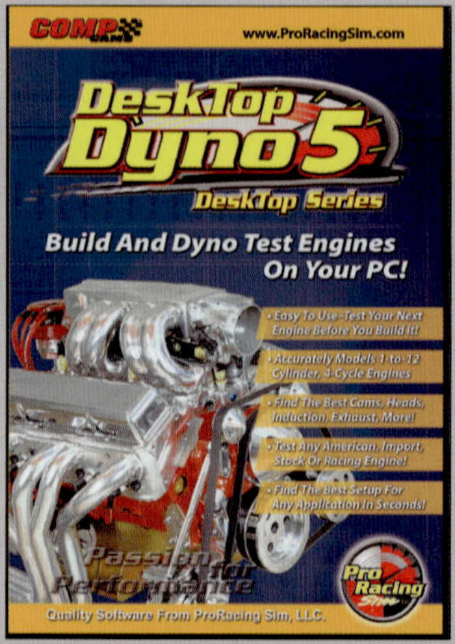

The DesktopDyno5 offers a host of simulation features. (Photo Courtesy ProRacingSim.com)

The Dynomation-5 Wave Action Engine Simulation is an advanced engine simulation software. It illustrates live pressure waves and airflow through the cylinders and engine passages. In addition to providing horsepower data for a given combination, the wave action program provides a 3D cutaway engine view that illustrates airflow, intake and exhaust port pressures, and velocity according to the cranskshaft angle.

This program provides an aid in analyzing intake runner length, volume, and shape in conjunction with cylinder head flow and variations in camshaft profile. Windows 7 compatible.

Challenger Engine Software

Challenger offers three programs suitable for engine design and analysis: Virtual Engine Dyno Professional, Dynamic

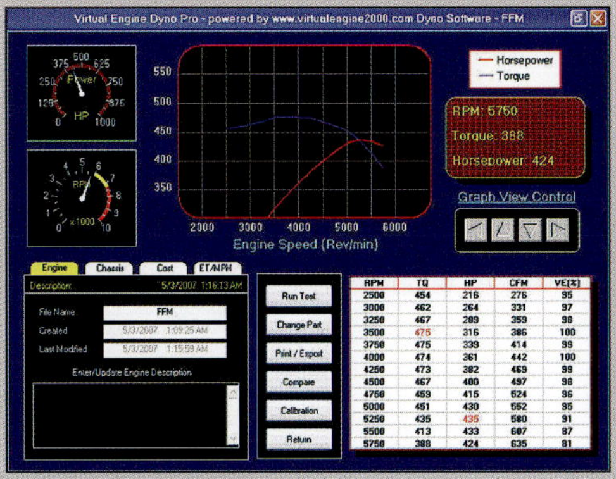

A sample screen display from Virtual Engine Dyno Professional.

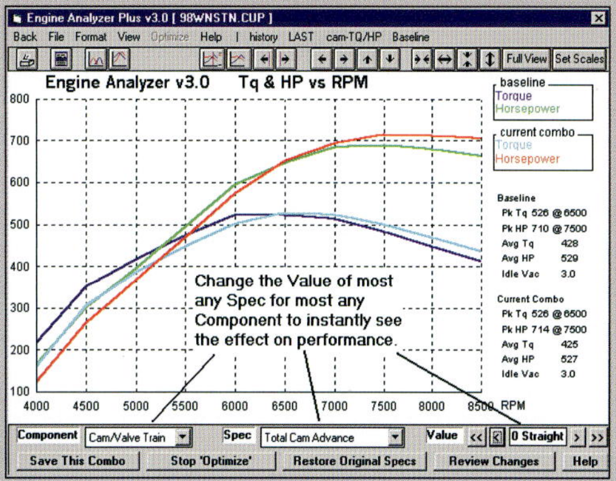

A sample screen display from Performance Trends' Engine Analyzer Plus.

Compression Ratio Calculator and Camshaft Selection Utility, and Engine Builder 3D. They are only available by purchasing website downloads. All are compatible with Windows 95, 98, 2000, NT, XP, or Vista.

Virtual Engine Dyno Professional provides component selection menus based on brand name and part number. This program provides relevant information for race engine builders and for blueprinting procedures. It also has a very useful engine math calculator. The input data includes horsepower curves, torque curves, and maximum engine speed (based on variables that include compression ratio, bore and stroke combination, piston design, rocker arm ratio, cylinder head combustion chamber volume, and more). Another feature is the ability to select and compare specific engine block types and engine block material.

The Dynamic Compression Ratio Calculator and Camshaft Selection Utility is a less elaborate software package offered as a stand-alone program designed strictly to aid in camshaft profile and compression ratio.

The Engine Builder 3D program is specifically applicable to the early (Gen-1) small-block Chevy engine platform, designed to experiment with a variety of component combinations. Test data compiled using this program can be saved and exported to the Virtual Engine Dyno program for simulated dyno testing of your build parameters.

Performance Trends

Engine Analyzer Plus is a professional-grade software program, which is designed for engine builders to predict and analyze stock or racing engine dyno performance. This program includes the ability to test various alternate fuels (alcohol, E85, etc.) in addition to variables in engine component selection. It also assists with camshaft profile selection. Compatible with Windows 98, Me, XP, 2000, NT, Vista, and Windows 7. (Note that in order to purchase the Plus version, you must first buy Engine Analyzer v3.2.)

Audie Technology

This company offers a variety of professional-level engine development software programs. Audie also offers additional testing and measuring hardware systems that are computer interfaced.

Cam Pro Plus is a high-end software and sensor package specifically designed for measuring and analyzing lifter- and valve-motion profiles. This program enables the camshaft engineer or advanced builder to measure camshaft profiles on a dedicated test stand or with the camshaft installed in an engine. This level of software is likely more suited to camshaft development applications.

Flow Pro is a complete line of flow bench software and electronics. It includes multiple options for a variety of flow benches. Every major flow-bench brand as well as most homemade benches are supported. The available system components (available separately or as a package) include software, pressure and temperature sensors, digital dial indicator, electronic depression controller, automatic valve opener, pitot tubes, and swirl meter. Of special interest to the cylinder head blueprinter are special modes for port mapping and carburetor testing.

CROWER CUBIC INCH CHART

STROKE

BORE DIAMETER	3.000	3.125	3.250	3.375	3.480	3.500	3.562	3.625	3.750	3.760	3.875	4.000	4.125	4.250	4.375	4.500	4.625	4.750	5.000	5.125
3.8750	283.0	294.8	306.6	318.4	323.0	330.2	336.1	342.0	353.8	354.7	365.6	377.4	389.2	401.0	412.8	424.6	-	-	-	-
3.9375	292.2	304.4	316.6	328.8	339.0	340.9	347.0	353.1	365.3	366.2	377.5	389.7	401.8	414.0	426.2	438.4	-	-	-	-
4.0000	301.6	314.2	**326.7**	339.3	**349.8**	351.9	358.1	364.4	377.0	377.9	389.6	402.1	414.7	427.3	439.8	452.4	-	-	-	-
4.0300	306.4	318.8	331.9	344.7	355.1	357.4	363.5	370.2	**383.0**	383.6	395.7	408.5	421.3	434.0	446.4	459.2	-	-	-	-
4.0625	311.1	324.1	337.0	350.0	360.8	362.9	369.4	375.9	388.9	389.8	401.8	414.8	427.8	440.7	453.7	466.6	-	-	-	-
4.1250	320.7	334.1	347.5	360.8	372.0	374.2	380.9	387.6	400.9	401.9	414.3	427.6	441.0	454.4	467.7	481.1	-	-	-	-
4.1550	325.6	339.2	352.8	366.3	**377.5**	379.9	386.4	393.5	407.1	**407.8**	420.6	**434.2**	447.8	461.4	474.6	488.1	-	-	-	-
4.1875	330.5	344.3	358.1	371.8	383.4	385.6	392.5	399.4	413.2	414.2	426.9	440.7	454.5	468.3	482.0	495.8	-	-	-	-
4.2500	340.5	354.7	368.8	383.0	394.9	397.2	404.3	411.4	425.6	426.6	439.8	454.0	468.1	482.3	496.5	510.7	524.8	539.1	567.5	581.6
4.3125	350.6	365.2	379.8	394.4	406.6	409.0	416.3	423.6	438.2	439.3	452.8	467.4	482.0	496.6	511.2	525.8	540.4	555.0	584.3	598.8
4.3750	360.8	375.8	390.9	405.9	418.5	420.9	428.4	436.0	451.0	452.1	466.0	481.1	496.1	511.1	526.2	541.2	556.0	571.0	601.3	616.3
4.4375	371.2	386.6	402.1	417.6	430.5	433.0	440.8	448.5	464.0	465.2	479.4	494.9	510.4	525.8	541.3	556.8	572.0	587.7	618.6	634.1
4.5000	381.7	397.6	413.5	429.4	442.7	445.3	453.3	461.2	477.1	478.4	493.0	508.9	524.8	540.7	556.7	572.6	588.5	604.4	636.1	652.1
4.5625	392.4	408.7	425.1	441.4	455.1	457.8	466.0	474.1	490.5	491.7	506.8	523.2	539.5	555.9	572.2	588.6	604.9	621.3	653.9	679.5
4.6250	403.2	420.0	436.8	453.6	467.6	470.4	478.8	487.2	504.0	505.3	520.8	537.6	554.4	571.2	588.0	604.8	621.6	638.0	671.9	688.7
4.6875	414.2	431.4	448.7	465.9	480.4	483.2	491.8	500.5	517.7	519.1	535.0	552.2	569.5	586.7	604.0	621.3	638.5	655.7	690.2	707.5
4.7500	425.3	443.0	460.7	478.5	493.3	496.2	505.0	513.9	531.6	533.0	549.3	567.1	584.8	602.5	620.2	637.9	655.6	673.4	708.8	726.5

As an example, here's a handy bore and stroke chart. Cross-referencing the cylinder bore diameter and the stroke shows the engine displacement. If your particular bore and stroke isn't listed on a chart such as this, it's easy to calculate by multiplying bore diameter x bore diameter x stroke x 0.7854 x number of cylinders. For example: an LS2 with a stock bore of 4.000 inches and a stroke of 4.125 inches has a final displacement of 414.7 ci. (Chart Courtesy Crower Cams and Equipment)

Piston Compression Height

Also referred to as piston compression distance (CD), this is the distance from the centerline of the piston's wrist-pin bore to the piston deck. Performance piston manufacturers offer pistons in a variety of piston compression heights to match your crankshaft stroke, rod length, and block deck height.

Changing crankshaft stroke requires changing other parts of the rotating assembly. This includes altering the connecting rod length and the piston height.

Your point of reference is the block's deck height. Depending on the nature of the specific build, you may want the piston at zero deck (piston deck flush with the block deck at TDC), or you may want the piston to be slightly below or above deck. Once you determine where you want the piston to be placed relative to the block deck, you can "work backward" to choose the best stroke, rod length, and piston height combination.

Before you begin, the block decks must be finish-machined. This means that you cannot assume that the block deck is at a specified height. The block, whether OEM or aftermarket, may have a block deck height that is taller than the specification. For example, even though an OEM block is specified at a block deck height of 9.240 inches, it may actually be 9.320 inches, or 9.248 inches, etc.

Piston compression distance (also called compression height) is measured from the centerline of the wrist pin bore to the flat (or quench) surface of the piston top. This does not include any protruding dome.

Most performance aftermarket blocks intentionally have extra material (thickness) at the decks, which allows you to cut the decks to the preferred dimension. Also, a new OEM or used block may have variances in deck height (the decks may be twisted, taller on one end, or differ from bank to bank, etc.). You must know exactly what the finished deck height is in order to obtain a useful deck height reference.

For illustration, let's assume that you want a zero deck, and use a big-block Pontiac as an example. The stock stroke is 4.210 inches, stock rod length is 6.625 inches, stock piston CD is 1.480 inches, and the stock bore diameter is 4.151 inches. Use this formula for finding block deck height:

Block Deck Height =
1/2 stroke + rod length + piston CD
(4.210 ÷ 2) + 6.625 + 1.480
2.105 + 6.625 + 1.480
10.210 inches

This block deck could be refinished in order to "square" the decks. That means the decks are flat, parallel to the main bore centerline, and the same distance from the main centerline to the decks from front to rear. This is done by reducing the deck height to 10.205 inches, which makes the pistons theoretically stick up out of the deck by .005 inch.

You also consider head gasket crushed thickness, cylinder head chamber design, valve diameters, and valve lift in order to determine valve clearance.

Compression Ratio (part 1)

Engine compression is a major factor in building horsepower and tailoring an engine for a specific range of fuel octane. This section explains which factors affect compression, and how compression ratio (CR) is determined.

Compression ratio is the volume of the cylinder with the piston at BDC compared to the piston at TDC. Or, thinking of it another way, it's the relationship between the combined capacities of a cylinder and combustion chamber with the piston at BDC and the piston at TDC. A two-part formula is used to determine compression ratio using five related measurements: combustion chamber volume (C), piston dome volume (P), head gasket volume (G), deck height volume (D), and cylinder swept volume (V).

The first part of the formula is the BDC Factor, which is comprised of C, P, G, D, and V. The second part is the TDC Factor, comprised of C, P, and D. To get the compression ratio, divide the BDC Factor by the TDC Factor.

$$CR = \text{BDC Factor} \div \text{TDC Factor}$$
$$(C - P + G + D + V) \div (C - P + D)$$

As you can see, the only differences between the first and second half of the formula are the cylinder swept volume and the head gasket volume. This is the actual displacement of the cylinder, which is controlled by the bore diameter and the crank stroke.

Before we get to an example of determining an actual compression ratio, let's examine the individual factors.

Combustion Chamber Volume

We don't use a formula to calculate chamber volume; it is measured using a burette.

The first step is to install spark plugs and valves. Place a smear of lithium grease on the seats to aid in sealing the valves. With the head deck facing up, position a flat piece of plexiglas (about 6 inches square) on the deck, centered over the chamber. Drill a single 1/4-inch chamfered hole through the plexiglas near the edge of the chamber. Angle the head slightly so that the hole in the plexiglas plate allows air to escape as fluid is added.

Place the hole over the deepest area of the chamber side. Apply a light smear of lithium grease around the perimeter of the chamber and install the plexiglas plate by giving it a small twist motion to improve the seal. If grease enters the chamber, it displaces liquid and creates an incorrect measurement.

If the valve lips protrude beyond the deck, they hit the plexiglas plate and prevent a flat seal. If that's the case, grind small reliefs in the bottom of the plate and seal them with grease.

Prepare a glass burette on a stand. Fill the burette with colored liquid (such as solvent or rubbing alcohol tinted with food coloring). Open the petcock and allow some fluid to drain, until the fluid sits exactly at the zero mark. The fluid creates a slight cup shape at the top because it's trapped in this small-diameter glass tube. Use the bottom of the cup as your index at all times, instead of using the highest point of the fluid where it meets the sides of the glass. Make sure there are no trapped air bubbles in the fluid.

Place the burette over the chamber you're measuring, and insert the burette's tip through the hole in the plexiglas plate. Open the valve and begin to slowly fill the chamber. Check to make sure there are no leaks (at the spark plug, plexiglas plate, valves). Before the chamber is full, close the petcock so the fluid just drips into the hole and then shut the petcock completely when the fluid reaches the bottom of the hole in the plexiglas plate. Note the level of the fluid in the burette and write this down. This shows how many cubic inches (or cubic centimeters, depending on how your burette is graded) of fluid were required to fill the chamber.

Repeat the procedure for the rest of the head's chambers.

If all chambers match (unlikely, unless they have been equalized via grinding or milling), this is your chamber volume figure. If they don't match, you can take an average of the measured volumes (or you can slightly relieve the smaller chambers to match the largest chamber). Unless you're building an all-out killer race engine, taking an average reading is likely fine.

Piston Dome Volume

Even though some piston manufacturers provide dome volume numbers for their various piston part numbers, it's still best to measure this yourself. In some cases the manufacturer may not have accounted for valve relief pockets. By the same token, if you have modified the piston domes by smoothing dome edges or changing the profile in any way, you must measure the dome volume yourself.

The procedure for finding piston dome volume begins with connecting rod, piston, and rings installed. Bring the piston to BDC, and coat the cylinder wall with grease. Raise the piston to exactly 1 inch BTDC. Using a sealed plexiglas plate with a small fill-hole on the deck, fill the cylinder area with fluid from a burette until the fluid touches the bottom of the fill-hole (tilt the engine slightly so air can escape from the fill-hole in the plexiglas plate). Record the volume of liquid required, which is Actual P.

After the cylinder area is filled with liquid, check the underside of the bore to make sure that the liquid is not leaking past the rings. If you find a leak, no matter how slight, your measurement is wrong, so you have to start over by resealing the cylinder with grease.

Next, calculate how much volume exists in theory (Theoretical P), for that size bore at 1 inch below deck. Here is the formula (in which Piston Dome Volume = P):

Theoretical P = bore radius^2 x 51.48

P = Theoretical P – Actual P

For instance, if the bore is 3.50 inches in diameter, the radius is 1.75 inches.

$1.75^2 \times 51.48$
3.06×51.48
157.53 ci

This is the volume if the piston were a true flat-top with no dome or reliefs. Compare this theoretical number to the actual volume you measured using the burette. If you had actually used, say, 151 ci to fill the space, the dome volume is the difference between the actual measurement and the theoretical volume, which in this case is 6.53 ci.

$157.53 - 151 = 6.53$

So the piston dome volume is 6.53 ci.

If the piston dome is flat, with no dish, but with valve notches, you can simply measure the volume of the notches instead of measuring the dome volume in the cylinder. The easiest way to do this is to clean the notch and fill it with a small piece of modeling clay.

Slide a sharp, flat ruler across the dome to cut the clay flush with the dome and carefully remove the clay without distorting it (use an Exacto-type blade or other small, fine instrument, and avoid gouging or mashing the clay). Drop the clay into a burette and notice how much volume the clay displaces.

If the clay caused the fluid to rise 2.5 ci, then the volume of that one notch is 2.5 ci. If the piston has four notches of the same size, the total notch volume is 2.5 x 4, or 10 ci. If the notches are of different sizes, take a measurement of each notch, one at a time, and add them together to find total volume used by the notches.

Head Gasket Volume

You don't often need a procedure to determine compressed head gasket volume on your own. Many of today's gasket manufacturers provide the volume on the spec sheet that is packaged with the head gaskets. But just in case, here is the formula (in which Head Gasket Volume = G):

G = gasket hole radius^2 x compressed gasket thickness x 51.48

For example, let's say the gasket hole diameter measures 4.040 inches, which makes the gasket hole radius 2.02 inches, and the compressed gasket thickness is .015 inch. Therefore:

$G = 2.02^2 \times .015 \times 51.48$
4.08 .030 x 51.48
3.15 ci

When determining the gasket hole diameter, be sure to measure the actual gasket hole. Do not use the cylinder bore diameter; the gasket hole is likely to be larger than the bore. Also, because many gasket holes are irregular in shape (not a perfect circle), you have to estimate, as best you can, the hole diameter. If the gasket manufacturer has provided the gasket volume, simply use that number instead of measuring it manually.

Deck Height Volume

If the deck height is zero, deck height volume is zero. If the deck height is positive (with the piston quench surface above the block at TDC), the deck height volume is is subtracted from the gasket volume when used in the formula to obtain final compression ratio.

Here is the formula (in which Deck Height Volume = D):

D = bore radius^2 x deck height x 51.48)

For example, the bore diameter is 3.50 inches, the radius is 1.75 inches, and the deck height is negative (piston quench surface is below the deck with the piston at TDC), at .014 inch. Using the formula:

$D = 1.75^2 \times .014 \times 51.48$
$3.06 \times .014 \times 51.48$
2.20 ci

Cylinder Swept Volume

Here is the formula for finding cylinder swept volume (in which Cylinder Swept Volume = V):

V = bore radius^2 x stroke x 51.48

For example, the bore diameter is 3.5 inches, making its radius half that, or 1.75 inches, and the stroke is 3.30 inches. Using the formula:

$V = 1.75^2 \times 3.30 \times 51.48$
$3.06 \times 3.30 \times 51.48$
519.85 ci

Compression Ratio (part 2)

Now that you know how to compute the volume for all of the individual components you'll be able to determine the compression ratio of your engine. Here's an example where: C = 56.00 ci, P = 6.53 ci, G = 3.15 ci, D = 2.20 ci, and V = 519.85 ci:

$CR = (56.00 - 6.53 + 3.15 + 2.20 + 519.85) \div (56.00 - 6.53 + 2.20)$
$574.57 \div 51.67$
11.12

The actual compression ratio is 11.12:1

Crankshafts

The crankshaft is the heart of the engine. Therefore, it must be durable enough to withstand the engine's dynamic demands, and that means it must handle the specific horsepower and torque loads. It must also cope with the crank speed and the deflection forces imposed by cylinder firing. The main and rod bearing clearances must be correct so the crankshaft's main journals and rod big ends are supported. Also, the crank must be straight to eliminate rolling resistance and prevent bearing wear, and it must be properly balanced with the rotating and reciprocating assemblies.

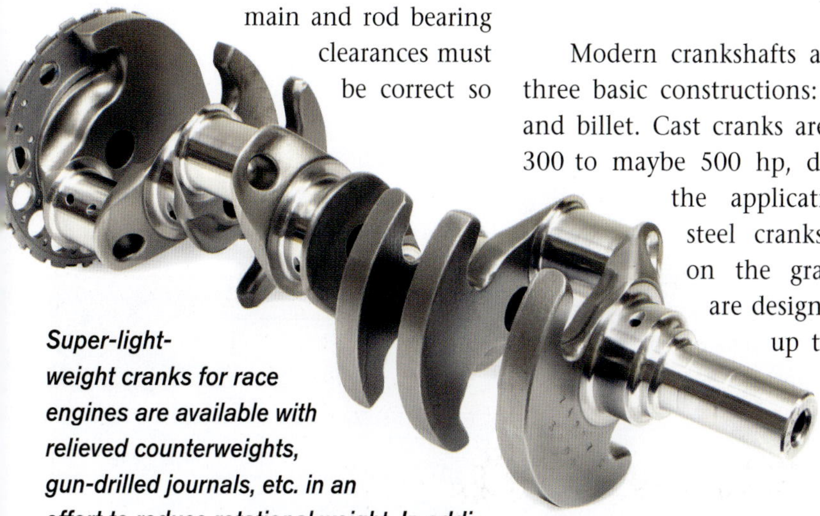

Super-light-weight cranks for race engines are available with relieved counterweights, gun-drilled journals, etc. in an effort to reduce rotational weight. In addition, a popular modification involves bull-nosing and knife-edging the edges of the counterweights. Bull-nosing (radiused profile) and knife-edging (narrower chamfer profile) together create an "airfoil" cross-section of the end of the counterweight. The bull-nose profile is at the leading end of the counterweight while the knife-edging is at the trailing end, similar in concept to the cross-section of an airplane wing. In theory, drag factors, such as air resistance and oil cling, are reduced. For a street engine, it's not worth the time and effort. Reserve this for race applications where (in theory) you'll gain benefits with decreased air resistance and increased oil slinging. (Photo Courtesy Callies)

Crankshaft Types

Modern crankshafts are offered in three basic constructions: cast, forged, and billet. Cast cranks are suitable for 300 to maybe 500 hp, depending on the application. Forged-steel cranks, depending on the grade of steel, are designed to handle up to (and often beyond) 1,000 hp. Billet crankshafts represent the ultimate in strength for high horsepower and are predominantly used in professional-level racing applications.

Cast Crankshafts

The first step in the casting process involves pouring a molten mix of iron and other alloys into a two-piece mold. The casting cools and solidifies and is released from the mold. At this stage the finish machining takes place: machining all journals; finalizing counterweights, flanges, and nose; drilling/tapping flywheel bolt holes; and drilling critical oil passages. The casting process creates a random grain structure and the material is relatively porous, so the cast crank is susceptible to fracturing and failure under high stress. A distinct "parting line" from the mold halves identifies cast cranks.

Forged Crankshafts

A forged crank begins as a dense, forged chunk of steel. Although specific procedures may vary among aftermarket crankshaft manufacturers, forged crankshafts are commonly made by starting with a steel ingot that is heated to about 2,200 degrees F, placed into its

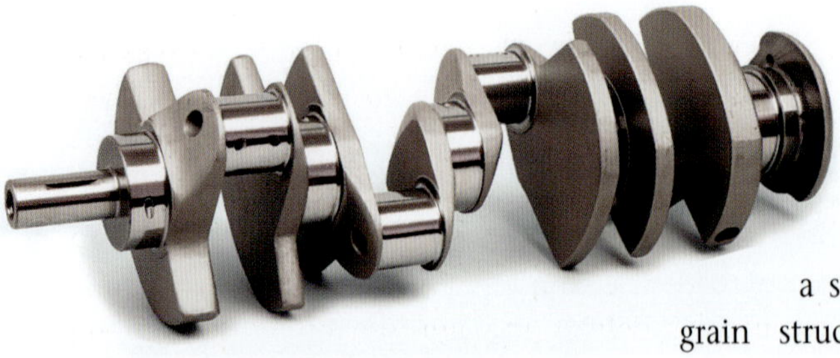

Forged crankshafts are significantly stronger than cast crankshafts. Grain structure is more uniform and dense, making a forging much less prone to embittlement and cracking.

forming die, and press/hammer forged to rough shape. The enormous pressures involved (around 240,000 psi with each hit) compacts the steel molecules into a very dense grain structure, providing increased strength.

Any excess steel that is forced out of the die is then trimmed off, usually in a shearing process. Next the rough forging is heat treated and tempered. This is followed by finish-machining and stress relief. Stress relieving is done to eliminate any internal stresses that may have occurred during machining. Finally, surface hardening is performed. Using dense steel ingots that are forged under heat and pressure results in a much stronger crankshaft that is much more resistance to cracking than is a casting.

The pressure and heat of this process makes a stronger crank with a tighter grain structure. A forged-steel crank may exhibit a wider parting line, but it is not the result of a mold. Rather, this is evidence of the excess material that has been pushed out of the die and then snipped off while still hot. High-quality aftermarket forged cranks are often so finely finished that you may not see any evidence of a parting line.

Manufacturers use one of two methods to forge crankshafts: twist or non-twist. In the twist approach, the die is shaped to orient two pairs of rod throws 180 degrees apart. This eases removal of the raw forging from the die. While the raw forging is still hot, the crankshaft is twisted in order to place the rod throws in the desired clock positions. Although this process creates correct geometry, the twisting procedure tends to interrupt the internal grain structure, which can result in potential stress risers.

A non-twist forging involves a more complex die and eliminates the need to twist the crank in order to place the rod throws in the proper positions.

A non-twist crankshaft is more expensive than a twist-forged crankshaft.

Billet Crankshafts

A billet crank starts off as a chunk of dense billet steel with a very uniform grain structure. It is dense steel and is machined to its final shape rather than

Aftermarket performance cranks are commonly laser-etched with important information. This Scat crank is labeled for internal balance. The forged crank is also identified as being made from 4340 steel. The top number in this example (45545002200) indicates that the application is for a Pontiac 455, the stroke is 4.500 inches, and the rod journals are 2.200 inches in diameter.

Oil passages at rod pins and mains are drilled using calibrated-angle fixtures.

Computer-aided machining control places oil passages at required intersecting points.

The journal and fillet shapes are finalized during the grinding process.

Main journals are ground with the crankshaft centerline positioned concentrically. Rod journals are ground with the crankshaft centerline mounted eccentrically, placing the centerline of the rod journals in a centered axis.

using a forging press. The true benefit of steel billet crankshaft manufacturing is that, through CNC machining, custom crankshafts can be produced in virtually any configuration without the need for a die. Prices are around $3,000 for these robust and durable crankshafts and they are worth it when you need ultimate strength.

Surface Hardening

Several processes can be used to increase the crankshaft's strength, particularly the main/pin fillet area, and each has benefits and drawbacks. The popular processes include induction hardening and nitriding.

Induction Hardening

The process of induction hardening creates a harder crankshaft journal surface. This involves exposing the crankshaft to an alternating magnetic field, which heats the component to a temperature within or above the steel's transformation range. Then it is immediately quenched. Induction hardening does not affect the core of the component.

Crank manufacturers don't widely use induction hardening because it isn't easily controllable. Thus, the depth of hardening can vary, potentially creating isolated hard spots, which can result in the creation of stress points. Induction hardening can also excessively migrate much deeper into the crank, potentially affecting the strength of the core. The crankshaft also has a variation of material thickness. If the crank is not cooled in a controlled manner, major stress areas are likely to be created. Induction hardening for crankshafts has many detractors. The potential exists for unwanted harmonics and/or stress-related cracking issues.

Nitriding

This process involves the absorption of nitrogen into the steel. Prior to nitriding, the crankshaft is machined, stress relieved, and tempered. Nitriding takes place in a heated container where the crankshaft is exposed to ammonia and nitrogen gas. Nitrides are formed as the gas reacts with the carbon on the surface of the steel. This results in a superior surface hardness that is more resistant to abrasion and stress-related failure. Another aspect of nitriding is that the process treats the entire crankshaft surface instead of only journals and fillets, creating a much more uniform surface hardness.

Nitriding process time is also a factor, since the hardening takes place deeper into the steel as nitriding exposure is increased. Typical nitriding treatment takes about 24 hours. The goal is to harden the surface deep enough to offer the required wear and surface strength, but not so deep as to create a potentially brittle crank that might snap under extreme loads. Depending on the manufacturer's process, hardness depth on a high-performance crankshaft is likely to be about .020 inch.

The depth of hardness dictates how far a crankshaft can be reground in the future. If ground beyond the initial hardness depth, a crankshaft may need to be renitrided after reconditioning.

Reducing Rotating Mass

Building a lightweight race car frees horsepower and can aid in better handling and braking. When you consider the weight of the engine as a system, removing weight benefits overall vehicle performance. When you also consider how heavy specific engine components are, removing weight not only affects vehicle static weight, but engine response and durability as well.

The idea of making something lighter may initially seem attractive, but you must remember that the crankshaft must remain balanced. If excessive weight is removed from the counterweights (exceeding the bobweight factor) some tungsten (heavy metal) balancing weight must then be added to the counterweights. Unfortunately, this defeats the purpose of lightening the crankshaft.

Centerline Drilling

One common method to achieve weight reduction during manufacturing is "gun-drilling" the centerline of the main, which is performed by a qualified machine shop. This entails drilling a hole through the centerline of the main to remove static weight from the engine assembly. Gun drilling can easily remove as much as 3 to 6 pounds (depending on the individual crankshaft), without compromising strength. This can also help to equalize crank case pressure and vacuum, transferring air back and forth to improve scavenging.

Another method to reduce weight involves drilling the centerline of the rod pins. This not only reduces weight, but also helps to balance the cranksharft when lightweight connecting rods and pistons are used. Commonly, the material removed by drilling the rod pins can easily equal the weight of one or two slugs of heavy, metal tungsten.

Counterweight Reduction

You can also reduce crankshaft weight by reducing the weight of the counterweights (see Chapter 6 for more details). This can be done by undercutting, a machining process that cuts the counterweights to a thinner width in specific locations. Counterweights can also be reprofiled by knife-edging or bull-nosing their outer edges. This not only reduces weight but theoretically improves the aerodynamics of the counterweights, enabling them to quickly sling-off parasitic oil.

External Balancer

You can spend substantial money for a lightweight crankshaft, but you can easily negate the benefits by installing a large-diameter and heavy crankshaft balancer. For example, an 8-inch balancer might weigh 12 pounds, and a 6-inch balancer might weigh about 8 pounds.

Here's a formula that you can use for reference:

$$\text{Rotating Mass} = \text{distance from centerline}^2 \times \text{weight}$$

For example, you have a 12-pound dampener with an 8-inch diameter (4-inch radius). Using the formula:

$$\text{Rotating Mass} = 4.00^2 \times 12$$
$$16.00 \times 12$$
$$192 \text{ pounds}$$

By substituting a 6-inch-diameter (3-inch-radius) dampener that weighs about 8 pounds, you reduce the rotating mass to 72 pounds.

Light Rods and Pistons

By selecting lightweight connecting rods and pistons, you gain quicker acceleration (the engine revs quicker). In addition, you reduce the dynamic stress on the crankshaft. By reducing the reciprocating mass, less force is exerted on the crankshaft as the rod tries to stop the piston at the end of the exhaust stroke. When you reduce the stress experienced by the crankshaft, you potentially extend crankshaft life and reliability.

A common practice among race engine and some street builders is to use a smaller diameter rod journal on the crank, which allows lighter rods. Common examples include 1.88-inch rod pins on small-block Chevy engines, allowing the use of Honda rod bearings. Big-end rods are available in a variety of lengths from leading rod manufacturers.

Inspection

Take time to inspect the crankshaft. With a new crankshaft, check for dimensions and runout. With a used crankshaft, check for flaws (cracks). Inspecting the crankshaft before installation verifies its condition and allows you to avoid problems and/or return visits to the machine shop.

Flaws and Cracks

Clean the crankshaft before you perform a thorough inspection. Always use a magnetic particle inspection process to identify any flaws or cracks in the crankshaft. Magnaflux is one brand name for the process; other companies offer similar products to perform the same process.

First, mount the crank in a fixture on a work bench. Pass the crank through a large-diameter circular magnet. Under ultraviolet (black) light, look for any cracks or damage. Cracks are readily apparent and appear as white lines. If you discover cracks, your crank needs to be recycled and a new one must be purchased. This should be your first step in the inspection process. So don't bother taking any measurements until after a flaw and crack inspection.

Using an ultraviolet light, any cracks or pores are easily visible. Once the inspection is complete, the crank passes through the ring again to remove the magnetic charge.

Crankshaft Runout

Next, check crankshaft runout. Place the crankshaft on a pair of level V-blocks and make sure it is level. Rest the crank's front and rear main journals on the V-block and mount the dial indicator at the center main journal. The indicator's probe needs to be slightly offset so it does not come in contact with the journal's oil feed hole. The first step is to preload the

Crank runout is performed by resting the crankshaft on smooth V-blocks or rollers. A dial indicator is located at the center main journal and the crank is slowly rotated while monitoring the gauge.

indicator by about .050 inch and then the dial needs to be returned to zero. Gradually turn the crankshaft over and at the same time keep an eye on the gauge.

Once the crank has made one complete revolution, write down the maximum runout. For example, the maximum OEM-spec for allowable runout for a given crankshaft may be listed as .000119 inch. A runout limit of .001-inch or less is typically acceptable. If runout exceeds these limits, straighten the crankshaft or get a new one. However, a qualified technician must complete the crankshaft straightening at a machine shop. Since the crankshaft's material likely has a "memory" in its grain structure, the crankshaft may need to be forced a bit beyond the desired amount to allow for spring-back. For instance, if a crankshaft has .0015-inch runout, the crankshaft may need to be pushed by about .002-inch, anticipating a .0005-inch spring-back.

Journal Measurements

Use a micrometer to measure the main journal diameter and verify that it's to spec. In addition, take the measurement at the center of each main journal. Write down your findings and check to make sure it's within spec for that particular engine. For many crankshafts, the manufacturer provides a tolerance specification of about .001 inch. For example, a typical spec could be 2.558 to 2.559 inches. A reconditioned crankshaft uses main journals that are reground to a smaller diameter so it's within tolerance. Take the smaller diameter journal into consideration when selecting bearings.

Measure journal diameters in several locations (different clock positions), and perform a backup measurement with a different micrometer.

Also measure the taper of each main journal near the front and rear of each journal. The limit of journal taper is usually between .0002 and .0004 inch. When measuring the journal concentricity or roundness, take measurements at many radial locations on each main journal, not just the center. The maximum-allowable out-of-round tolerance is usually about .000110 inch. Each rod journal needs to be measured for diameter, at several radial locations, and check the measurements against spec. The allowable tolerance range is approximately .0007 inch.

Be sure to measure each rod journal width. Take this measurement from the base of the front fillet to the base of the rear fillet on any given journal. Check this against the stacked width of two connecting-rod big ends. If rod journal width is too tight, the rod big ends have insufficient sideplay. Within general parameters, rod sideplay (the distance the rods can move from front to rear on a journal) should be at least .014 inch.

If journal diameter, taper, or width is out of round or beyond tolerance, regrind the crank on a dedicated crankshaft grinding machine. In order to correct journals, you end up moving to an undersize diameter, in which case you can easily purchase a set of undersized-ID main and/or rod bearings.

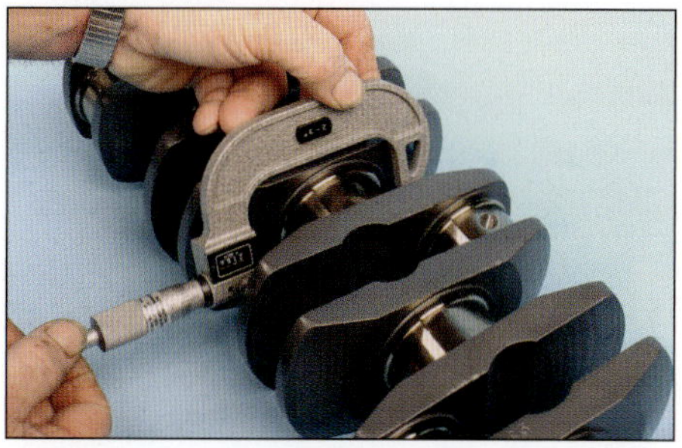

When measuring crank journals, be careful to avoid contacting the micrometer's anvils to the journal's oil feed hole.

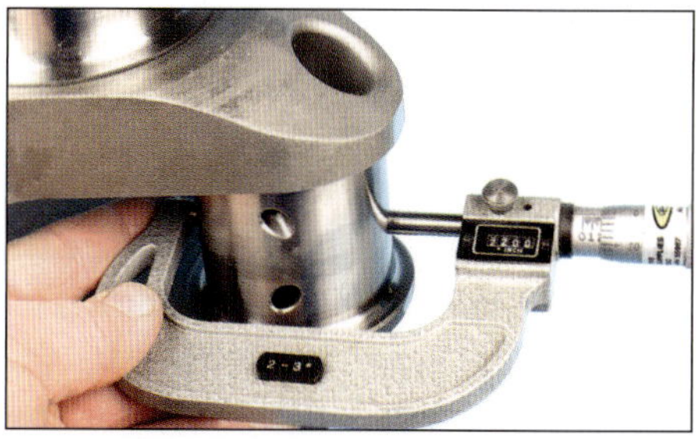

Use a quality calibrated micrometer to measure main and rod journal diameters. Avoid cheap gauges, as accuracy is critical. Also, check your mic for calibration before each use, using the calibration checking standards (these should be included in a quality micrometer kit).

Also remember that bearing size needs to be uniform. If one main journal must be reground to accept an undersize main bearing, then *all* of the main journals should be ground to that same size. The same holds true for rod bearings. If even one rod journal needs to be undersized, then *all* rod journals must be ground to the same diameter. Always check with your bearing supplier to first find out what undersize bearings are available (-.0005, -.005, -.010 inch, -.020 inch, etc.). This determines the diameter of the regrind.

If a used crank passes inspection and you intend to reuse it, each journal can be polished on a crankshaft belt polisher, using 400 grit, then stepped up to 600-grit. Polishing usually eliminates small surface scratches. Different equipment manufacturers may specify different grit-grade abrasives for polishing. The journals should not be "mirror" polished, since microscopic scratches are needed to provide oil cling.

Stroker Clearance

Always check any crank for block clearance, but this is critical when using a stroker crank that has a longer stroke than the original. Test fit the crank to the block. Install the upper main bearings to the block saddles. Make sure the saddles and rear side of the bearings are clean and dry. Install lower bearings in the main caps as well but don't install the caps just yet. Lube the exposed surfaces of the bearings with clean engine oil or assembly lube. Make sure that the crank is clean. *Carefully* lay the crank onto the upper bearings, observing the counterweight positions. If you do have a clearance issue between counterweights and the block, you don't want to crash the crank into the block.

Before test fitting your crank, install the bearings that you intend to use, and apply a coating of clean engine oil or a quality assembly lube.

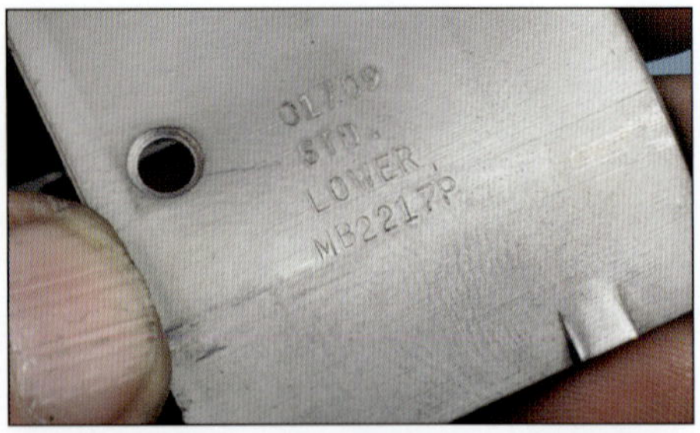

Pay attention to the bearings; some bearing sets have dedicated upper and lower bearing inserts when you're dealing with the block upside-down on a stand, upper main bearings install to the block saddles and lower bearings install to the main caps.

Test Fit

With the crank resting evenly on all of the upper bearings, rock it back and forth on the upper bearings just a bit to make sure that all journals are evenly resting on all upper bearings. Slowly begin to rotate the crank, while observing each counterweight as it approaches the block surfaces. If the crank dead-stops, find the contact point and mark it on the block.

If you can't rotate the crank 360 degrees due to contact, mark the first contact spots found, then rotate the crank in the opposite direction to find more interference locations. Mark all problem spots where the counterweights are actually interfering with crank rotation.

Remove the crank and the upper bearings. Use a hand grinder (pneumatic or electric) with either a radiused milling bit or abrasive stone, to minimally relieve the contact areas on the block (don't remove more than needed).

Clean the block and re-install the upper bearings and the crank. As you rotate the crank without the main caps, watch the upper bearings. If a bearing moves a bit, push the protruded end back down with your finger. Once you verify clearance to rotate the crank a full 360 degrees, perform another clearance check

Undersize Grinding

As a result of journal damage such as scoring, main or rod journals may be ground undersize, which is performed on a dedicated crankshaft grinder, using specific-width abrasive stone wheels. When main journals are ground, the crankshaft is mounted and rotated straight with zero runout. When rod journals are ground, since they are offset from the crank centerline, the crank is adjusted on the machine to run at an offset, with the rod journals positioned at zero.

Oil feed holes only require a deburring/softening of the edges if the hole edges are currently sharp. Don't get too carried away. The edges should be softly radiused to promote easy oil flow.

Cooling fluid is applied during grinding to cool and clean the journal surfaces. Regrinding a journal to an undersize in order to save the crankshaft requires undersize bearings. An undersize bearing has the original outer-diameter dimension but has a smaller inside diameter. As a result, the bearing wall thickness is increased.

Be aware that the depth of surface hardening on journals can vary, depending on the specific brand or model of crankshaft. In some cases, the journals may be ground to a .020-inch undersize without losing surface hardness; some crankshafts allow as much as .030-inch undersizing. If surface hardness is compromised as a result of regrinding, the crankshaft can be sent out for renitriding to restore the required hardness.

During crankshaft test fitting, be sure to lube the threads of the main cap bolts (always follow the bolt manufacturer's lube and torque specs!). Installing bolt threads dry provides a false torque reading (excess friction results in lower clamping load).

at each counterweight. Rotate the crank and observe the counterweight-to-block clearance. Minimum clearance between a counterweight and any area of the block (on the rotation plane) should be around .060 inch. Mark and relieve the block as needed.

After test fitting the crank a second time, if more material needs to be removed, repeat the process. You can always remove more metal, but it's very difficult to replace. You don't want to remove more than needed to avoid weakening any area of the block.

Check Rotation

After the block has been align-honed and bearing clearances have been measured, verify crank rotation. First, install the upper bearings and the crank. Install the main caps with bearings installed and lubed. Tighten the cap fasteners to manufacturer's torque spec with oil or moly. Tighten all main cap fasteners in the proper sequence and to the spec'd torque value. Also tighten the main cap fasteners in stages.

For instance, if the spec calls for a final value of 110 ft-lbs, first tighten to 10 ft-lbs, followed by another pass at 25 ft-lbs, followed by another pass at, say, 50 ft-lbs, then 80-, then 100-. After the first tightening phase, gently rotate the crank just a bit to make sure that it isn't jammed up. After each tightening step, rock the crank just a bit after snugging each main cap. If you run into a bind all of a sudden, you may have a problem with that specific main bore.

It's important to understand that an OEM aluminum block, such as the GM LS aluminum block (LS1, LS6, LS2, LS7, LS3, or LS9), is very susceptible to distortion. Don't expect the main bores to arrive at final alignment until *all* main cap fasteners are tightened to specification, which includes the main-cap side bolts. If the block has main-cap side bolts, tighten the side bolts only after tightening all primary cap bolts to their full spec.

Here's an example of a stroker crank in a Honda race build. Notice where the counterweight hits the black wall. Remember: When inspecting for stroker crank clearance, rotate the crank slowly as you monitor clearances.

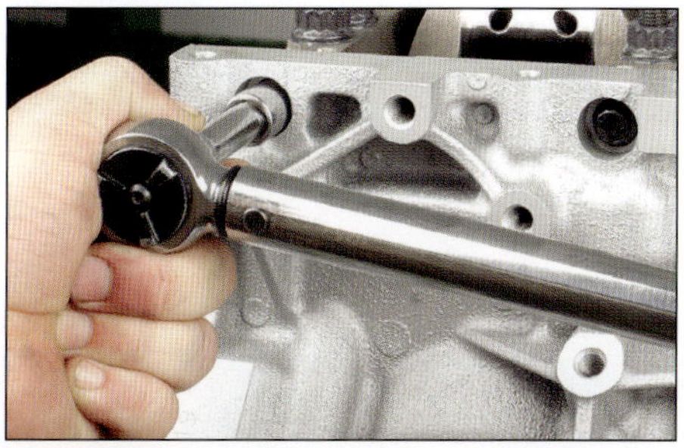

If you're dealing with a block with additional main cap side bolts, always tighten these last; tighten after the primary main cap fasteners have been tightened to spec. In some cases, such as with the GM LS engines, the main bores don't properly align until all of the main cap fasteners are fully tightened. If you tighten only the main (vertical) fasteners, you may experience crank bind, which leads you to mistakenly think that you have a problem. Only rotate the crank after all main cap fasteners have been fully tightened to spec.

Measure Thrust

Once the crank has been fully installed in the block, measure crankshaft thrust. Thrust is the amount of movement available for fore/aft movement of the crankshaft in the block.

Securely mount a magnetic-base dial indicator at the front of the block. Use a magnetic indicator base for iron blocks and mount the magnetic base to the face of the number-1 main cap on an aluminum block. Using a long and clean flat-blade screwdriver, gently pry between a counterweight and a main cap to move the crank fully rearward. Place the dial indicator's plunger onto the crank snout

With the crank installed, place a dial indicator at a fixed position on the block, with the plunger contacting the crank nose or snout base. Using a flat-blade screwdriver (between a main cap and a counterweight), push the crank fully rearward. Preload the gauge to about .050 inch, then zero the gauge. Push the crank fully forward and note the gauge reading. This allows you to measure crank thrust. Repeat the procedure several times to verify your readings. Compare your thrust reading with specs. Crank thrust is usually somewhere around .006 to .007 inch.

Bearing Design and Clearances

When you install a main or rod bearing shell into its saddle, you may notice that each end of the bearing protrudes slightly beyond the mating surface. This is intentional and provides a slight crush area. Take the main bearings as an example, with the upper bearing installed in the block saddle and the lower bearing installed in the main cap.

When the cap is installed and the main cap bolts are tightened to specification, this slight crush area (where the edges of the bearings make contact with each other) forces the bearing shells to try to expand, which applies radial pressure, seating the bearings to the round main bore. This locks the bearing package in place and provides the bearing inside diameter to achieve its proper geometric shape.

Although many assume that the bearing tangs (which register into saddle- and cap-tang relief slots) are responsible for holding the bearings in place, they are actually present to aid in assembly. The installed (crushed) radial force is actually responsible for locking the bearings in place, preventing movement.

Although the bore (main bore or rod big-end bore) must be perfectly round, and the outside of the bearing package must conform to this round shape, the inside diameter of the bearing package is actually not perfectly round. Rather, the end of the bearing shells are slightly tapered. This creates a ramp effect for the oil, with the wider gap at the parting-line areas allowing oil to expand and fill this void. At the same time the oil is being squeezed and forced around the rest of the bearing surface. This creates oil pressure to properly support the crankshaft on a film of oil.

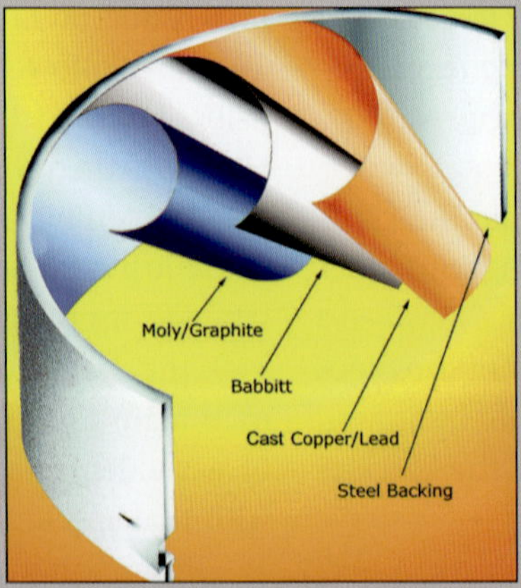

High-performance main and rod bearings often have multiple layers of material. This view shows the bearing construction of a Clevite TriArmor bearing, including the steel backing, cast copper/lead layer, babbit outer layer, and a moly/graphite surface coating. (Photo Courtesy MAHLE Clevite)

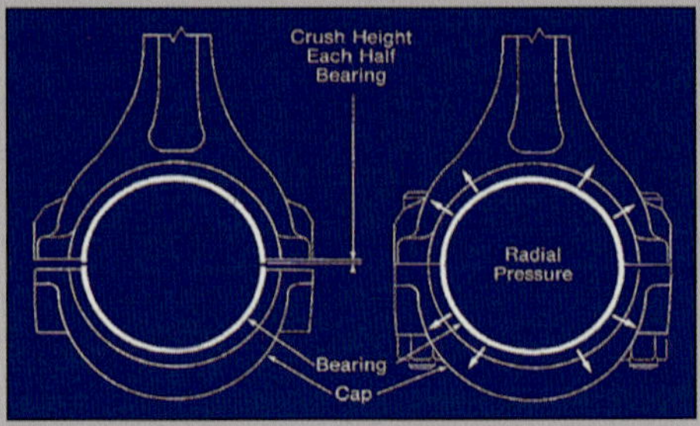

Bearing crush is applied when the cap is tightened to the rod. This crush effect, where the upper and lower bearings meet, provides the final bearing profile and secures the bearing shells in place. (Illustration Courtesy MAHLE Clevite)

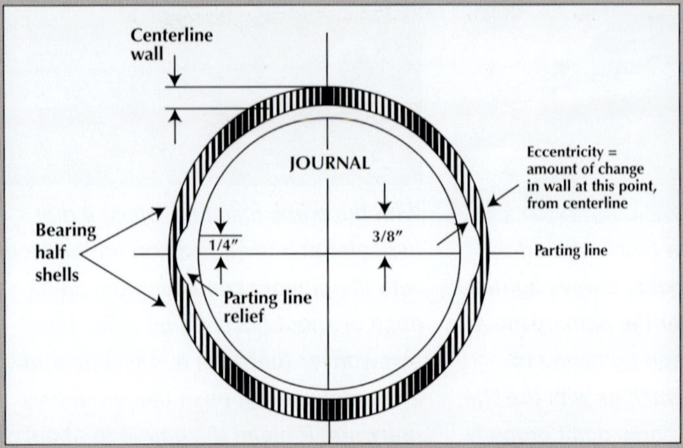

The inside diameter of main or rod bearings is not perfectly round; there is a slight taper on the bearing walls. This creates a ramping effect that builds oil pressure as the journal rotates. The pressurized oil film supports the bearing on the journal. (Illustration Courtesy MAHLE Clevite)

Pay attention when fitting and installing bearing shells because the set includes dedicated upper and lower bearings in the majority of applications. Upper always refers to the block main saddle or rod saddle locations. Lower bearings are to be installed in main or rod caps. Don't make the mistake of mixing them up when your block is positioned upside-down on your stand. This is an upper main bearing. Note the oil feed holes that must align with the oil passages in the block's main saddle.

Rod bearings may or may not be labeled for position so be sure to check. This is an upper rod bearing. Upper/lower bearing inserts may have slightly different profiles, so don't mix them up.

This is a lower main bearing. It installs to a main bearing cap.

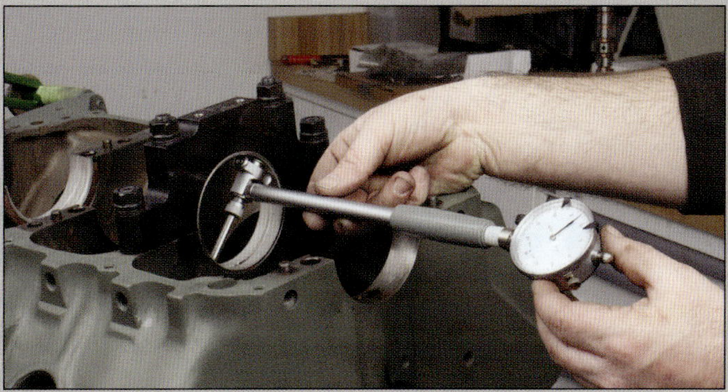

With main bearings installed, bearing diameters must be checked and compared to the outside diameter of each crank main journal to determine and verify oil clearance. Use a caliper micrometer to check this measurement. The main caps must be fully torqued to specification before attempting to measure installed bearing ID.

face or snout base (a flat surface). Adjust the indicator to slightly preload the gauge about .050 inch. Zero the gauge. Pry the crank backwards again to make sure that it's still fully rearward. Zero the gauge again if needed. Then pry the crank fully forward while observing the dial gauge.

Refer to the recommended crankshaft thrust specifications for your particular application. Thrust specs vary depending on the engine, but a fore/aft movement of approximately .006 to .007 inch is a commonly acceptable range.

Shot Peening

Shot peening is a process that involves pounding the surface of the crankshaft with steel or stainless steel round shot (tiny balls). This is not a blasting process intended for surface cleaning, where abrasive media is used. Shot peening is not abrasive; it's a material-compression process. Shot peening hammers and work-hardens the surface while compressing the steel under the surface. This increases below-surface density, serving to make the crankshaft stronger and less prone to distortion or cracking.

Shot peening compresses the metal and reduces the chance for surface galling, cracking, or corrosion fatigue. Shot peening, especially in the journal fillet areas (the edges of the journals where they meet the counterweights) can be beneficial in reducing the possibility of stress cracking.

Stress Relieving

The purpose of any stress-relieving process is to eliminate internal metallurgical stresses in the steel, which reduces the chance for fracturing by allowing the molecules to align more uniformly. Shot peening is not really a stress-relief process. Because it compacts the surface of the metal, it can actually induce additional stress inside the crank. True stress relief should treat the entire crankshaft all the way through, not only the surface.

Stress relieving a crankshaft is performed as part of the manufacturing process, to reduce the internal stresses of the metal that occur during the forging and initial rough-machining steps. This is commonly done by heating the crank

Bearing Cavitation

Bearing damage in the form of surface deterioration is referred to as cavitation erosion. This can happen if oil clearance is a bit too great. As the journal rotates in the bearing, small vapor bubbles can be created in the oil film between the journal and bearing. These small bubbles can pop, with resulting force that can eventually erode the bearing's soft surface layer. This leads to eventual bearing failure. Many of today's race engines run slightly tighter rod bearing clearances. In some high-speed/high-load racing applications, a change of as little as .0008 inch has been reported to stop cavitation erosion.

High-performance bearings today often have specialized surface layers and coating treatments to resist cavitation erosion. A note concerning crankshaft journal oil holes: Although many builders tend to drastically radius the oil holes, chamfering and blending the hole edges, this isn't always a good idea. It is very possible to create too much oil bleed-off by excessively chamfering these holes. If in doubt, it's safer to simply deburr the holes and remove all sharp edges. Again, this is an area where an engine builder's years of experience comes into play with what works best for specific applications.

In all-out racing builds where the crankshaft has a small-diameter rod journal (in an effort to reduce mass), the bearing load increases as you decrease journal size. In order to deliver more oil to the rod bearings, a popular modification is to create a teardrop from the crankshaft's main oil holes, where the leading side of the oil hole is slightly extended in a teardrop ramp. This provides a slight cavity that fills with oil, boosting oil pressure through to the rod bearing oil feeds. The width, depth, and length of this groove is critical. If machined too wide or too long, you run the risk of decreasing oil delivery to the main bearings. This custom task should only be performed by an experienced builder who has experience with this modification.

To further scrutinize bearings, consider suggested clearance, anticipated load of the bearing surface, and bearing speed, which is based on journal circumference.

In high-end engines, where you plan to run smaller journal sizes, you must consider the load-carrying capabilities. Some high-end race engine builders sometimes drill extra oil holes to provide adequate oil delivery in the bearings and partial-radius grooves in the housing or saddle area of the mains to create multiple oil supply points. This is especially important in engines that use smaller bearings and experience higher loads. But don't do this at home.

Especially in race engine applications that experience constant high-speed operation and high loads, avoid machining bearing bores to their maximum allowable specification. This can reduce bearing crush. Although you may have sufficient bearing crush to lock the bearings in place, you may not have enough crush to allow sufficient heat transfer from the bearing to the bore contact.

Position each main bearing in the block saddles with the bearing ends protruding by the same amount at each end. Verify that the oil holes in the bearings align with the oil feed holes in the saddles. If a slight misalignment is found, you can carefully elongate the hole in the bearing, being careful to remove any burrs.

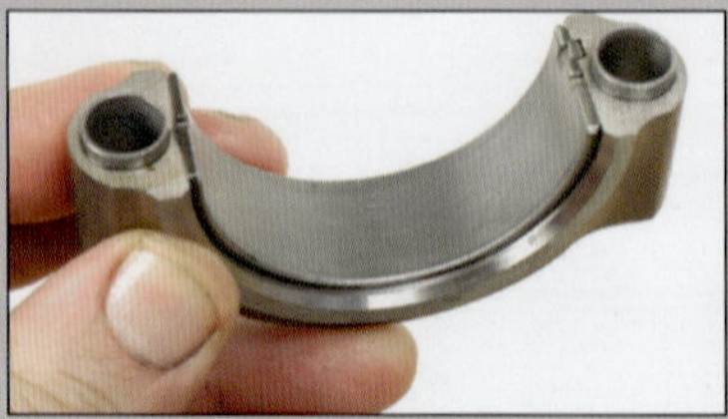

Install rod bearings by aligning the bearing tang with the locating notch and press down each end with your fingers. Each end of the bearing shell should protrude by the same amount.

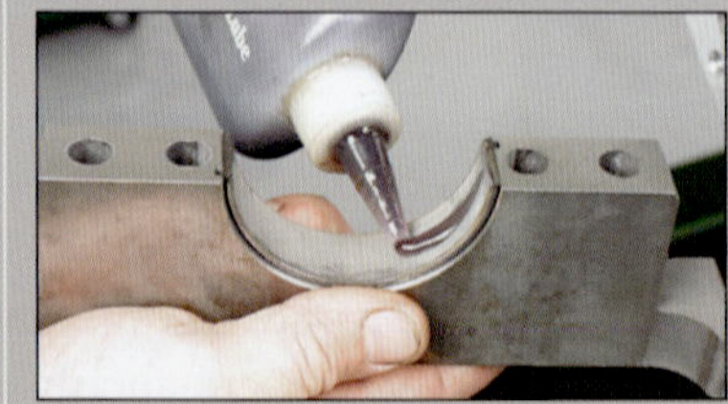

Once the bearings have been installed dry, apply a film of assembly lubricant to the exposed bearing surface. If you don't plan to fire the engine for a while, be sure to use an assembly lubricant that provides good "cling" to prevent a dry start. Several synthetic assembly lubes provide excellent retention and super-slick friction reduction. Here, I'm applying Royal Purple's Max Tuff synthetic assembly lube.

Apply a generous amount of assembly lube to the installed bearing. When applying lube to a rod bearing, also apply a bit to both sides of the rod's bore, which is at the chamfered side and the opposite side. This aids in rod positioning at the crank fillet side and between pairs of installed rods.

in a controlled, heat-treatment oven. Initially grinding journals to rough size with abrasive wheels tends to generate substantial heat, which can induce stress. CNC machining the journals (as opposed to grinding) results in less distortional heat. The crank needs to be stress relieved after either type of rough machining.

When a crankshaft is repaired, by regrinding journals and/or weld repairs, stresses are induced within the crankshaft. Stress relief "relaxes" and realigns the molecules. If a crankshaft has some runout and the crank is straightened but not stress relieved, the memory in the material may allow the crank to once again bend and return to the former runout state. Basically, whenever a crankshaft has been exposed to high heat and/or mechanical stress during repairs, it should be stress relieved.

Additional benefits of stress relief include the long-term stability of the crankshaft for performance use. Both cryogenics and vibratory stress relief (see Chapter 18 for more details) are treatments that can improve the molecular structure and can improve durability and reduced harmonics.

Counterweight Modifications

Crankshaft counterweight modifications, either by design during manufacturing or during later modification, are performed to reduce counterweight mass (weight reduction) or to address oil windage concerns. Specialty lightweight racing crankshafts often have counterweights that are "carved" or "skeletonized" to reduce weight. A lighter crankshaft means less rotating mass, which means faster crankshaft acceleration (quicker revs). However, you can't simply hack off counterweight material. The crankshaft must still be balanced by matching rod and piston weight.

Counterweight modifications that address windage concerns typically involve bullnosing and/or knife-edging. Altering the profile of the counterweights is primarily done to reduce parasitic oil cling and drag. During engine operation, parasitic oil (oil that slings around and clings onto the counterweights) can result in slight drag and minor imbalance conditions as the oil clings and then is slung off. This is only a concern for racing engines, where the goal is to obtain maximum performance.

Bullnosing involves radiusing the leading edge of the counterweight. By rounding-off the leading edge of the counterweight (eliminating a flat face), any parasitic oil that might cling onto the leading edge is more easily diverted off the leading edge.

Knife-edging involves tapering the trailing edge of the counterweight. By bullnosing and knife-edging, the profile of the counterweight is then shaped similar to an airplane wing. The radiused leading edge theoretically allows oil to slip over the leading edge. The reduced thickness of the trailing edge then causes the oil to more easily be pulled off the counterweight.

Modifying the counterweights to become more aerodynamic may free some power in a high-revving race engine, but it's really a waste of time and money to perform this alteration on a street-performance crank. As with many modifications, changing the profile of crankshaft counterweights is for the race engine builder who is trying to extract as much power and longevity as possible, where every ounce counts. Any modifications to a crankshaft that involves material removal requires rebalancing.

Cross Drilling

Don't do it. Cross drilling is an old-school approach intended to feed more oil to the rod bearings for high-revving applications. Cross drilling refers to drilling the oil holes completely through the journals, followed by drilling an intersecting hole from the main to the rod journal. The theory is that this provides a more continuous flow of oil to the rod bearings. However, cross drilling can potentially weaken the crank because it introduces more stress riser points. Also, oil then needs help to overcome crankshaft centrifugal force. Even increasing oil pressure may not help.

Today's high-quality aftermarket performance crank manufacturers don't offer cross drilling for these reasons. Although some folks stick with this outdated approach, you're taking a chance with the potential for severe rod bearing damage. Play it safe and just don't do it.

Chamfering Oil Holes

Oil holes in the main journals and rod journals should be slightly chamfered, eliminating any sharp edges at the perimeter of the hole. Softening these oil-feed hole edges reduces the chance for stress risers and provides a smoother path for oil travel. Some builders get carried away by severely enlarging the chamfer area, which in most cases is not necessary. However, removal of sharp edges aside, additional radiusing/blending on the trailing side of the oil holes can produce a more efficient path of travel for oil to the bearing.

Blending a softy radiused and slightly extended "teardrop" path for oil can promote more efficient oil travel. The blend area should not be wider than the diameter of the oil hole, and should only be about .200 inch long at the most. Teardrop blending is not needed in all applications. Defer to the crankshaft manufacturer's design or an experienced engine builder for a specific application.

CHAPTER 7

Connecting Rods

Connecting rods are placed under more stress than any other component in the engine. Selecting the strongest rod for a performance target is absolutely essential. You have a variety of materials to choose from: powdered metal, forged steel, aluminum, titanium, billet steel, and aluminum. Connecting rods are offered in H- and I-beam configurations, and you have weight, balance, and dimensional factors to consider. Connecting rod bolts are also placed under enormous stress and they must not fail. As we all know, if a connecting rod fails, the engine can be turned to junk in a fraction of a second. So choose wisely and do not select inexpensive rods.

Rod Types

Connecting rods are available in a variety of materials and design processes. Older production rods for passenger car engine applications were typically made of cast iron. For high-performance production in select engines, rods were commonly made of forged steel. Today, the majority of OEM production rods are made of powdered metal (often referred to as PM rods), with select applications using forged-steel rods.

Powdered-Metal Rods

Powdered-metal rods are made in a similar way as casting or forging. A specialized powdered mixture of alloys is placed into a mold, heated (to melt and flow), and then pressurized. The process results in a surprisingly strong product that requires only big-end and small-end honing and bolt thread tapping (no additional exterior finishing is required). Instead of having separate rod body and cap (as with cast or forged rods), PM rods are manufactured as one piece.

A series of machining steps creates the pin bore, crank bore, and bolt faces. The cap is also separated by a cut, faced and attached in order to finalize the big-end dimension.

A powder-forged rod eliminates most of the machining and forming steps, since the initial forged shape is extremely close to all final dimensions. Fracturing the rod end creates the cap. The rod is held in a fixture, the parting line area is scored, and the cap is literally snapped off.

Unlike a rod and cap that have each been machined flat at their mating surfaces, PM rods have uneven surfaces at the mating areas. The benefit of this is that it creates a perfect mating between rod and cap, since no material is lost

A billet connecting rod provides far greater strength than a cast-iron rod.

After the powder mix is formed in a die, the mix is heated to more than 1,500 degrees F and forged with more than 750 tons of pressure. The powdered-metal forging is then heat treated and finish-machined. Rods are seen here on their way into the heat-treating oven. (Photo Courtesy Howards Cams)

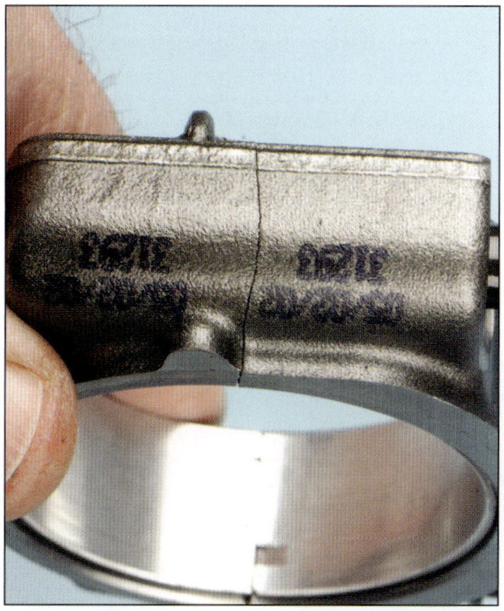

Once a PM connecting rod is formed in a mold, the cap is created by breaking it off in a holding fixture. This is also referred to as a snapped-cap design. While this process creates a very precise cap-to-rod mating, a PM rod cannot be resized by traditional methods. If resizing is required, it possible to simply hone the big ends larger and select an oversize-OD bearing.

This close-up of a powdered-metal "cracked cap" rod's cap mating surface shows the irregularity of the surface. The two fractured surfaces align perfectly when the cap is installed, providing precise cap register.

during the separation. The cap fits to the rod precisely (mirror-image surfaces). The cap is now dedicated to its original rod, and when mated and the bolts are tightened the parting line is invisible to the naked eye. This provides perfect cap-to-rod alignment, with no wiggle room and no need for locating slots or keys.

Aside from saving on production costs, the irregular mating surfaces created when the cap is severed provides a precision mating of cap to rod that perfectly aligns the cap during assembly. The metallurgy of the rod prevents elongation of the bore during the fracturing process, so there's no concern for creating an out-of-round, big-end bore. Also, due to the compacted powdered metal construction, no stress cracks or weak areas are created when the fracture takes place.

The irregular mating fracture surfaces provide an "intimate interlock" between rod and cap. This virtually eliminates cap shift (rotation of the cap relative to the rod) and lateral movement of the cap relative to the rod. Cap shift can lead to accelerated wear of bearing surfaces and, in extreme cases, bearing seizure. Lateral movement can result in high shear stress on connecting rod bolts at high engine speed (RPM).

Close-up of a powdered-metal cracked-cap rod-mating surface. When dealing with powdered-metal rods, be careful. Do not disturb the irregular mating surfaces. If you blast them or touch them with a file, or drop the rod onto the floor, you ruin the mirror-image mating surfaces, and the rod becomes worthless. The profile of these mating surfaces (on rod saddle and cap) must remain absolutely intact and undisturbed.

OEM PM rods are usually okay for up to about 400 hp. Beyond that, move up to forged rods. The downside of PM rods is that they should not be reconditioned in the traditional manner. In fact, resurfacing the mating ends reduces big-end diameter and creates an out-of-round condition, and then they need to be rehoned round to the specified diameter. PM rods don't have enough material to cut in this manner. However, depending on the engine application, you can simply hone the big ends to an oversized

Because the cap-to-rod mating surfaces are precise mirror images, once the cap is fully tightened, the parting lines virtually disappear.

bore dimension and then install oversize-OD bearings. If a PM rod is damaged, you often need to pitch it and buy another. They're not designed to be reconditioned.

Although PM rods have traditionally been used only in OEM engines, you're now also starting to see high-performance aftermarket PM rods. Howards Cams, as an example, teamed up with GKN and now offers forged, powdered-metal rod technology with an extremely dense and non-directional grain structure.

A high-tech base powder is blended with select alloy elements. Melting, atomizing, and annealing are controlled to exacting standards. The metal mix is compacted (in dies) under tremendous pressure, at more than 1,500 degrees F. Hot forging with a 750-ton press finalizes the structure of the metal.

This new generation of PM rods represents a hybrid of PM and forging. Although OEM-level PM rods typically withstand about 400 hp, Howard's new PM rods with 5/16-inch ARP 2000 rod bolts are capable of withstanding more than 585 hp, but they require L19 rod bolts. These rods have even survived at more than 800 hp.

The OEM PMs are a practical choice for a street engine up to about 500 hp. I recently built a 5.3L iron-block LS engine that I overbored to 327 ci and assembled using stock OEM powder-metal rods on a stock crank. In carbureted form (single 650-cfm), the engine pulled 425 hp and had no issues whatsoever. If you're planning on any high-performance build, forged aftermarket rods from Scat, Eagle, Lunati, Callies, Crower, Oliver, etc. are good choices. If you're going to use PM rods, you shouldn't expect to pull more than 400 hp and your engine will be reliable.

Reconditioned PM Rods

A major concern when trying to recondition rods with a cracked cap in the traditional manner relates to the irregular mating surfaces of the rod and cap. This irregular surface provides an accurate locating of the cap to the rod, preventing any misalignment of the cap during assembly.

However, if the mating surfaces are machined or ground flat to reduce the rod's large-end bore in preparation for rehoning to size, all centering ability is lost. This destroys the unique cap-to-rod interlock. Since no interlock mating remains because there are no positioning tangs to use, it's possible to install the cap slightly off-center, due to the small tolerance range of the bolts to the cap's bolt holes. As a result, the cap might be installed off-center left-to-right (laterally), or at an angle relative to the axis of the large-end bore. If only for this reason, it is not advisable to reface the rod and cap mating surfaces.

Another reason that resizing these rods can create a problem is the relatively thin cap material. Once the mating surfaces are ground flat, the new smaller and non-round large-end bore may require so much enlargement in order to create a round hole that the cap material may be reduced enough to create a potential weak area. Note that in the process of creating flat mating surfaces, it may be necessary to reduce the mating surfaces by .040 inch or more, which could result in a combined reduction of the hole by .080 inch or more. Precious little cap material may be left after resizing.

If resizing is necessary, do not disturb the irregular cracked mating surfaces. Instead, hone the big end to an oversize dimension to accommodate fitting oversized-OD rod bearings. However, oversized bearings are not available for all cracked cap applications. You may find them with a standard-size ID and a .010 inch larger OD; or in an undersized ID (to accommodate an undersized ground crank) and an oversized OD, to accommodate an enlarged connecting-rod big end.

If an oversized OD bearing is not available for your particular application, replace it with a new rod. In some cases a new rod may not be available as a separate piece. A case in point is the Dodge/Plymouth 2.0 SOHC and DOHC engines,

TECH TIP

Design Advantages of Powdered-Metal Forging

- Even distribution of stress over the entire side of the I-beam area
- Increased I-beam rigidity (100-percent increase in stiffness over conventional forged)
- Improved noise/vibration/harshness level, which improves piston and ring action
- No bore offset between rod and cap
- Increased bore stiffness and geometry
- Increased bearing reliability
- Reduced weight
- Crankshaft mass reduction
- Single weight grade (+/- 2 g on a 400-g rod)
- Reduced mass variation for improved balance
- Full density across entire rod
- Lower cost carbon-steel
- Proven fracture split method (no crank bore elongation during cracking)
- High machinability

High-quality aftermarket connecting rods, whether forged or billet, commonly are supplied with high-grade, high-tensile-strength rod bolts. Generally, rod manufacturers supply a torque specification as well as a rod bolt-stretch range, providing a choice of method for you to follow. Even when using OEM cast, forged, or powder-metal rods, if you're using quality aftermarket rod bolts that are supplied with a stretch recommendation, tightening by stretch (rather than torque or torque-plus-angle) is possible.

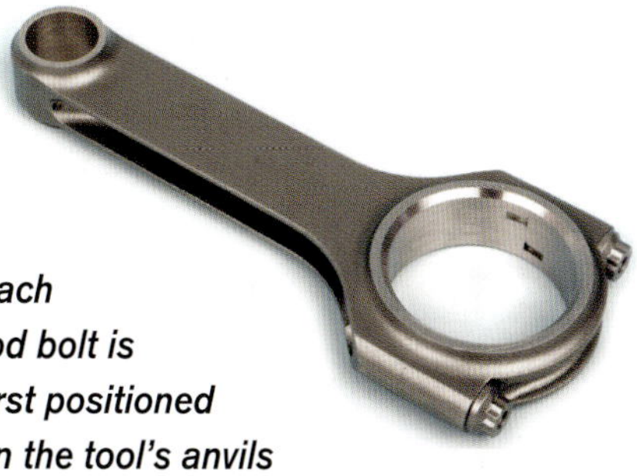

Each rod bolt is first positioned on the tool's anvils and the gauge is zeroed. This provides a reference point (while the bolt is in its relaxed state). A GearHead stretch gauge is shown here. Notice the convenient thumbhole on the billet housing. During test fitting, final assembly, and any future teardowns, always keep the rod bolts with their original rod (keep them organized), so that you can remeasure each bolt to determine if it's still usable and hasn't lost some of its original elasticity.

I-beam rods have a recess on the beam faces. The strength of H-beam versus I-beam rods is often debated, as well as the theoretical characteristics of each beam design to sling parasitic oil during engine operation. For extreme engine speed and for increased cylinder pressure applications, such as nitrous/forced induction, H-beam designs are often recommended. Major rod manufacturers usually offer both styles.

where a complete rod/piston/pin assembly must be purchased from the automaker because individual rods or pistons are not offered. So, in a situation where a rod's big end has been distorted and an oversize OD bearing is not available, you are forced to pay for a complete rod/piston/pin assembly, even if the original piston is perfectly serviceable.

Forged Rods

Forged steel rods start as an ingot, or billet, of alloy steel. In fact, performance aftermarket forged rods are usually made from 4340 chrome-moly steel. A steel ingot is usually heated in an oven to about 2,200 degrees F during conventional steel forging, at which point the steel is very formable. The ingot is then placed in a forging die and squeezed into the approximate shape of the desired profile. A hammering or pressing process performs this squeezing with as much as 240,000 pounds of pressure. This increases the strength of the alloy by compacting, tightening, and aligning the molecules.

The size of the ingot is much larger than required in the die, so it starts with an ingot that weighs about twice as much as the desired final product. During the forging/compacting process, the excess material is forced out of the die at its mating lines. This excess is later sheared off in a trimming die. Depending on the manufacturer, the rods may be induction hardened, shot peened and/or cryogenically stress relieved, and heat treated. Individual manufacturers often employ their own proprietary formulas.

The forged rods are rough shaped and must be trimmed then quenched and tempered. Before being machined, they should be tempered because the process can alter the shape of the part. The rods can deform by as much as .060 inch. Although the process may differ from manufacturer to manufacturer, the rod is typically quenched in a glycol solution. Once it has been quenched, the rod is finish machined to attain its final shape. Then the rod is put through stress relieving so it is resistant to the formation of stress cracks. These parts are baked in an oven at 400 to 600 degrees F to remove stresses that occurred during the machining process. During this process, carefully controlled heating and cool-down cycles properly cure the metal. The rod is then final-machined for small- and big-end bore size. And last, the final surface hardness is set.

When examining forged rods, you may find what looks like a large parting line. In fact, this is not a parting line. This line was established when the excess steel was pushed out of the die, and then it was cut off after forging. Ultimately, final machining was performed but this line remained. In some cases, hot-hammering or pressing squeezes the malleable steel out of the die, and then machining

Unlike OEM cast or forged connecting rods, high-quality aftermarket performance rods are manufactured to a much higher level of precision and uniformity. Note that this H-beam rod's cap has strengthening ribs but no weighty "balance pad." Today's quality performance rods are made to such tight dimensional and weight tolerances that it's extremely rare that any balance-correction work is needed, hence the absence of extra metal (that is, the pad) on the rod cap. Balance pads are common on older OEM rods, providing an area of mass that can be reduced in order to properly balance a set of rods.

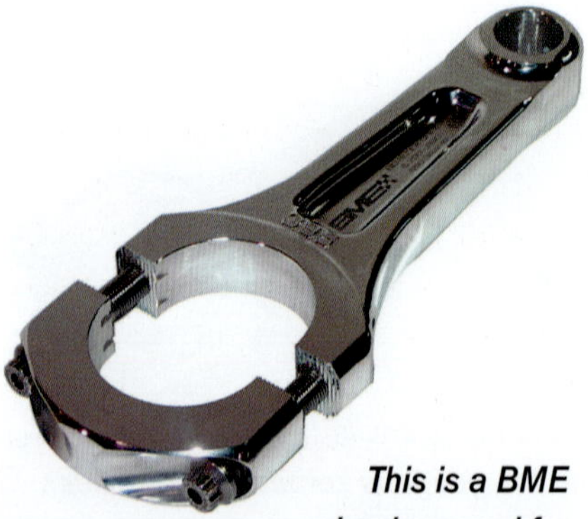

This is a BME aluminum rod for a small-block Chevy. Note the precision polished surfaces (this eliminates stress risers and aids in tossing-off parasitic oil-cling). Also note the precision-cut mirror image serrated mating surfaces for cap-to-rod registering. (Photo Courtesy Bill Miller Engineering)

removes the excess. As a result, no evidence of a trim line remains.

In other cases, the trim area may not have been machined as closely, so you can faintly see the line in the trim area. Forged parts may slightly show that a die was used during the manufacturing process. Some evidence of a trim area is common and does not create any issues. Most quality aftermarket forged rods are machined with high precision and, in the process, any parting line traces are eliminated.

Aluminum Rods

Forged aluminum rods are usually made from 7075 or 7075-T6 aluminum alloy. They may be made from forged flat blanks or forged aluminum that has been extruded. A common belief is that aluminum rods have a relatively short lifespan (due to fatigue) and are not suitable for street applications where routine teardown doesn't take place. But that is not true. Aluminum rods can be used for the street.

Some of today's premium aluminum rods are die-forged instead of being cut from a plate of aluminum. Die-forged rods begin as high-density aluminum bar stock and tend to have a denser grain structure. During the die-forging process, the aluminum is heated to around 700 degrees F and then pressurized with about 2,200 tons of pressure. This enhances the grain flow and increases grain density. This also forces the grain around the rod bearing bore area for additional strength. If you want to spend the money and if the aluminum rods clear the block and crank, this is a lightweight alternative to steel.

Billet Rods

Billet rods come in alloy steel and aluminum. With the capabilities of today's CNC machining, it's now possible to machine connecting rods from raw stock. However, the stock is actually a dense-grain steel that has been made by a forging process. So in reality, billet steel rods use forged steel that is then CNC machined to final state.

Billet rods are more expensive than forged rods due to the higher cost of the steel alloy and the machining time. Since these rods are CNC machined, they are manufactured to very precise weight, dimensions, and specifications.

Titanium Rods

Titanium has an incredible strength-to-weight ratio. It is billet-machined from Ti6AL4V stock, and is about 33 percent lighter than a comparably sized forged steel rod. As an example, a complete titanium rod may be lighter than only the big end of a comparably sized steel rod. Lighter reciprocating weight translates into quicker revs and more power because of reduced parasitic mass. Titanium rods are much more expensive than forged steel.

These rods reduce rotating mass, which is a discernible advantage at engine speeds in excess of 5,000 rpm or so. In a race engine that has real benefits, but in a street engine, it's really a waste of money. Also, titanium is a fragile material that is sensitive to scratches. Small scratches in the surface can lead to stress cracks that can lead to rod failure.

Titanium from a friction/machining standpoint is rather "gummy" and often galls when rubbed. The concern comes at the rod's big-end sides. To prevent this condition, titanium rods must be polished and/or coated with a hard-surface coating, such as chromium nitride (see Chapter 6 for more details).

Relative Material Cost

Forged or billet steel is suited for the vast majority of street and race applications. Where further weight reduction is desired, forged/billet aluminum rods are available at a higher cost.

For high-RPM engines where weight savings are really critical, titanium rods are often the best option but they tend to fatigue more than steel over time. In racing use, they need to be replaced more frequently compared to steel rods.

Titanium rods are also very expensive, which is a major factor when you're on a real-world budget.

Aluminum rods are lighter than steel and nearly as light as titanium but cost less than titanium and more than steel. Aluminum rods tend to be more on the chunky side and generally require more block clearance.

Center-to-Center Length

Center-to-center (CTC) length is the actual distance from the center of the rod pin bore to the center of the rod bearing bore. (See Chapter 5 for details on calculating this dimension.)

Rod length is a factor in determining the combination required to achieve a specific stroke, relative to the block deck height.

When blueprinting, you want an equal TDC location for each piston to obtain an equal compression ratio in each bore location. During test fitting, install the crank, rods, and pistons with bearings, but without rings. Slowly rotate the crank to bring each piston to TDC and measure the distance from the top of the piston's compression deck to the block deck surface.

Slight deviations in tolerances of the crank, rods, and pistons may result in differences in TDC height. By swapping rods to other cylinder locations, you can optimize the components. For example, mix and match the rods/pistons until you obtain the most equalized dimensions. Yes, this is time consuming and nit-picky, but that's part of blueprinting: the attempt to optimize all dimensions, weights, and clearances.

Piston-to-Rod Clearance

The small end of the rod must not touch any part of the piston. On the bench, test assemble each rod to its piston with wrist pin and check the clearance between the top of the rod to the underside of the piston. This generally isn't an issue unless you're running OEM stock rods with big balance pads at the top of the rod and non-stock pistons. Pivot the rod small end on the wrist pin and make sure that there is adequate clearance at the underside of the piston. Even considering thermal expansion, you should have at least .080-inch clearance.

Also check the clearance between the rod small end and the piston pin bosses (where the rod slides on the wrist pin). Even if you have clearance on the bench, that's no guarantee that you will have clearance when the package is installed. In many engine designs, the rod beam is not centered under the piston (slight offset) when the rod is positioned onto the crank.

Test install the rod/piston onto the installed crankshaft. Rotate the engine block on your stand upside-down. Work the rod back and forth to see how close the small end of the rod gets to the pin bosses. You should have at least .060- to .080-inch clearance. If the small end touches the pin boss, mill a bit from the pin boss or narrow the rod small end to accommodate this.

Clearance issues more commonly occur when using aluminum rods because they are thicker. Never have the crankshaft balanced until all prefitting and clearancing has been accomplished.

A rod bore gauge (the gauge shown here is on a Sunnen rod-honing machine) allows accurate and quick measurement of the rod-bearing bore diameter as well as allowing checking for bore runout. After calibrating the bore gauge to the desired bore diameter specification, the rod's big end is placed onto the gauge to verify bore diameter. This reveals whether the bore is correct, undersized, or oversized.

Rod-to-Block Clearance

This should only be a concern when using a stroker crankshaft (or possibly when using thicker aluminum rods). With the crank installed and the pistons/rods test installed, slowly rotate the crank to inspect rod big-end clearance at the bottom of all cylinders. If rods touch or if the clearance is too tight, mark the block and grind material to obtain clearance. Generally speaking, rod-to-block clearance should be at least .080 inch.

Rod-to-Camshaft Clearance

As the crank rotates and the rods travel toward TDC, the big ends of the rods approach the camshaft. Especially with a long-stroke crank, high-lift cam, and/or thicker steel or aluminum rods, there is a concern that the rods might hit the cam lobes. During test fitting, slowly rotate the crank and watch for rod-to-cam interference. If you feel resistance, stop. Using a slim flashlight, you may be able to visually check for clearance.

The camshaft rotates at half the speed of the crankshaft, so be sure to check through at least four crankshaft revolutions. An aid in checking clearance is to apply a strip of .125-inch-thick clay to the side of the rods, at the lower beam to about the rod cap parting line. Be sure to clean the rod surface to remove any oils before attaching the clay.

When you rotate the crank, any area tighter than the thickness of the clay imprints and provides a witness mark. If you find any contact marks, use a razor blade to cut out the section of clay and carefully measure the compressed clay thickness. You want about .060 inch of clearance. If the rod dead-stops against cam lobes, remove rod material at the contact area, re-install, reclay, and recheck clearance.

If more clearance is required, you have two choices: remove material from the rod or change to a camshaft with a smaller base circle. If you change to a cam with smaller base circle, you need longer pushrods. Obtaining such a cam was once a real hassle, requiring much expense and time. But with today's CNC capabilities, most cam manufacturers can produce what you need relatively quickly.

If you opt to relieve the rods, do it carefully to avoid weakening the rod bolt's female threaded areas. Also, the rod needs to be free of sharp edges and grind/scratch marks. After grinding the necessary relief, reassemble and recheck clearance. The ground area must be carefully polished to remove any potential stress risers.

Again, never perform crank balancing until after all clearancing has been verified. Remember to mark all rods and their caps with bore location numbers. Once a rod has been fit and clearance verified to a specific bore location, keep it in that location.

Rod Bearing Clearance

Measure the crankshaft rod journal diameter with a micrometer. Don't just rely on the published specs that came with the crank. Record this diameter for *every* rod journal. Never assume that all journals were ground identically. Install a pair of new rod bearings to the rod and cap. Make sure that the rod and cap saddles are clean and dry first—no oil should be between the bearing and saddle. Install the rod cap and, in a rod vise, tighten both rod bolts to the specified value (whether you're using torque or tightening by stretch).

Measure the crankshaft's rod journal using a micrometer and record the reading. Install the rod bearings in the connecting rod and fully tighten the rod cap to spec. Calibrate a dial bore gauge to the recorded diameter of the rod journal. Use the bore gauge to check the installed bearing diameter in the rod. Any difference in the bore gauge (plus or minus) reveals the existing oil clearance. Be sure to measure the bearing diameter at the 12-o'clock and 6-o'clock positions, at 90 degrees to the parting line.

Assuming that you've already verified that the rod big-end bore is within specification, if oil clearance is too tight or too loose, undersize or oversize bearings are available for most applications.

Be sure to inspect rod-bearing edge clearance to the crankshaft fillets. Rather than trying to view this with the crankshaft installed to the block, position the crankshaft on a pair of clean V-blocks on a bench. Install the rod bearings to the rod saddle and cap. Before installing the rod to the crankshaft, use a felt-tip marker to paint the edge of the bearing that faces the journal fillet.

Assemble the rod to the crankshaft's rod journal. Position the rod fully against the fillet and rotate the rod back and forth against the fillet. Remove the rod from the crankshaft and inspect the marked edge for witness marks that indicate whether the bearing was rubbing against the fillet. If so, the bearing edge can be carefully relieved with a bearing scraper tool.

Rod Side Clearance

Side clearance refers to the front and back clearance of a pair of rod big ends on the crankshaft's rod journal. With the crank installed in the block, with a pair of rods and pistons installed, and with the rods on a shared journal, push the rod big ends apart (shoving both against their fillets and creating a gap between them). Use a feeler gauge to measure clearance. An acceptable range for side clearance is around .014 to .020 inch for steel rods, and perhaps a bit more for aluminum rods (considering theoretical expansion

Never assume that rod side clearance is correct, whether you're dealing with new or used cranks and/or rods. Always check side clearance by inserting a clean feeler gauge between the two rod big ends. Be sure to pull the rods apart by hand while inserting the gauge. Always refer to side clearance specifications for the engine at hand, but in general terms, you should have at least .012 inch or so clearance.

rates), around .017 to .022 inch. Always check to see what the manufacturer recommends.

Beam Design

Connecting rods are usually of I- or H-beam construction. The designations refer to the shape of the rod beam cross section. I-beam rods have a smooth, solid surface at the beam sides and a recess along each side of the beam faces. H-beam rods have flat, solid surfaces on the beam faces and grooves (recesses) along the beam sides.

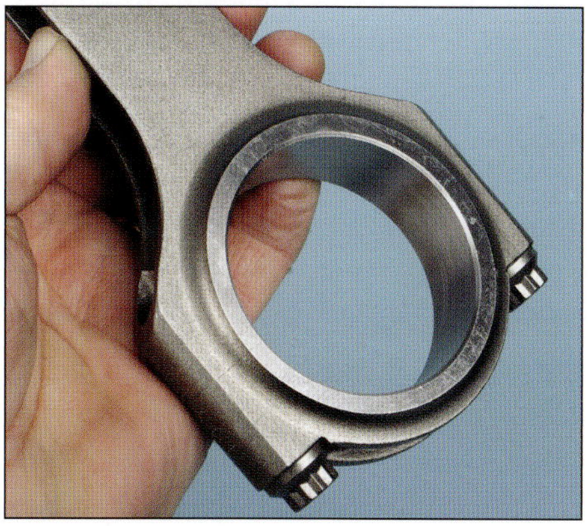

All connecting rod big ends feature a flat-machined surface on one side and a chamfered edge on the opposite side. A flat side is shown in this example. The chamfered side always faces the crankshaft journal fillet, and the flat sides always face each other (rod to rod) on the shared journal. Also, never swap rod caps among your rod set. The original cap that was mated to the rod by the manufacturer must always remain with the same rod. Rod manufacturers usually help by laser-etch matching numbers on both the rod and the cap to ease identification.

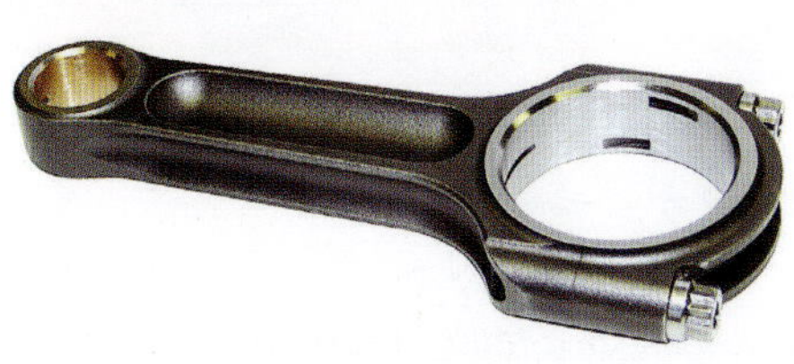

Example of an X-beam rod. This is one of Howards Cams new Extreme Duty X-beam rods for import turbo applications. Notice the lightening groves on both the beam faces and sides. This provides reduced reciprocating weight while maintaining rod strength. (Photo Courtesy Howards Cams)

Which style is better? In theory H-beam rods are stronger but, in reality, an H-beam rod can be lighter while being as strong as an I-beam rod. From a windage standpoint (oil clinging to the rod during engine operation) the I-beam is theoretically better. However, there are exceptions because specialty or "oil shedding" coatings can be applied to promote less parasitic oil cling to any style rod. In many cases, choosing between I-beam and H-beam boils down to manufacturer availability and/or engine builder preference.

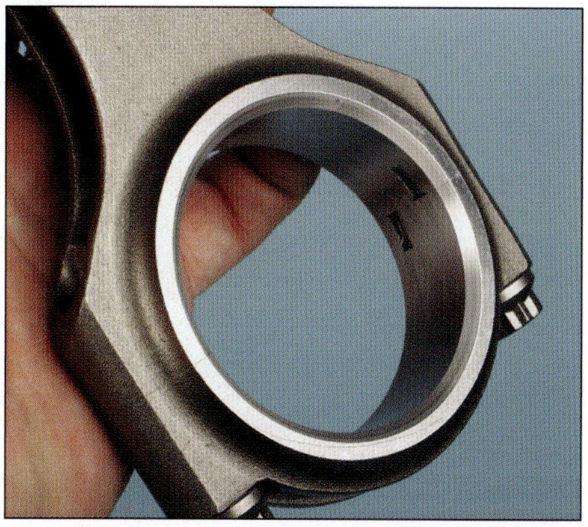

Shown here is the chamfered side of the same rod big end. Notice how pronounced the chamfer is on this side. The chamfered side must face the crank fillet (the area where the rod journal meets the crank's counterweight base).

Most performance connecting rods feature floating pin designs, which allow the wrist pin to freely rotate in both the piston pin bore and the rod small-end bore. The rod big end has a bronze bushing.

Another style of rod is the X-beam, which is available for various automotive gas-engine applications as well. The X-beam is sort of a mix of both I-beam and H-beam, and it has weight-saving grooves on the beam faces and the beam sides. This provides a substantial weight savings, while also increasing the beam surface area, offering lighter weight while retaining strength.

Rod Bolt Tightening

One of the easiest and least expensive ways to ensure longevity of connecting rods and rod bearings is to use only the highest-quality rod bolts. That means that you should buy ARP, A1, or another performance brand.

Rod bolts can be tightened in one of three ways: using a torque-plus-angle method, torque application, or monitoring bolt stretch.

Regardless of which tightening method you choose, it's a good idea to take advantage of a stretch gauge. Regardless of what type of rod bolts you have (OEM or aftermarket), first measure and record each rod bolt's overall free length (when new and uninstalled). Be sure to record which rod each bolt will be installed to (cylinder number-1, -2, etc.). During any future engine teardowns (or when the opportunity arises), remeasure each bolt's free length and compare it to the original (new) free length that you recorded. If the bolt has elongated (stretched) by more than .0005 inch, replace the bolt because it has begun to lose its elastic properties. Never assume that a used rod bolt is still serviceable.

Rod and Cap Numbering

Connecting rods *must* be kept matched with their rod caps. When rods are manufactured or reconditioned, the big end is honed with the cap attached and fully tightened to specification. *Never* mix up your caps; this guarantees a diameter and out-of-round problem that reduces rod bearing clearance.

To keep rods matched to their respective caps, each rod and its cap should be marked with matching numbers. Quality performance aftermarket rods are usually laser-etched with matching numbers. These may be factory-assigned sequential numbers, such as matching 3-, 4-, or 5-digit numbers. The numbers may not mean anything to you, but each rod and its cap has the same number. This makes it easy to match up your caps and rods if they become separated during cleaning.

If your rods and caps are not already numbered (usually with OEM rods), you can do it. Bore location numbers are the easiest to use. Simply follow the engine's firing order to identify each rod for its location, by placing the numbers on one side of the rod big end (just above the parting line) and the side of the cap (just below the parting line on the same side). For example, for number-1 rod, place a "1" on the rod and "1" on the cap, etc.

However, it is *not* advisable to stamp the numbers by using a steel number punch and a hammer. If you don't have a very skilled feel for this, you can easily distort the rod bearing bore, depending on the amount of force applied, which may be difficult to control. The safe method is to etch the numbers. If you or your shop does not have access to a laser etcher, a simple and inexpensive electric etching pen (available at most hardware stores) can be used.

Keep in mind that PM rods have a cracked-off or fractured rod-to-cap parting surface. These surfaces are irregular and permit a perfect mating of each cap to its original rod. With that said, don't rely on your eye to identify this matchup. Even PM rods and caps need to be match-numbered to avoid destroying the mating surfaces if you accidentally attempt to install the wrong cap to a rod.

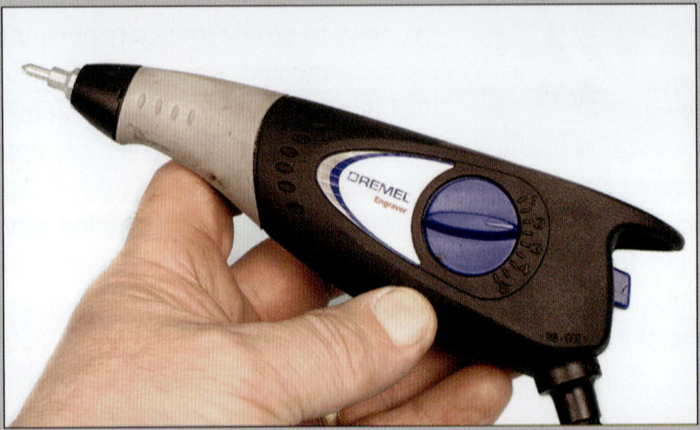

If you don't have access to a laser-etching machine (and most of us don't), an inexpensive hardware-store electric etching pen works fine to etch identification numbers on the rods and rod caps. Even this inexpensive amateur-level etching pen does the job.

During the entire build process (test fitting, measuring, cleaning, prepping for final assembly, etc.) keep all components organized on a clean work surface.

Etch the bore location number on both the rod and the cap side, next to the cap parting line.

Torque-Plus-Angle Tightening

The torque-plus-angle method is only to be used on OEM rod bolts, when the specifications call for torque-plus-angle, for a specific engine application. If you're using aftermarket performance rod bolts, in the majority (if not all) cases, the manufacturer provides a torque specification and a stretch range (giving you a choice of tightening methods).

Torque Spec Tightening

If you intend to use the torque spec method, bolt manufacturers usually provide two torque values: one with engine oil as a lubricant and one with a specific moly lube. Torque values are always a bit lower with moly because moly decreases thread and underhead friction (if you use moly but tighten to the oil spec, you run the risk of overtightening). Moly is preferred because it greatly reduces friction and provides a more accurate (and consistent) torque value. When you lubricate the rod bolt prior to installation, be sure to apply lube on the bolt threads and to the underside of the rod bolt head.

Bolt Stretch Monitoring

Remember that bolts are designed to stretch when they enter their elastic state (you want this "alive" elastic state in order to provide sufficient clamping force). However, if the rod bolt is overtightened and exceeds its designed elastic state, the bolt immediately weakens. If a rod bolt has been stretched beyond its elastic range, it must be replaced.

If you're blueprinting an engine, you are most likely building a high-performance or race engine and should use aftermarket high-performance connecting rod bolts. This tightening process is used by all major race engine builders.

You should not just tighten the rod bolts to reach a particular torque value. Instead use a precision-tightening technique for the connecting rod bolts. This

Each rod bolt is first positioned on the tool's anvils and the gauge is zeroed. This provides a reference point (while the bolt is in its relaxed state). A GearHead stretch gauge is shown here. Notice the convenient thumbhole on the billet housing. During test fitting, final assembly, and during any future teardowns, always keep the rod bolts with their original rod (keep them organized), so that you can remeasure each bolt to determine if it's still usable and hasn't lost some of its original elasticity.

How to Measure Bolt Stretch

Here are the steps: Measure the static bolt at the bottom of the shank to the head. When installing the fastener, use a high-quality moly lube to lubricate the bolt threads and bottom side of the bolt head. Position the rod cap on the big end of the rod and, by hand, thread in both bolts to the connecting rod.

Once both bolts have been finger tightened, use a clean plastic hammer to carefully (gently!) tap the cap into the piston. Using a torque wrench, tighten the bolts to the manufacturer's specification (using the moly spec).

Next, use a stretch gauge to check the bolt's amount of stretch. Refer to the manufacturer's specs for stretch. If the bolt did not stretch as far as needed, *do not* continue to tighten it. Instead, back the bolt off, apply more lube, and re-install it.

Tighten to the listed torque spec and again check bolt length with the stretch gauge (you may add a few pounds of torque in order to achieve the specified stretch). The difference in length indicates the amount of stretch the bolt experiences in its installed state.

Even when you plan to use a stretch gauge to monitor bolt stretch, you still tighten by torque (to the bolt-manufacturer's spec!), and then measure the bolt to find out how much it has stretched.

To accurately measure bolt stretch, use a specialty fixture and a dial indicator. You can use a micrometer to measure bolt stretch but this is not the most accurate method.

ARP, Gear Head Tools, and Goodson all sell high-quality fixtures for taking this measurement. A Goodson Shop Supplies rod bolt stretch gauge (PN RBG-4) has spherical points for consistent and accurate readings. ARP sells a bolt stretch gauge, which has a sturdy spring and ball tips and can take measurements with .0005-inch increments. ARP and Gear Head (P/N 13002) offer stretch gauges that also have a finger-hole in the body, which is extremely handy in holding the gauge steady while taking the reading.

Today's high-quality aftermarket performance connecting rod bolts always have a dimple at both ends of the bolt (at the head and at the tip of the shank). These dimples provide precise locating contacts for a stretch gauge.

requires you to use a bolt stretch gauge during the process. The torque values for aftermarket connecting rod bolts may vary by as much as 10 ft-lbs from batch to batch because of differences in heat treating and materials. The bolt needs to be measured so you attain peak bolt strength and big-end roundness. Measuring bolt stretch allows you to accurately attain the target clamping loads rather than just using a torque wrench.

Granted, when dealing with production engines that utilize torque-plus-angle specifications (for example, 20 ft-lbs followed by a 90-degree additional turn) follow the OEM procedure. Tightening connecting rod bolts using the stretch-monitoring method really only applies to performance engines that have aftermarket high-performance rod bolts and the bolt manufacturer has provided a stretch value reference.

There is some debate among engine builders regarding the validity of measuring rod bolt stretch, due to potential compression of the rod material as the rod cap is clamped to the rod. Although this may occur, a stretch gauge remains the optimum method of accurately determining connecting-rod bolt clamping load.

Tightening Method Variations

Think of a bolt and other similar fasteners as a high-resistance spring. To get the best clamping force from a bolt, you must not overtighten it or you exceed the yield point. When tightened below the yield point, the bolt provides consistent and accurate clamping force that you need for a high-performance engine build. You can damage the rod threads through overtorquing, and bolts can have unequal clamping force. This can lead to failure of the bolt and make the bolt bore out of round.

OEM rod bolts commonly provide a tensile strength of approximately 150,000 to 160,000 psi. Variances in bolt production can greatly affect the tolerances, and as a result, peak bolt stretch travel can occur from .003 to .006 inch. Use a bolt-stretch gauge and the torque spec to reach the ideal bolt stretch for an application. If you simply use the torque spec, you may end up with unequal rod-bolt clamping force.

Manufacturers produce premium-quality rod bolts to achieve far tighter tensile-strength tolerances. ARP calculates stretch and yield point for all bolts and its bolt packages include all specification data so you can safely and correctly attain the correct clamping force. The instructions state the particular bolt stretch for each bolt. ARP states that base tensile rating is 190,000 psi. Actual ratings are significantly higher for some specific products.

You must always strive for consistent and accurate torque of the connecting rod bolts. You must use the same method for tightening bolts at all stages of the engine build process. Different or inaccurate tightening methods can produce unequal or inadequate bolt tightness. In turn, this can damage the big-end bore shape. For instance, if one technician uses one torque value to recondition the connecting rods using torque value alone while another mechanic employs the bolt-stretch method, out-of-round bores can be the final outcome. The two methods produce different frictional variances. Thus, a higher clamping load may be reached using the stretch method, in contrast to using only bolt torque (without regard to actual bolt stretch). Also, only 80 percent of the torque can

Once the gauge has been zeroed (on a relaxed rod bolt), the bolt is lubed and tightened with a torque wrench to the rod or bolt manufacturer's torque spec. Then using the stretch gauge (still in its zeroed state from the original bolt free-length check), the bolt is checked to find out how far it has stretched. In this example, the rod bolt has stretched by .005 inch. Although a more time-consuming procedure than simply tightening to a specified torque value, achieving precise bolt clamping load is worth the trouble, especially for a race engine that represents a huge investment. It's important to understand that bolts stretch as they are tightened. By tightening the bolt to its optimum clamping load while staying within its elastic range, many race engine builders prefer to tighten by monitoring bolt stretch, rather than a specific torque value. Monitoring bolt stretch eliminates the potential variables of tightening-by-torque, such as thread friction, bolt underhead contact friction, and torque wrench calibration/accuracy.

This rod cap has been drawn from its rod. Before removing the rod from the tool, I temporarily and loosely replaced the rod bolts, simply to ease the task of keeping the cap with its rod (makes it easier to avoid cap mismatch mistakes).

A connecting rod vise is definitely worth the investment. When removing a rod cap from a new rod, or when tightening rod bolts with the rod off the crank you need to secure the rod. Using a common bench vise can cause serious damage to the rod; the creation of burrs, gouges, or worse can lead to stress cracks and failures. A dedicated, purpose-designed rod vise allows securing of the rod without damage.

A dedicated "rod cap splitter" is another specialty tool to consider. With the need to remove a rod cap on a previously assembled rod, the all-too-common approach is to first loosen the rod bolts, and then smack the rod bolt heads with a plastic hammer to dislodge the cap. That method is crude and always has the potential for dropping the rod. A rod cap splitter such as this one makes rod cap separation easy and precise. The rod big end is placed onto the tool's split collar. The spindle is then rotated, moving the collar halves away from each other. This smoothly draws the cap from the rod.

be exerted on a bolt without lubricant because of increased friction.

As you can see, variances can be caused by how the rod bolts were stretched (basing the results on torque value or bolt stretch), and how thread lubrication varies (if the bolts were tightened first with one type of lube then tightened subsequently with a different type).

When engines are hand-built, this close attention to bolt stretch and torque values is indispensable for achieving the highest-quality build. A methodical engine-build that is aiming for the highest level of accuracy should use the stretch method, rather than the torque method.

The Friction Factor

If a bolt is tightened by addressing only torque value, you may not necessarily achieve the desired preload due to the variable of friction. Consider the material of the connecting rod itself, which is beyond your control. As you tighten the bolt, the head of the bolt tends to embed itself into the rod, which slightly compresses the material of the connecting rod cap. The hardness of the rod cap material varies among OEM rods and aftermarket-forged billet steel, titanium, and aluminum rods. By monitoring bolt stretch, you eliminate this variable in the pursuit of achieving the required clamping force.

Simply put, by using the stretch-monitoring method, you eliminate the variables associated with thread friction and rod material hardness. If the manufacturer's specified amount of stretch for a given rod bolt is .005 to .006 inch, and you tighten by a measured stretch of .0055 inch, for example, you know that you've achieved the needed clamping force while neither undertightening nor overtightening.

Torque Requirements

As you tighten any threaded fastener, you fight friction in thread engagement. The underside of the bolt head or nut rubs against its mating surface. As a result, a significant amount of torque energy is lost. Although you may apply the specified amount of torque, according to the torque wrench, you really have no idea exactly how much clamping force is being generated. This is another example of why it makes sense to tighten critical rod bolts by using the stretch method.

When you rely on torque to achieve clamping load, you maximize your efforts by minimizing frictional losses. It's critical to make sure that the threaded holes in the block are clean and free of debris, burrs, and other contaminants.

In order to clean female threads in the engine block, do not use a cutting tap.

Recommended Torque for Optimum Clamping Force

The torque values below state clamping loads reached with a particular kind of fastener. This chart should be used as guide and not a definitive reference because it does not necessarily reflect the torque value specification for a given engine application. For that, you must follow the engine manufacturer or the fastener supplier specifications.

Always lubricate threads prior to torquing to ensure accurate readings. This chart is courtesy of ARP.

Fastener Tensile Strength 170,000 psi

Thread Size	*Torque (ft-lbs)*	*Torque Preload (psi)*	*W30 Oil with Moly*
5/16-24	28	22	6,948
3/8-24	50	39	10,512
7/16-20	80	62	14,220

Fastener Tensile Strength 220,000 psi

Thread Size	*Torque (ft-lbs)*	*Torque Preload (psi)*	*W30 Oil with Moly*
5/16-24	35	28	8,685
3/8-24	63	49	13,140
7/16-20	100	78	17,775

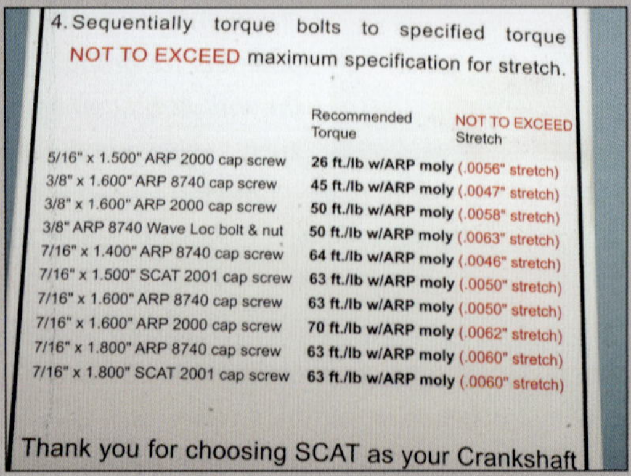

High-quality, high-performance connecting rods are usually packaged in boxes that include instructions regarding bolt torque and stretch specifications based on the particular bolts being used. (This varies depending on bolt diameter, bolt length and specific model of the bolt. This box has torque information (with torque, with moly, as well as bolt stretch specifications).

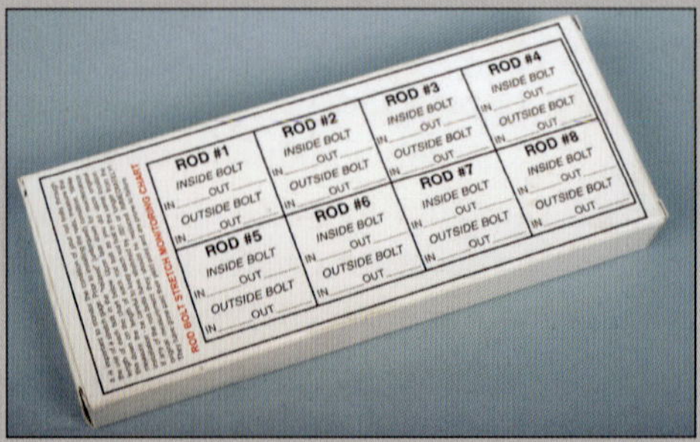

The opposite side of this rod box also includes a handy data-recording chart that allows you to record torque or stretch data on a per-bolt basis, as well as the number of times the bolts have been tightened.

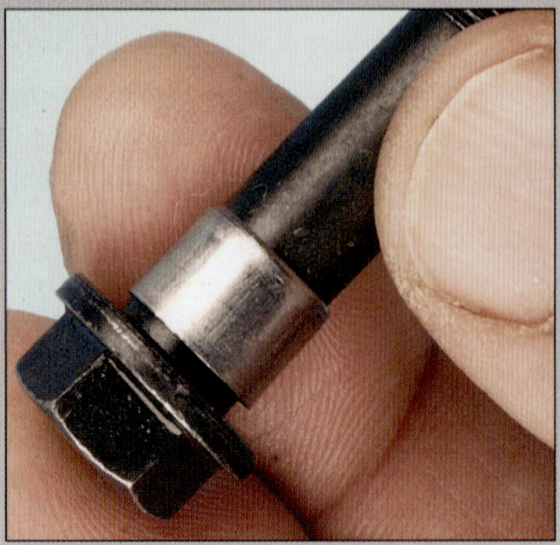

Some OEM rod bolts (such as this stock LS 9-mm bolt) have a thin metal dowel sleeve that centers the bolt shank in the rod cap. If you're using OEM rods that have this style, toss the bolts and dowels into the trash and replace them with performance aftermarket rod bolts (from ARP, for example) that offer not only superior tensile strength, but have shoulders that provide proper bolt centering. Incidentally, most OEM rod bolts are torque-to-yield style and should not be reused anyway.

They are designed to create threads by removing material. In order to clean existing threaded holes, start with a cleaning procedure (such as hot-tanking the block). Never assume that all threaded holes are clean by simply washing the block.

Hand-clean each holes with solvent, a rifle brush, and compressed air (always wear safety glasses). If thread condition is suspect, you may use a forming tap to recondition the threads. A forming tap is designed to reform the threads, in contrast to a cutting tap, which removes material. Special forming taps are available through machinist specialty supply sources and engine rebuilding tool sources.

When using high-performance aftermarket cylinder head bolts and main cap bolts, always use the thread lubricant specified by the bolt manufacturer, and follow the torque value recommendations. There is a great difference in lubricity between engine oil and moly-type thread lubricants. A moly lube provides much less frictional resistance.

If a bolt is specified for, say, 60 ft-lbs of torque with oil, but the threads are lubed with moly, you overtighten the

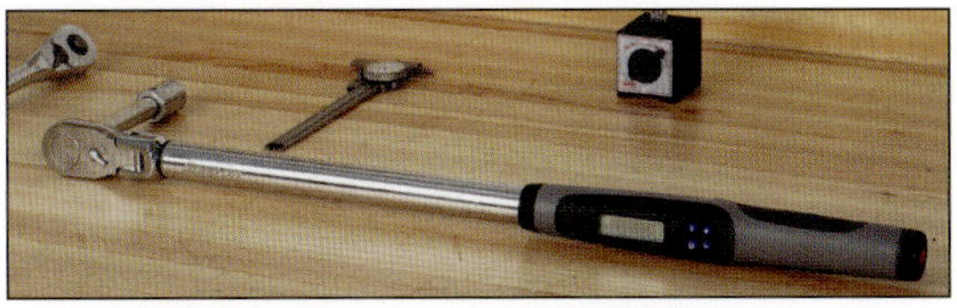

While a fairly expensive tool, a digital torque/angle wrench provides accuracy and work time efficiency; there's no need to stop and change to an angle wrench. If you perform OEM torque/angle tightening on a regular basis, consider buying one of these.

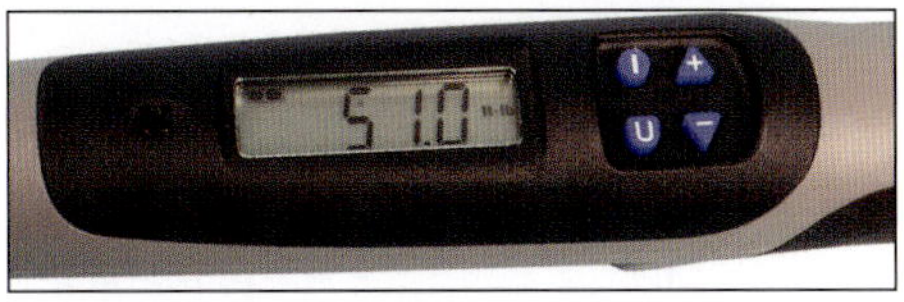

With the push of a button, you can quickly set your torque value (in ft-lbs, in-lbs, or Nm).

With the push of another button (changing mode), you quickly set your angle adjustment. One nice feature of this type of wrench (Snap-On shown) is that you can actually ratchet-tighten even in the angle mode without losing the reference point. Pretty slick.

If the rod bolt-tightening specifications call for a torque-plus-angle (typical for many of today's OEM rod bolts), a digital torque/angle wrench makes the job easy. Press a button, set the torque, and tighten until the wrench beeps. Push another button to set the required angle, and tighten again until the wrench beeps. Shown here is rod bolt tightening (with new OEM rod bolts), on a GM LS engine, using a Snap-On digital torque/angle wrench.

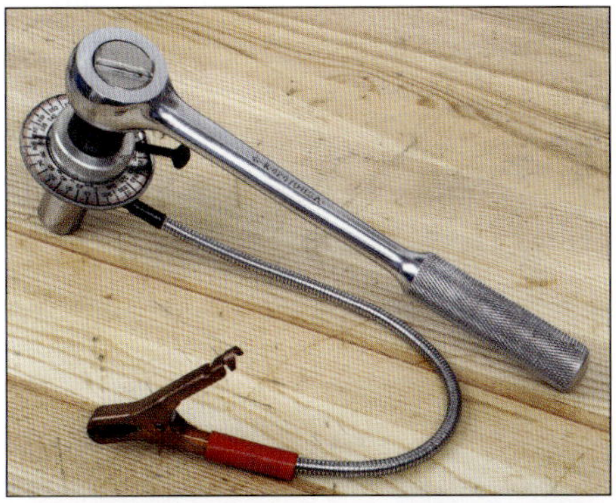

If your OEM rod bolts call for torque plus angle tightening, first use a torque wrench to tighten to the torque specification. Then continue to rotate the bolt for the specified number of degrees. Inexpensive metal or plastic angle gauges are available that attach to the head of a ratchet or breaker bar. This simple adapter, attached between the ratchet and socket, allows visual monitoring of tightening angle. With the tool engaged to the fastener, zero the gauge. As you continue to tighten, simply observe the gauge needle. Once it reaches the desired angle, stop.

If you tighten with a torque wrench, always make sure that you grip the handle at its center. Also, do your best to keep the wrench as level as possible (90 degrees to the bolt), and brace the wrench head with your other hand. Tighten slowly/gradually until you reach the desired torque value. Avoid quick pulls and yanks that can result in slight over tightening.

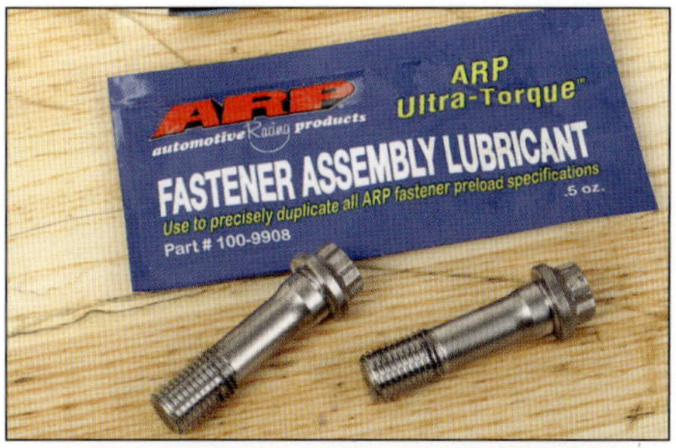

Specially developed fastener assembly lubricant is designed to drastically reduce friction (at the threads and bolt head underside) during torquing or tightening via stretch. Do not use a different type of lubricant. Rod bolt tightening is a critical aspect of engine assembly, and it's no place to cut corners!

When applying assembly lube to the rod bolts, apply an even coating to the bolt threads.

bolt. If the bolt is specified at 60 ft-lbs with moly and you use oil instead, you undertighten. Pay strict attention to the instructions included with any set of aftermarket head or main bolts.

Connecting Rod Inspection

In almost all circumstances, I recommend installing new connecting rod bolts for high-performance builds or stock rebuilds. It's simply not worth taking the risk with used and possibly overstretched bolts. Whether you're installing new or used connecting rods, you should always inspect each one for center-to-center length, big-end bore diameter, small-end bore diameter, and any out-of-round big-end bore. For used rods, also check for rod bend, rod twist, and cracks.

Also apply a coating of assembly lube to the underside of the bolt head. Remember, friction occurs between threads and between the bolt head and the rod cap.

When measuring the big-end bore for diameter and out-of-round, the cap must be fully installed, with rod bolts fully tightened to the torque (or bolt stretch) specified by the manufacturer. Also, always use the specified bolt thread lubricant. This is especially important if you're tightening to achieve torque value.

Torque value (and clamping load) differs between oil and a moly-type lubricant. If you're using OEM rod bolts that are specified for tightening by a torque-plus-angle method, the following procedure must be followed.

After measuring the big-end bores for out-of-round (and if the deviation is less than .002 inch), the bores can be reconditioned by grinding material from the mating surfaces of the rod and cap, then reassembled and honed round to size.

PM rods have a unique, irregular mating surface that must not be disturbed. If a PM rod is out of round by no more than .002 inch, the big-end bore may be honed to oversize, providing that oversize-OD rod bearings are available for the application. Otherwise, the rods cannot be reused and must be replaced.

The same rod bore gauge is used to check the rod's pin bore. If using aftermarket performance full-floating rods, diameter is based on recommended oil clearance to the wrist pins.

Certain aftermarket performance rods may have a serrated mating surface that registers the cap to the rod. This design also cannot be reconditioned by grinding material from the mating surfaces. Honing to a slight oversize and using oversize bearings is required. If any rod big-end bores are out of round by more than .002 inch, the rods should be discarded and replaced.

Each connecting rod (especially used) should always be checked for rod deformation, using a rod alignment stand. Here a rod is checked for bend. Note that the upper anvil is resting on top of the wrist pin. With the rod placed onto the tool-centering fixture, use a feeler gauge between the wrist pin surface and the upper checking base.

The same fixture is used to inspect the rod for twist, with the upper anvil contacting the side of the rod pin. Any measured bend or twist beyond .003 inch is cause for concern.

CHAPTER 8

Pistons

Pistons must withstand tremendous heat and cylinder pressures during their four-stroke cycle. Pistons take quite a beating, especially in high-performance or racing applications. High-compression ratios and/or cylinder pressure boosting with nitrous injection, turbocharging, or supercharging places greater demands on the piston. Add to this the possibility of detonation, and you're asking too much from these slugs. For any serious build, it's imperative to use forged or billet pistons. Cast pistons, even today's hypereutectic pistons, simply do not hold up to extreme demands.

Performance Pistons

Current-day OEM pistons are most commonly hypereutectic, which means they are precision-cast pistons with a high silica content. Hypereutectic pistons are much stronger and more stable than the cast pistons of old, and are generally good for up to about 400 hp.

A step up is forged or billet pistons for much-improved strength. Forged aluminum pistons start as a dense forging and are then CNC machined to final shape. Billet pistons begin as a dense-alloy billet block and are fully CNC machined to the final product. Billets are more expensive, mostly due to the waste of material that's machined away during manufacturing.

If you're building a performance engine from scratch, going to forged or billet simply makes sense. They're stronger and you have a much wider range of sizes and configurations available. If you're upgrading an OEM engine with hypereutectics, or if you're overboring and/or changing compression ratio, altering stroke, etc., forged and billet pistons are readily available in any configuration. Hypereutectic pistons are primarily available as direct replacements for rebuilds.

If you plan to boost cylinder pressure with nitrous or forced induction, you must upgrade to forged or billet pistons. They withstand combustion temperatures better than cast or hypereutectic pistons.

The following are features commonly found on performance pistons.

Vertical Gas Ports

These small, vertical holes in the piston deck (around the perimeter area of the piston deck) allow combustion pressure to directly enter behind the top ring (on the power stroke). This pressurizes the area behind the top ring for improved ring sealing (the pressure pushes the top ring against the cylinder wall). During operation other than the power stroke,

You can use an inside micrometer or telescoping gauge to measure the pin bore diameter. This is an important measurement when blueprinting the pistons.

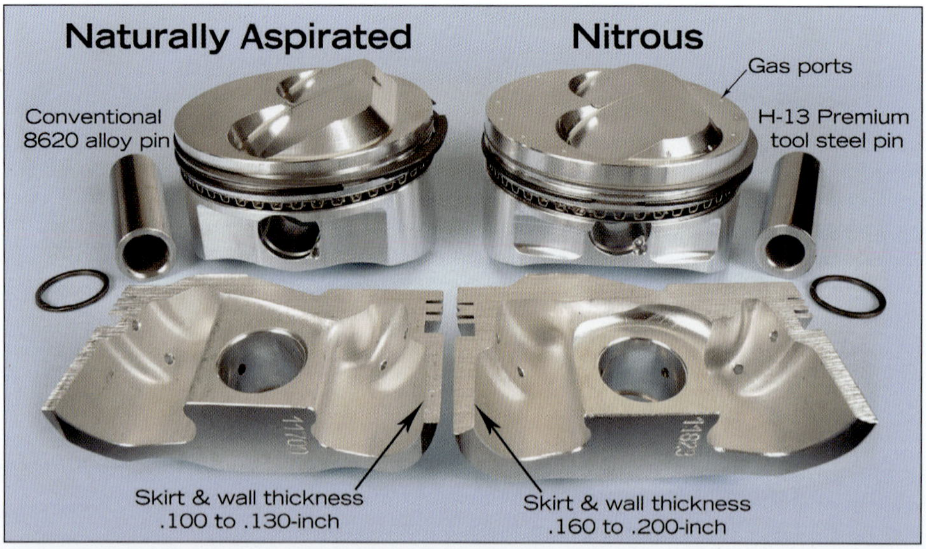

These cutaways provide an example of material thickness in strategic areas between a normally aspirated application (left) and a piston designed for use with heavy nitrous injection. (Photo Courtesy Diamond Racing Products)

the top ring experiences normal pressure and less drag.

Drag racing is the most common application.

Lateral Gas Ports

These very small slots are milled into the top (roof) of the top ring grooves, providing a path for combustion pressure to push the top rings out against the cylinder wall for improved ring seal.

This type of gas porting is most commonly used in circle track applications.

Vertical gas ports allow combustion pressure to push the top ring outward to provide better sealing.

Contact Reduction Grooves

Some pistons have a series of narrow circumferential grooves machined around the piston area between the top ring groove and the piston deck. According to JE Pistons, the purpose is to reduce the amount of contact area against the cylinder wall when the piston "rocks over" during the transition from TDC. Contact reduction disrupts the flame travel, reducing the chance of detonation.

Accumulator Groove

This radiused groove is machined into the piston land between the top and second ring grooves. This groove provides added volume for residual combustion gas that blows past the top ring. This reduces pressure between the top and second rings. The benefit: It reduces ring flutter and helps top ring seal.

Notice the shallow radiused groove (accumulator groove) between the top and second ring lands. This provides a small volume area for combustion pressure that sneaks past the top ring and reduces second ring flutter.

TECH TIP

Spin Boss and Window Milling

Spin boss refers to machining material from the bottom of the pin boss in areas where strength is not compromised.

Window milling removes material and weight from the inside of the piston skirt area while retaining skirt strength. This is basically this is a process to reduce weight where possible without sacrificing piston integrity. ■

Want a lightweight piston? When every gram counts, some piston manufacturers offer highly skeletonized pistons for optimum weight reduction. (Photo Courtesy Diamond Racing Products)

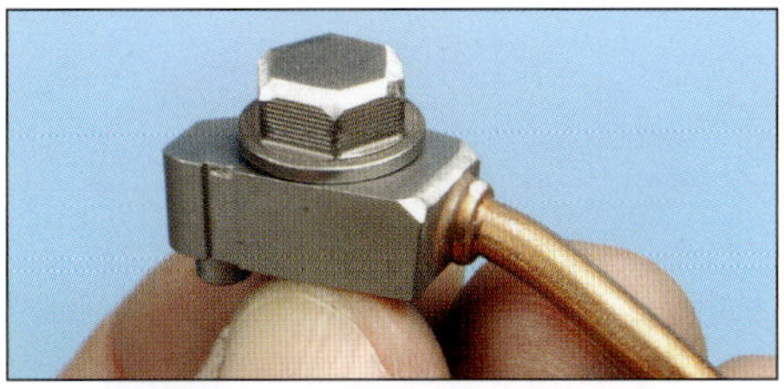

Oil squirters (OEM on some engines and added to some race engines) provide added oil cooling to the underside of the pistons.

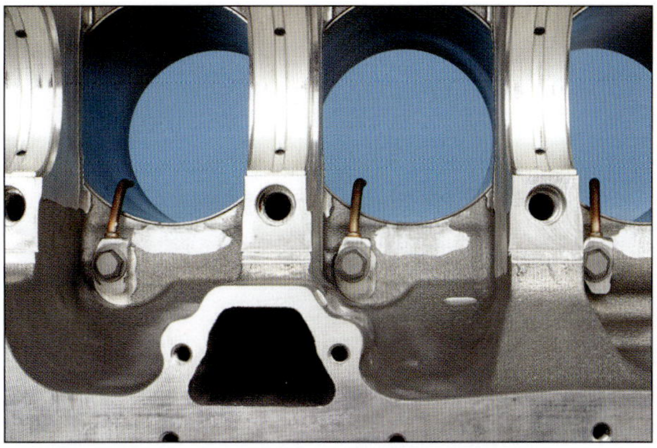

Oil squirters (stock OEM) on a B18C1 Honda block. When using a stroker crank, check for clearance between the squirters and the counterweights.

Double Pin Oilers

These large oil holes are milled into the floor of the oil ring groove. They deliver additional oil to the piston's wrist pin.

Oil Squirt Notch

Oil squirters are small nozzles secured to the block that squirt pressurized oil from the galley to the underside of the pistons for additional cooling. They have a notch machined into the bottom of the piston skirt to provide clearance for the squirter nozzle to avoid contact between the nozzle and piston.

Bottom Oilers

These small oil holes are drilled into the underside of the pin boss (on each side of the boss) and provide additional splash oiling to the wrist pin.

Pin Lock Grooves

Any piston designed for full-floating wrist pins must be secured so the wrist pin doesn't walk out of its bore. Floating pins are secured with pin locks, which are inserted into the pin bore after the pin has been installed. A groove is machined into the entrance of each side of the pin bore to accept wire locks or flat-wound spiral locks.

Piston-to-Wall Clearance

Piston-to-wall clearance specifically refers to the clearance between the widest area of the piston skirt to the cylinder wall. If clearance is too tight, the pistons can scrub against the cylinder wall and possibly seize. If clearance is too great, the pistons rock back and forth excessively, which degrades ring seal and can result in piston skirt cracking as the piston skirts slap the walls.

All pistons expand when operated at full temperature. This must be taken into account in order to arrive at the ideal spec. Desired clearance is based on the piston and bore diameter under operating temperature. That's why you may hear piston slap noise when the engine is fired cold. The noise goes away once the pistons warm up.

Factors that affect wall clearance include actual cylinder wall thickness, including how the cylinder bore changes geometry during engine operation. Piston compression distance (CD), piston material (specific alloy mix and density), piston dome thickness, and the operation of the engine are all factors contributing to bore shape. This is critical if the engine is subjected to short runs in drag racing, run for long periods at lower RPM on the street, or run for extended periods in oval track or road racing.

In very general terms, .001 inch of clearance per inch of bore diameter is accepted. But this always depends on the specific alloy and density of the piston material. Performance and racing piston manufacturers use their own alloy formulas, so always follow the piston manufacturer's specified wall clearance.

Typical small-bock engines have a wall clearance of about .004 inch. Typical big-block engines call for about .005-inch clearance. Whatever clearance is listed for naturally aspirated engines, you add about .001 inch if the engine uses a high-pressure, forced-induction system because the pistons run hotter and therefore grow in diameter under operating temperatures.

Again, follow the piston manufacturer's clearance specs. JE, Diamond, Ross, Probe, Wiseco, and others have performed the research and development to determine how their material reacts to temperature and pressure. Don't make the mistake of thinking that you know more than they do.

Piston diameter is always measured at the widest area of the skirts. The piston manufacturer specifies the exact location for measuring (for instance, .500 inch from the bottom of the skirt).

If you look closely, you'll notice a series of circumferential machining lines on the skirt surface. This "spun" finish isn't just for looks. The tiny groove lines promote oil retention for the skirts.

Another general rule is that the higher the anticipated engine speed (RPM), the greater the required clearance. In simple terms, if the manufacturer says that the wall clearance range for a particular piston is, say, .003 to .005 inch, the .003-inch clearance is better suited to a street application, while the higher end of the range is better for high-RPM use. In a race engine, where clearance is set a bit wide, any cold piston slap noise isn't the concern it might be in a street engine.

Piston Diameter

Before final-honing your cylinder bores for proper piston clearance, never assume that the pistons are manufactured to the specific diameter. In most cases, however, top-quality pistons are often precisely made to the correct specification because of today's CNC machining and tight manufacturing tolerances. Always measure each piston's diameter for two reasons: to verify consistency piston to piston and to determine the required honed-bore diameter in order to achieve desired wall clearance.

Never measure a piston diameter anywhere near the ring grooves. Always measure the widest area of the skirt. Manufacturer instructions may tell you to measure exactly .500 inch above the bottom of the skirt, while some pistons must be measured directly under the oil ring groove. It depends on the profile of the piston.

Moly anti-friction skirt coating has become commonplace for many street and race pistons. (Photo Courtesy Diamond Racing Products)

Pistons are not usually perfectly round from top to bottom. The upper area (ring groove area) is round, but the skirt is likely machined to an eccentric (oval) shape. This out-of-round condition might only be .020 inch or so, but the widest skirt area is located 90 degrees to the piston pin bore centerline (the cam profile). This oval profile exists in order to compensate for the piston's rate of expansion in specific areas of the piston. Due to varying material thickness in the cross-section, different areas of the piston are expected to react at different rates (determined by design). Always measure piston diameter exactly where the piston manufacturer tells you to measure.

Piston Skirt to Crankshaft Clearance

When using components with non-stock dimensions, check clearance, which includes stroker cranks and different-length rods. Also check clearance between the bottom of the piston skirts and the crank counterweights, primarily as the piston approaches BDC. Most high-performance piston bottoms have a relief to add clearance and to reduce weight.

High-temperature ceramic thermal barrier coating and moly skirt coating. (Photo Courtesy Diamond Racing Products)

During test fitting, slowly rotate the crank and perform a visual check for this clearance. As long as you have any clearance, you should be okay but, in general, a minimum of about .060 inch provides a safe margin.

Coatings

Specialty coatings and treatments aid in piston performance and longevity. See Chapter 18 for in-depth information about coatings.

Piston Pins

Piston pins (often referred to as wrist pins) provide the critical link between the piston and connecting rod. Here I discuss pin design, clearances, and locating methods for floating pins.

Pin Bore Design

Piston pin bores are commonly (but not always) slightly offset from the piston-diameter centerline. The reason for this is to compensate for friction resulting from the thrust side of the piston as it travels through the cylinder bore. The angle of the connecting rod changes as the crank rotates. When the rod is at its greatest angle, it pushes against one side of the piston harder than the other side.

By offsetting the piston pin bore toward the thrust side, friction is slightly increased but piston skirt slap noise is reduced. If you turn the piston 180 degrees, with the offset toward the minor thrust side (the side with less thrust force), friction is slightly reduced but noise increases. One of the main reasons that some pistons have an offset pin bore is to reduce slap noise.

The piston is offset from the small end of the rod slightly in order to soften the loading. Instead of one entire side of the piston abruptly smacking against the cylinder wall during the rocking transition before and after TDC, the offset reduces the impact, allowing the piston skirt to make contact first, followed by the rest of the piston side.

The thrust side is where greater force and pressure is placed on the piston and the cylinder wall. It's fairly common for some race builders to reverse the piston to reduce friction, since slap noise (cold start) isn't a concern.

Pin Bore Clearance

On a full-floating pin design in which the wrist pin floats on the rod and the piston pin bore, oil clearance from the pin to the piston bore is generally about .0008 to .001 inch. When you purchase performance pistons, the pins are usually included and already provide the necessary oil clearance, but it never hurts to measure for yourself just to verify.

Pin Locks

A retention device (or pin lock) is required to prevent the pin from walking out of its bore into the cylinder wall. Many pistons use a full-floating pin in which the pin floats in the rod's small end and in the piston's pin bore. There are three options, depending on the piston design: wire locks, spiral locks, or buttons.

Wire locks are tempered round-wire clips with an end gap. Spiral locks are made of tempered, flat-wound steel that is spirally wrapped (similar to a spring that is fully compressed). Pistons designed for wire or spiral locks usually have one or two small notches around the perimeter of the pin bore to provide access with a small, flat-blade screwdriver that makes removal easier. Wire locks and spiral locks function the same: After the pin is inserted in the piston, the lock snaps into a groove at the end of each side of the piston pin bore.

Wire locks are spring steel clips. Insert one end and use your finger or a small drift to walk the clip into its groove.

Here's a wire lock fully installed. Be sure to position a wire lock with the lock gap away from the small access notches. During future removal, you'll be able to insert a pick through one of the notches to grab the lock.

Specialty installation tools for spiral locks are easy to use. Spread the spiral, thread it onto the tool, enter the tool at a slight angle, and rotate counterclockwise. Straighten the tool out (parallel to the pin) and continue to rotate the tool counterclockwise until the lock snaps into the groove. These tools are available for both spiral locks and wire locks, in a variety of diameter sizes to match the pin bore size. Steel versions run about $75 each and plastic versions are about $25 (made by Lock-In-Tool).

Spiral locks can be tricky at first, but once you install a few it becomes second nature. After gently spreading the spiral apart, insert one end and "walk" the remainder in a counterclockwise direction into the groove until the final end snaps into place. Some installers use their fingers along with a small, flat-blade screwdriver, but specialty installer tools are available.

Depending on piston pin bore design, the piston requires one wire lock per side, one spiral lock per side, or two spiral locks per side. These provide a positive stop for the pin. Installing and removing these snap-in tensioned locks isn't that difficult, but dealing with spiral locks requires practice. After installing a couple of them, you get the hang of it.

When installing a spiral lock, first gently spread it apart so that it isn't mashed together. Insert one end of the spiral into the groove while pushing the rest of the lock into the groove in a counterclockwise direction. Once it's fully inserted, it snaps in place.

Buttons

The third method of piston retention is the use of buttons. These work well when the piston pin is short (still providing enough support through the pin bore but recessed farther at the ends). A soft material slug (softer than the cylinder wall) is finger inserted into each end of the pin bore. Once the piston is inside the cylinder bore, the buttons are captive.

Button materials include nylon, Delrin, and even aluminum. Buttons are great for racers who routinely perform engine teardowns at the track (no spring clips to mess with), but due to the eventual wear on the button faces, they're not intended for long-term or street use.

An example of a piston that has pin buttons rather than wire or spiral pin locks. (Photo Courtesy Diamond Racing Products)

Ring Grooves

Most pistons have three ring grooves: the top compression groove, a second compression groove, and an oil ring package groove. Ring grooves are machined to a very precise dimension and surface finished to allow the rings to seal properly and to slide in and out. It's extremely critical to avoid disturbing this surface finish. If you plan to glass-bead blast a piston to clean or to soften custom-machined, valve-pocket surfaces, carefully mask off the ring grooves. Ring "lands" refer to the outer-diameter surface of the piston at the ring areas.

The required CD of the piston pin bore can be such that the pin bore encroaches into the oil-ring groove area, and this condition is very common with short-CD pistons. Fortunately, the floor of the ring groove has an open area above each side of the pin bore. In order to provide uniform support for the oil ring package, a support rail is installed to the floor of the oil ring groove before the oil ring package.

This underside view shows the oil ring support rail properly positioned, with the male dimple centered in the opening above the wrist pin bore. The dimple prevents the support rail from rotating too far (this prevents the support rail gap/ends from moving to an area with no support).

This is an example of an oil ring support rail installed to the floor of the oil ring groove. The rod and piston pin must be installed before installing the rings, due to the wrist pin being captive behind the support rail.

The support rail has a small dimple that protrudes downward. When the support rail is installed, this dimple must be positioned at this open space, above the pin bore. The dimple prevents the support rail from rotating out of position. When required, support rails are supplied with the pistons. The end gap of these rails is fairly wide, so even if the compression rings require file fitting, you don't have to mess with the support rail gap.

Piston-to-Head Clearance

You don't want your pistons smacking into the head combustion-chamber quench area. Even slight contact between the piston and the head eventually (if not quickly) destroys the rod bearings, piston wrist pin bore, etc. Compensate for thermal and dynamic growth of the rod and piston to avoid this degradation. Other contributing factors are connecting rod material and its rate of expansion, piston mass, and piston speed.

If you're using forged steel connecting rods, minimum piston-to-head clearance should be around .035 inch in a typical small-block and about .045 inch in a big-block. If you're using aluminum rods, add about .010 inch more clearance. Remember to factor-in the compressed head-gasket thickness, which can vary widely depending on the gasket (composite, multi-layer steel, copper, etc.).

Piston-to-Valve Clearance

Piston-to-valve clearance isn't so much an issue at full lift when the cam lobe is at peak height because the piston has already moved into the bore when the valve is at full lift. Camshaft duration is the real issue, along with radial clearance of the valve head to the piston.

Use the clay-checking method or a degree wheel and a dial indicator or to measure piston-to-valve clearance. When you do, don't rely on just one measurement at just one cylinder location. Tolerance stackup could result in different piston-to-valve clearances from the front of the block to the rear, so take the time to measure your valve clearance at each cylinder.

In either case, assemble the block with the crank and its main bearings for at least one rod and piston. I recommend doing this for number-1 cylinder. Then install the camshaft and solid lifters. If you're using hydraulic lifters, the solids need to be the same length at the corresponding cylinder location, timing gear setup, a head gasket, and a cylinder head with valves.

After adjusting the degree wheel for the piston at TDC, rotate the crank to 10 degrees before and after TDC in order to bring the exhaust and intake valves closest to the piston.

Clay Checking

For the clay checking method, it's best to use the actual springs that you plan to use in the final assembly because the light checking springs may not be strong enough to fully compress the clay, which results in a false (too shallow) depression. Install the correct pushrods (the length that you plan to use) and rocker arms on the corresponding valve location. Make sure that the cam timing is adjusted where you plan to run the bumpstick, since this affects valve clearance.

For this method, lower the piston below deck (about an inch or so) before installing the cylinder head. Apply a light coat of assembly oil to the cylinder wall to prevent the clay from sticking to the wall. The piston dome should be clean and dry so the clay can stick to the piston. Apply a coating of oil to the intake and exhaust valves and the quench area of the combustion chamber so the clay does not stick to the valves or head.

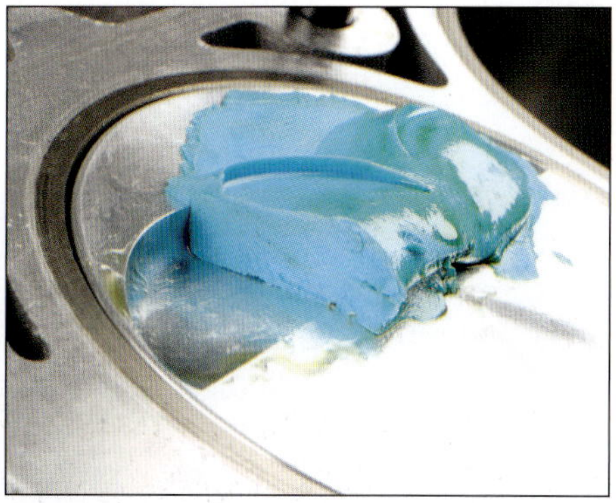

After rotating the crank to run the valves through a complete cycle, remove the head and, using a razor blade, carefully slice a cross section of the clay at the center of the relief. The clay thickness can be measured (carefully) with a dial caliper or small ruler that's incremented in one-thousandths of an inch.

Apply a thick slab of modeling clay (available at any craft store) to the piston valve pockets. Overlap a bit of clay beyond the valve pocket edges.

Install a head gasket (preferably one already crushed). If you don't have a used head gasket or the same type you plan for final installation, you can measure your new gasket and then check with the gasket manufacturer to determine the compressed thickness. The difference between the new and compressed figures can be subtracted from the total valve clearance.

Install the cylinder head and hand tighten at least four head bolts. Do not torque to final spec. Install the pushrods and rocker arms and zero the valve lash. Using a socket wrench on the crank snout, rotate the crank 360 degrees *at least two times* in the engine's direction of rotation. Remember, the crank needs to rotate twice in order to rotate the cam once.

Carefully remove the cylinder head. You should see clear valve impressions in the clay.

Using a sharp razor blade, carefully cut through the center of the exhaust-valve pocket area; be careful not to gouge the piston. Carefully remove the outer half of the clay. Now you can see a cross-section of the clay in the valve pocket.

Clay is also good for checking the angle of valve attack relative to the angle of the piston's valve pocket.

Using a depth mic or the end of a dial-caliper rule, measure the thinnest section of clay in the pocket (closest to the pocket "eyebrow").

Repeat this procedure at the intake valve. Record your findings.

Granted, the clay method is definitely old-school, but it gets you close enough. In order to obtain higher accuracy, it's best to measure with a dial indicator.

If you're using aluminum connecting rods, you likely need to allow an additional .030 inch or so for clearance, since aluminum rods tend to expand more when hot.

Dial Indicator

For the dial-indicator measurement method, install light checking springs on the intake and exhaust valves instead of using the actual valvesprings. This makes it easier to move the valves rather than fighting spring pressure.

With the degree wheel mounted to the crank, rotate the crank in the direction of common operation until your degree wheel indicates that number-1 piston is at 10 degrees ATDC on its intake stroke. Why? Because this is the position where valve-to-piston clearance is likely the tightest.

Without disturbing the piston location, mount the dial indicator onto the cylinder head. The indicator base must be mounted solidly. If you have an aluminum head and using a magnetic base, you may be able to solidly mount a steel pushrod guideplate at the adjacent cylinder to provide a flat contact point.

Position the dial indicator plunger onto the number-1 intake valve or its retainer. The dial indicator plunger needs to be parallel with the valvestem. Don't mount it at an angle.

Preload the dial indicator at about .050 inch, then zero the gauge. Carefully push down the rocker arm tip until the valve stops as it touches the piston. Note the amount of movement on the gauge and record this distance.

In order to measure exhaust valve clearance, smoothly rotate the crankshaft in the same direction as before until the degree wheel indicates 10 degrees BTDC on the exhaust stroke. This is the area at which the exhaust valve is likely closest to the piston.

As before, place the dial indicator on the exhaust valve, preload the plunger, zero the gauge, push down on the exhaust rocker tip until the valve hits the piston, and record the number.

As a general rule, you should have at least .080-inch valve-to-piston clearance at intake valves, and about .100 inch at the exhaust valves. The exhaust valves soak up more heat and expand more than the intakes.

Piston Compression Distance

Piston compression distance (CD) is the distance from the centerline of the piston pin bore to the flat deck of the piston. The height or this measurement is critical for establishing compression ratio and valve and head clearance. CD is one of the factors that determines where the piston deck is located relative to the block deck when the piston is at TDC. The centerline of the crankshaft main bore is a fixed position. The distance from the main bore centerline to the block deck is the reference for where the piston deck is at TDC. To determine this, with the crank rod journal at TDC, add together the following dimensions:

Block Deck Height = 1/2 stroke +
rod length + piston CD + deck clearance

The result of this combined measurement is then compared to the distance from the main bore centerline to the block deck.

As an example, let's say that your block deck height (main bore center to bock deck) is 10.2 inches. Let's say that your crank stroke is 4.500 inch and the rod length is 6.700 inches.

One-half of stroke (since you're only concerned with the max distance that the crank pushes the rod up at TDC) in this case is 2.250 inches. Adding this 2.250 inches to the 6.700-inch rod length results in 8.950 inches. Subtracting this 8.950 inches from the block deck height of 10.2 inches gives a dimension of 1.250 inches. This 1.250 inches is the available piston CD in order to place the piston deck flush with the block deck. If you want the piston deck to be below the block deck by, say, .015 inch, subtract this from the initial CD, which means that you want a piston CD of 1.235 inches. Remember that you still have to consider the crushed cylinder-head-gasket thickness (which gives us more volume above the piston) and required valve clearance.

Compression Ratio

Compression ratio represents the volume of the cylinder when the piston is at BDC compared to the volume when the piston reaches TDC. These factors are all affected by cylinder displacement: bore diameter, crank stroke, deck clearance, head gasket volume, and combustion chamber volume.

When you consider the piston's dome and dish/valve reliefs, you subtract piston dome volume and add dish or valve pocket volume.

See Chapter 5, page 46, for the formula and examples to find the compression ratio.

Deck Height

If the piston top quench area (not including a dome protrusion) is flush with the block deck at TDC, it is called zero deck and doesn't need to be included

If a piston deck is to be modified on a CNC lathe, each piston is first measured for existing dome height.

in the formula. If the top of the piston is below deck (negative deck) or above deck (positive deck), then this volume must be considered. If the piston is below deck, this volume is added to the formula (since it adds volume). If the piston quench is above deck, subtract this volume.

See Chapter 5, page 48, for the formula and examples to find the deck height volume.

Gasket Volume

The bore of the head gasket is larger in diameter than the cylinder bore. Even if the cylinder head gasket bore is not perfectly round, measure the diameter to obtain an average diameter.

See Chapter 5, page 48, for the formula and examples to find the head gasket volume.

Piston Dome Volume

Quality piston manufacturers provide dome volume figures for their pistons. You can rely on those figures or you can measure for yourself. Measure piston dome volume with the piston installed, with at least the top compression ring installed (to seal the bore). Rotate the crank to bring the piston down the bore near BDC. Then coat the cylinder wall with a thick lithium grease (to aid in sealing) at a point of about 1.50 inches below block deck.

Rotate the crank to raise the piston to a point exactly 1.00 inch below block deck (use a depth micrometer to measure this). Wipe off any excess grease that might be on the top of the piston.

Smear lithium grease onto the block deck around the bore and place a piece of stout, flat clear plexiglass or acrylic, with a small chamfered hole drilled in the

Once the program has been written, an entire set of pistons can be cut on a CNC lathe in a matter of a few minutes.

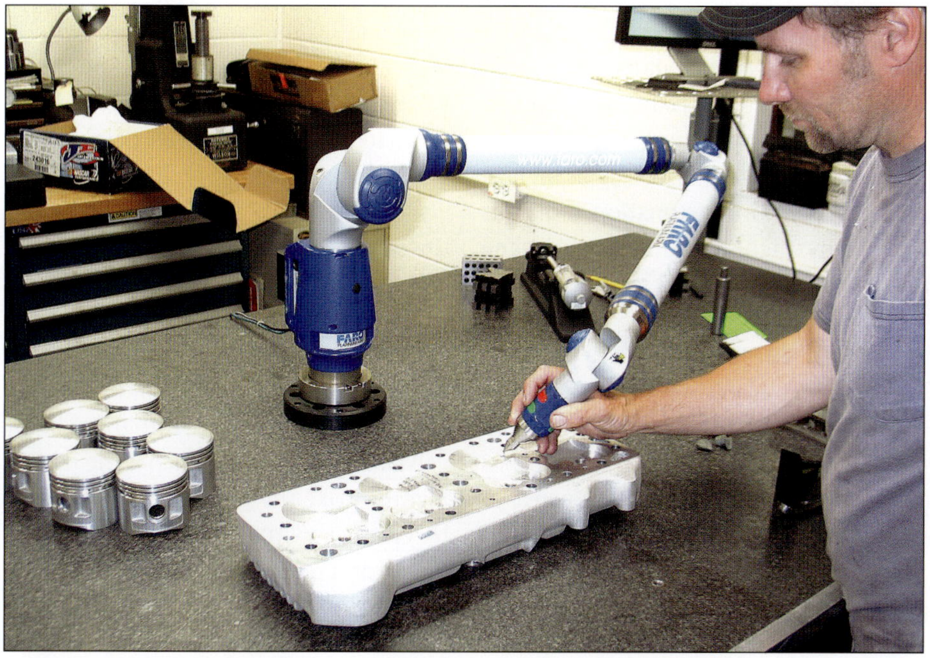

In order to provide a reference for the piston, a CMM (coordinate measuring machine) is used to create a profile of the cylinder head combustion chamber.

Once pistons have been machined on the CNC lathe, dome height is verified.

A dedicated precision piston vise allows the piston to be secured in the correct valve angle prior to pocket milling.

A milling bit adjusted to the proper valve head diameter (with proper clearance added) is used to mill the valve reliefs.

plate (about 1/4 inch in diameter) onto the block. Press the plate gently to obtain a good grease seal.

With the block deck level, you use a use a burette to add fluid (blue windshield washer solvent works well) until the cavity is filled, with no air bubbles trapped under the clear plate. Fill the burette to its top index mark. As you release fluid, monitor how much fluid it takes to fill the cavity (you may need to refill the burette several times).

Before adding fluid, place a clean piece of paper under the bore and watch for leaks. If fluid passes through the bore past the top ring, you gain a false reading.

See Chapter 5, page 47, for the formula and examples to find the piston dome volume.

Combustion Chamber Volume

You use the burette here also. With valves installed, place the head upside down on a workbench. Install a spark plug. Apply white lithium grease around the chamber and position your clear plate. Using the burette, add fluid and record how many cubic inches it takes to fill the chamber (make sure that no air bubbles are trapped.

Once you have recorded your data (combustion chamber volume, piston dome volume, head gasket volume, deck height volume, and cylinder swept volume, you're ready to do a bit of math to determine compression ratio.

See Chapter 5, page 47, for the formula and examples to find the combustion chamber volume.

Due to aluminum heads' faster heat dissipation, you are able to run a higher compression ratio than with cast-iron heads. Considering today's fuels, you should be able to get away with as much as 11.4:1 or so compression ratio on high-test pump gas (92 octane or higher) before pre-ignition/detonation problems occur. Any higher requires high-octane race gas or alcohol.

Valve Pockets

Whether you're using pistons that already have valve reliefs or flat-top pistons that require reliefs due to increased cam duration and/or oversized valves, you can perform your own check for valve diameter clearance.

The first step is to wipe the piston dome clean with a solvent and "paint" the dome with a broad, black marker.

Mount a degree wheel to the crank and adjust it to indicate TDC for the cylinder at hand.

Install a bare cylinder head (with no valves) to the block, using a crushed head gasket of the same type planned for assembly. Grab a junk valve that has the same stem diameter as the valves you intend to use and cut the head off the old valve. You can also use a piece of steel rod as long as it's the same diameter.

On a lathe, grind the cut end of the valve to a point. You must use a lathe in order to achieve a centered point. Don't try grinding the point by hand.

Apply a light smear of lube onto the stem to allow it to glide through the valve seal.

Rotate the crank to 10 degrees BTDC, where the valve is closest to the piston. Insert the pointed valvestem, which now serves as a center punch, into the exhaust valve location until it touches the piston. Lightly strike the punch with a hammer to create a witness mark. Remove the punch and rotate the crank to 10 degrees ATDC (this is where the intake valve is closest to the piston). Insert the punch into the intake valve location and tap a mark onto the piston.

Remove the cylinder head. You see both witness marks on the piston dome (one for intake and one for exhaust). These marks indicate where the center of each valve is in relation to the piston.

Next, carefully measure the diameters of the intake valve and the exhaust valve. Transfer these valve head diameters to the pistons, using the witness marks as your centers. You can use a sharp compass scribe to lightly scratch the valve head diameters.

In order to provide enough valve head clearance around the perimeter of the valve, add .080 inch to the measured diameter to compensate for piston rock and valve float. If your pistons already have valve notches, you are able to see if your valves clear the reliefs with about .080-inch clearance.

After you know where the outer diameter of each valve is in relation to the piston, determine valve-to-piston depth. Again, you're back to clay checking or measuring valve clearance with a dial indicator.

If valve reliefs need to be cut (or existing reliefs cut larger and/or deeper), this must be done on a vertical mill. An old-school method (which I don't recommend) is to take a junk valve, with the same valve head diameter and stem diameter as your good valves, and cut sawtooth notches onto the valve head. You then install the valve to the head, install the head to the block (with gasket), and turn the valve as you would turn a cutting tool with a hand drill. This is a crude and potentially disastrous method, so don't do it. Instead, pay a skilled engine machinist to handle this for you.

Before cutting the valve reliefs, the dome area must be measured for material thickness to make sure that there's enough material to handle the depth of the cut and still leave enough material to handle the combustion pressure. Generally speaking, the absolute minimum for a naturally aspirated engine is .150 inch of material thickness in the piston. For any forced-induction or nitrous application, around .200 inch is the bare minimum and .250 inch is better.

After the reliefs have been properly machined, re-install the head with valves and recheck valve-to-piston clearance. Once you've achieved the desired dimensional results, soften the sharp edges created by the pocket cutting: Remove the pistons from their rods. Using a hand-held die grinder or a Dremel with abrasive "tootsie-roll" sanding roll or drum, gently radius all sharp edges. This reduces the chance of having hot spots in the combustion process that can result in detonation. After softening these edges, thoroughly wash and rinse the pistons to remove all debris.

You want the valve relief to be at the same angle as the valve. A protractor or a valve angle gauge (from Goodson Tools for example) can be used to measure the angle of the installed valve. This is the angle between the cylinder head deck and the face of the installed valve. The same angle needs to be created at the valve pocket in the piston.

Ordering Custom Pistons

When an off-the-shelf piston isn't available to suit your needs, it's time to turn to custom sources such as JE, Diamond, Mahle, Ross, Probe, CP, Wiseco, and others. They offer CNC machining, which can provide about any dimensional piston configuration.

Custom pistons may be needed because of the combination of diameter, compression distance, valve clearance, compression ratio, skirt clearance, ring groove dimensions, height locations, etc. Or maybe you're restoring a vintage engine and you simply can't locate the pistons you need (discontinued or you've made changes so that a stock replacement doesn't work). Thanks to piston design and the increased use of CNC machining, today you can basically obtain any piston configuration. "That isn't available" no longer applies.

The cost of a custom piston was once astronomical and only within the budget of well-funded pro teams. Thanks to CNC machining, the cost has come down to the "still not cheap, but finally affordable" level, making custom pistons a viable option for just about any builder. Also, the turnaround time for custom orders has been reduced from months to weeks.

Ordering a set of custom slugs to work with your new heads and your bore/stroke combination requires you to provide the piston manufacturer with some detailed information. This includes

Custom pistons are available in any configuration, but knowing what's needed to order them requires careful measurement and sound decisions.

Required Information for Custom Pistons

You need to do your home work and provide accurate specs, so the piston manufacturer produces the required piston. Most custom piston builders guide you through the process, and are able to handle any calculations that might be needed. Be sure to provide detailed and accurate information to get the piston you need.

Proper planning also is required before ordering a custom piston because you don't want to make a mistake. After all, if you give the manufacturer the wrong information, you get an incorrect piston. The piston must be compatible and complementary with all the other internal parts. Lead or production time for a custom piston can vary greatly according to the time of year but common production time can be as little as three to five weeks but a wait of six to eight weeks or longer is possible. You need your new pistons before you perform a final hone of the bores because you must have the new pistons to measure skirt diameter. Ensure your cylinders clean-up straight at your new target bore size.

Here is the information you need to supply when ordering:

General Information

- Engine type (example: small-block Ford, LS1, big-block early-generation Chevy, 340 Mopar, etc.)
- Application (street, drag race, oval track, road race, marine, etc.)
- Desired compression ratio
- Estimated desired engine horsepower (an honest estimate)
- Maximum expected RPM

Crankshaft and Block

The piston manufacturer considers deck height, bore length, crank stroke, rod length, and crank counterweight radius. This is in part to make sure that the piston compression distance provides the desired compression ratio and deck clearance on the up-stroke, and also so that the piston skirt height clears the crank's counterweights on the down-stroke.

- Block deck height (main bore center to deck surface)
- Finished bore size (target cylinder bore diameter)
- Cylinder bore length (bottom of cylinder wall to block deck)
- Distance between cylinder bores (center-to-center)
- Crank stroke
- Crankshaft counterweight radius (distance from crank centerline to outer edge of the counterweight)

Cylinder Head

This information is needed to help determine compression ratio and valve pocket placement, pocket radius, and depth and angles.

- Cylinder head (make, type, model, part number, casting number)
- Cylinder head chamber volume (ci)
- Valve diameters (intake and exhaust)
- Valve angles (if non-OEM)
- Compressed head gasket thickness
- Head gasket bore diameter (assume a round gasket bore)
- Head gasket volume (ci)
- Desired minimum valve-to-piston clearance
- Ring dimensions (axial height and radial width)

Connecting Rod

Rod information is needed to help determine piston compression height and distance between pin bosses. Information regarding rod material may be considered as well, due to the theoretical elongation of the rod length when using steel, aluminum, or titanium.

- Length (center-to-center)
- Material (steel, aluminum, titanium)
- Small-end width and thickness over the top of the pin hole

Camshaft and Valvetrain

The piston manufacturer considers actual valve movement off the seat to calculate the needed valve-to-piston clearance. Rod ratio plays a major part in piston-to-valve requirements, due to varying piston speeds at TDC for a given rod/stroke combination.

Camshaft Type

This is needed whether you have hydraulic, solid/flat-tappet, or roller cams.

- Gross valve lift (intake and exhaust)
- Lobe separation angle
- Cam duration at .050 inch
- Valve lift at piston TDC (intake and exhaust)

- Piston valve pocket depth (intake and exhaust, from sample piston, if available)
- Intake valve angle
- Valve spacing

Piston

- Pin support (floating or fixed)
- Skirt shape (scalloped, full round, inboard, etc.)
- Desired piston-to-deck clearance (–above or +below)
- Compression height (distance from pin center to theoretical "flat" of piston)
- Type (flat-top, conical-dish, dome, reverse-image dish, round dish, or full 3D max dome)
- Dome volume (total effective, including valve pockets)
- Axial ring height (top, second, oil)
- Radial ring width (top, second, oil)
- Desired ring back clearance, if not standard (top, second, oil)
- Wrist pin diameter
- Wrist pin length
- Pin wall thickness

Power Adders

This information helps the piston manufacturer to determine piston material, crown, cross-sectional thickness required for the power level desired, and piston-to-bore clearance.

- Maximum boost (turbo or supercharger)
- Nitrous HP amount
- Number of nitrous stages
- Fuel type and octane planned

Optional Piston Features

- Double pin oilers with slots (if not standard)
- Gas ports (side/top)
- Internal weight-removal milling
- Chamfer pins (round wire lock)
- Offset pins (OEM noise control)
- Ceramic top coat
- Moly skirt coat
- Casidium coated pins (race, dry sump, high-pan vacuum applications)

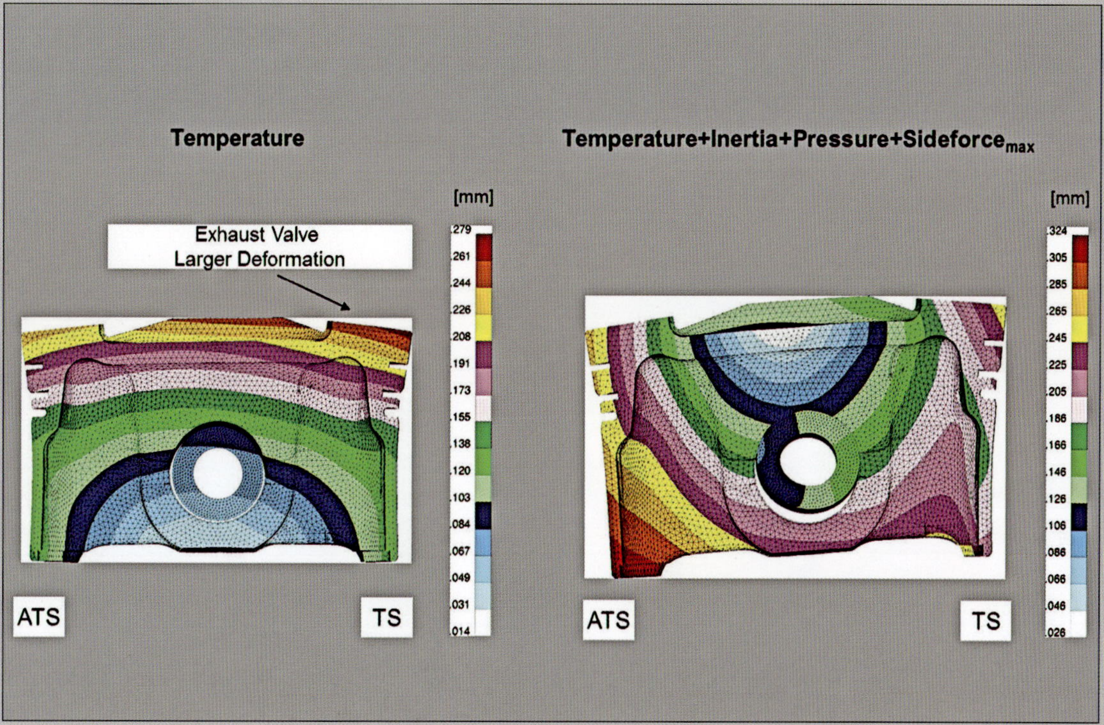

This is an example of how temperature and operating conditions can alter piston shape. At left, note larger deformation in the exhaust area. At right is an example of how temperature, inertia, pressure, and side force can result in shape changing during engine operation. (Illustration Courtesy MAHLE Clevite)

Manufacturers commonly utilize CNC machining for speed of product output and precision results. This shot shows a few of the production CNC machining centers in JE Pistons' facility.

data to achieve the correct sizing, compression ratio, and clearance.

Today's custom forged aluminum pistons are CNC machined from precision-forged 2618 or 4032 aluminum alloy slugs. Due to the precision and repeatability of CNC processing, piston dimensions (within a set or for future replacements that reference the original order) should be exact matches in every detail, including weight (usually +/- 1 to 2g).

When you place an order for a set of custom pistons, the manufacturer asks you to complete a custom piston data sheet, which can be requested from the piston manufacturer or downloaded from their website. The piston manufacturer's engineers examine the data sheet and may contact you to discuss any questionable areas, to confirm that the order is completed correctly.

It's important to provide the correct information For instance, if you say that your block deck height is 9.240 inches, when in fact it is 9.230 inches, your new pistons will be .010 inch taller than required, which can result in inadequate piston-to-valve clearance. It is your responsibility to provide accurate dimensional information on the data sheet.

When filling out your data sheet, avoid using nominal published dimensions for things such as block deck height, crank stroke, rod length, rod pin bore diameter, etc. Since *actual* dimensions can often vary (if even by a tiny bit) from published specs, having pistons made using *theoretical* measurements can create potential disasters on assembly day.

Verify Measurements

Eliminate potential variables by measuring, checking, and verifying *each* component. This can't be stressed enough. For instance, let's say that a certain OEM block is listed as featuring a 9.8-inch deck height: *Never* assume that your block meets that spec. Actually measure the distance from the main bore centerline to the block deck (on each bank, and front-to-rear), since your block might not match factory spec. Plus, your block may have been previously resurfaced for cleanup or it may have been cut for block squaring to a specific dimension for a previous build.

The piston manufacturer needs actual dimensions for your specific application. You want to make the piston to fit your application, rather than have something that doesn't work properly because no one measured the actual stack up of the parts beforehand. In other words, take the time to measure block deck height, check the block decks for squareness, measure crank stroke, measure

The CNC process continues with dome finishing (including valve reliefs) and skirt notching. When an extremely short compression height is needed, the pin bore may intersect the oil ring groove, in which case a set of oil ring support rails are needed (available from the piston manufacturer).

The valve pocket placement, radius, and depth on this custom slug are based on a range of input, including valve angle and valve plunge (lifter rise and rocker ratio).

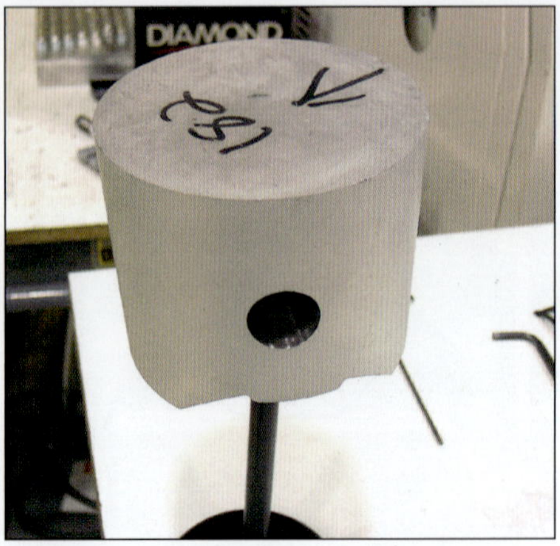

Custom forged pistons start off as billet blanks, followed by placement of the pin bore. (Photo Courtesy Diamond Racing Products)

connecting rod length, and check for piston skirt and pin boss clearance issues. Measure combustion chamber depth and volume, and measure valve-open locations, spark plug depth, etc.

Once the piston program has been written, the machining process starts with skirt profiling, ring groove machining, and port drilling. Here the dome has been machined to a raw stage. With ring grooves machined relative to the pin bore and anticipated dome height, finish machining creates the final skirt profile, dome, and valve pockets. (Photo Courtesy Diamond Racing Products)

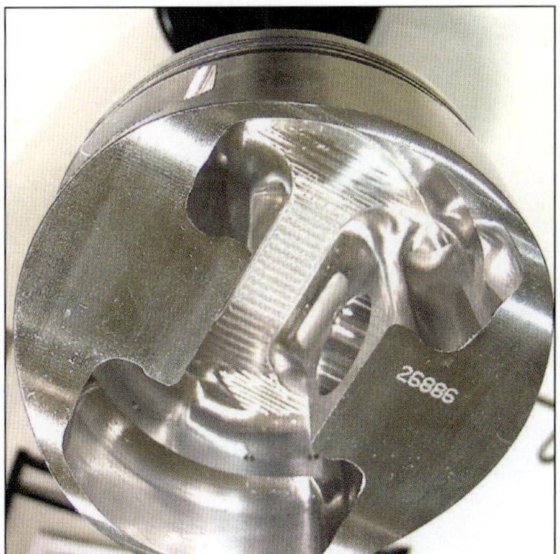

The underside of the piston is machined to remove unwanted weight and to finish the pin bosses. Thanks to today's CNC machining techniques, custom piston mass is extremely close in terms of weight. (Photo Courtesy Diamond Racing Products)

If you provide the wrong information and the piston is made incorrectly, don't blame them if you have a fitment problem on assembly day. Most manufacturers work with an engine builder to correct a problem; They have tractability built into their product via an engraved job number, so errors can be traced all the way back to the point of order, if necessary.

Bore Diameter

When specifying your bore diameter, don't assume that you can have any piston diameter that you want. Yes, the piston manufacturer can produce whatever diameter you request, to fit almost any bore size, but if you can't obtain the proper ring set for your new bore size, you can't use the pistons. Always select your rings before you order the pistons to ensure you can get the custom rings to fit the pistons. The piston manufacturer may be able to provide matching ring sets for many popular custom requirements. You can always have custom pistons made, but you can't have custom ring sizes made unless you have enormous amounts of money.

Once you know that you have the correct rings, you can then specify your finished bore size. Some piston manufacturers can supply the needed rings, in which case they machine the ring grooves

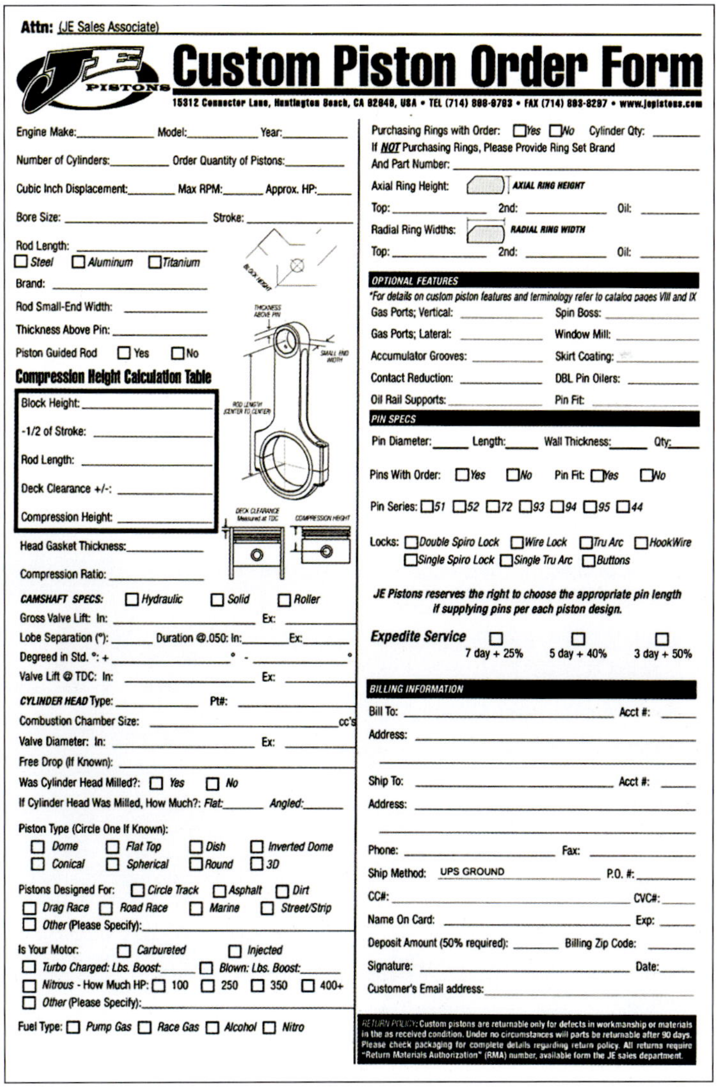

Attn: (JE Sales Associate)

JE PISTONS **Custom Piston Order Form**

15312 Connector Lane, Huntington Beach, CA 92649, USA • TEL (714) 898-9763 • FAX (714) 893-8297 • www.jepistons.com

Engine Make: ____ Model: ____ Year: ____

Number of Cylinders: ____ Order Quantity of Pistons: ____

Cubic Inch Displacement: ____ Max RPM: ____ Approx. HP: ____

Bore Size: ____ Stroke: ____

Rod Length: ____
☐ Steel ☐ Aluminum ☐ Titanium

Brand: ____

Rod Small-End Width: ____

Thickness Above Pin: ____

Piston Guided Rod ☐ Yes ☐ No

Compression Height Calculation Table

Block Height: ____

-1/2 of Stroke: ____

Rod Length: ____

Deck Clearance +/-: ____

Compression Height: ____

BLOCK HEIGHT / THICKNESS ABOVE PIN / SMALL END WIDTH / ROD LENGTH (CENTER TO CENTER) / DECK CLEARANCE Measured at TDC / COMPRESSION HEIGHT

Head Gasket Thickness: ____

Compression Ratio: ____

CAMSHAFT SPECS: ☐ Hydraulic ☐ Solid ☐ Roller

Gross Valve Lift: In: ____ Ex: ____

Lobe Separation (°): ____ Duration @.050: In: ____ Ex: ____

Degreed in Std. °: + ____ ° - ____ °

Valve Lift @ TDC: In: ____ Ex: ____

CYLINDER HEAD Type: ____ Pt#: ____

Combustion Chamber Size: ____ cc's

Valve Diameter: In: ____ Ex: ____

Free Drop (If Known): ____

Was Cylinder Head Milled?: ☐ Yes ☐ No

If Cylinder Head Was Milled, How Much?: Flat: ____ Angled: ____

Piston Type (Circle One If Known):
☐ Dome ☐ Flat Top ☐ Dish ☐ Inverted Dome
☐ Conical ☐ Spherical ☐ Round ☐ 3D

Pistons Designed For: ☐ Circle Track ☐ Asphalt ☐ Dirt
☐ Drag Race ☐ Road Race ☐ Marine ☐ Street/Strip
☐ Other (Please Specify): ____

Is Your Motor: ☐ Carbureted ☐ Injected
☐ Turbo Charged: Lbs. Boost: ____ ☐ Blown: Lbs. Boost: ____
☐ Nitrous - How Much HP: ☐ 100 ☐ 250 ☐ 350 ☐ 400+
☐ Other (Please Specify): ____

Fuel Type: ☐ Pump Gas ☐ Race Gas ☐ Alcohol ☐ Nitro

Purchasing Rings with Order: ☐ Yes ☐ No Cylinder Qty: ____
If NOT Purchasing Rings, Please Provide Ring Set Brand
And Part Number: ____

Axial Ring Height: AXIAL RING HEIGHT

Top: ____ 2nd: ____ Oil: ____

Radial Ring Widths: RADIAL RING WIDTH

Top: ____ 2nd: ____ Oil: ____

OPTIONAL FEATURES

*For details on custom piston features and terminology refer to catalog pages VIII and IX

Gas Ports; Vertical: ____ Spin Boss: ____

Gas Ports; Lateral: ____ Window Mill: ____

Accumulator Grooves: ____ Skirt Coating: ____

Contact Reduction: ____ DBL Pin Oilers: ____

Oil Rail Supports: ____ Pin Fit: ____

PIN SPECS

Pin Diameter: ____ Length: ____ Wall Thickness: ____ Qty: ____

Pins With Order: ☐ Yes ☐ No Pin Fit: ☐ Yes ☐ No

Pin Series: ☐ 51 ☐ 52 ☐ 72 ☐ 93 ☐ 94 ☐ 95 ☐ 44

Locks: ☐ Double Spiro Lock ☐ Wire Lock ☐ Tru Arc ☐ HookWire
☐ Single Spiro Lock ☐ Single Tru Arc ☐ Buttons

JE Pistons reserves the right to choose the appropriate pin length if supplying pins per each piston design.

Expedite Service ☐ 7 day + 25% ☐ 5 day + 40% ☐ 3 day + 50%

BILLING INFORMATION

Bill To: ____ Acct #: ____

Address: ____

Ship To: ____ Acct #: ____

Address: ____

Phone: ____ Fax: ____

Ship Method: UPS GROUND P.O. #: ____

CC#: ____ CVC#: ____

Name On Card: ____ Exp: ____

Deposit Amount (50% required): ____ Billing Zip Code: ____

Signature: ____ Date: ____

Customer's Email address: ____

RETURN POLICY: Custom pistons are returnable only for defects in workmanship or materials in the as received condition. Under no circumstances will parts be returnable after 90 days. Please check packaging for complete details regarding return policy. All returns require "Return Materials Authorization" (RMA) number, available form the JE sales department.

A blank custom piston data sheet. (Photo Courtesy JE Pistons)

Creating a Piston Mold

One method of supplying the piston manufacturer with useful reference information is to make a mold of your combustion chamber with the cylinder head installed to the block. Epoxy chamber mold kits are available that include an A-B pair of epoxy cans, mixing stick, nitrile gloves, and spark-plug hole plug. This eliminates the need to ship one of your heads to the piston manufacturer for scanning.

The kit is easy to use. With the loaded head (with valves) installed onto the otherwise bare block, turn the block upside-down and level the deck surface front-to-rear and side-to-side. Install the cylinder head and apply a release agent (such as WD40) to the bore and the chamber. Pour the mixed two-part epoxy into the bore, from the bottom of the cylinder and the top of the cylinder. Allow it to puddle and fill the combustion chamber.

The epoxy sets in about 30 minutes. Scribe a line on the back side of the mold parallel to the piston pin centerline, and pop it out. As the epoxy hardens, it also shrinks about 2 percent; so removing the mold is easy after it dries.

With this mold, the manufacturer doesn't need to guess or wonder about chamber shape, valve placement, etc. This is especially handy if the cylinder head is an oddball, unknown, or highly modified head.

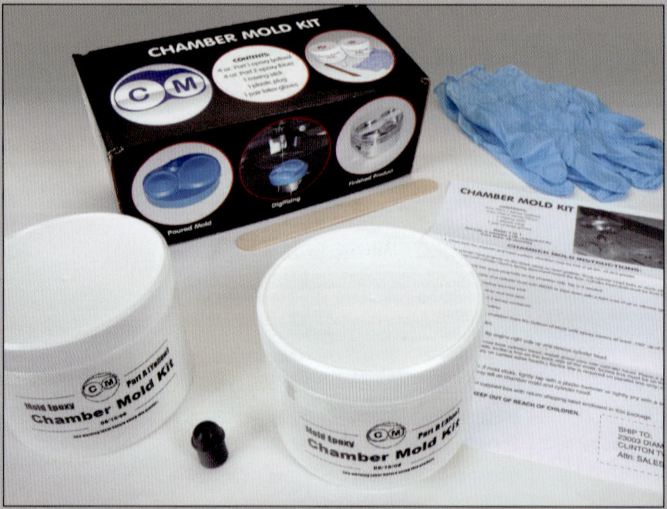

In order to provide the piston manufacturer with an actual (measurable) image of your cylinder head combustion chamber, a two-part mold kit may be used by builders. Diamond's chamber mold kit includes A-B epoxy, spark-plug hole plug, gloves, and a mixing stick. Using a bare block (no crank, no rods/pistons, no cam), a release agent is applied to the bore and to the combustion chamber. The cylinder head is then bolted to the block (with the specific type and size of head gasket used during final assembly in place and with the cylinder head fully torqued). The block is rotated to orient the block deck facing the floor. Next, the epoxy mix is poured into the cylinder bore (from the bottom) and allowed to cure. Finally, the head is removed and the mold popped out. (Photo Courtesy Diamond Racing Products)

Once the epoxy dries, it's easily popped out, aided by the release agent. The epoxy material maintains the actual size. The mold is then shipped to the piston manufacturer. (Photo Courtesy Diamond Racing Products)

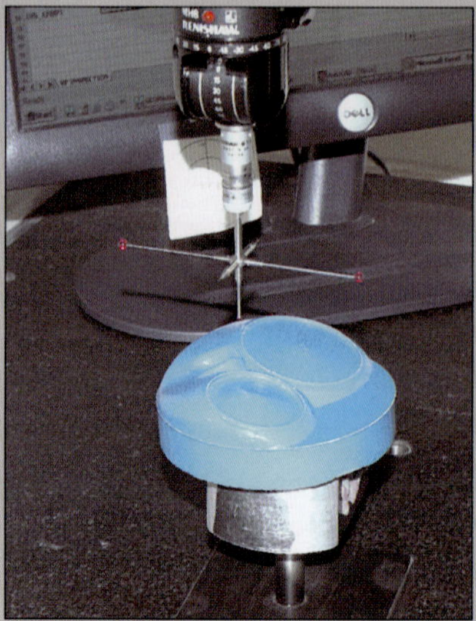

Once in the piston manufacturer's hands, the mold is digitally scanned on a CMM (capability maturity model) system. The custom piston program may then be created, factoring in the customer's camshaft lift, stroke, desired compression, valve clearance, etc. (Photo Courtesy Diamond Racing Products)

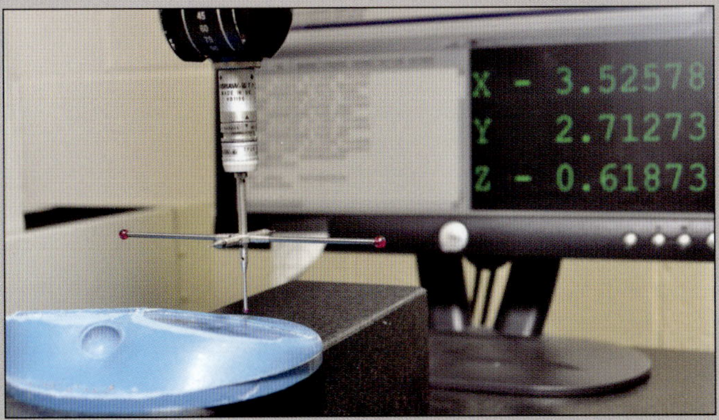

During CMM processing, a digital probe travels across the entire combustion chamber model, providing all dome information to the system database. (Photo Courtesy Diamond Racing Products)

Once the mold profile data has been input, piston engineers study the model and are then able to modify to compensate for desired compression ratio, head clearance, and valve clearance based on bore, stroke, and camshaft profile. The result is a truly custom piston that precisely fills the user's requirements. (Photo Courtesy Diamond Racing Products)

to match the style and radial depth of the rings on your new pistons. But remember, only certain specific ring diameters are available. Regardless of where you obtain your rings, make sure that the correct rings are actually available. This dictates your choices for desired piston size and finished bore diameter.

Valve Pocket Depth

Custom piston manufacturers have layouts for most popular OEM and aftermarket cylinder heads, but every engine combination is different so valve pocket depths vary. Valve-pocket depth directly relates to an engine's valve timing events, and actual valve position on the seat. The depth of the valve pockets absolutely dictates top ring placement and the total effective volume of the piston.

Flat-top and dished pistons are not quite as sensitive to ratio calculations as domed pistons. Domed pistons for some engines can be a real challenge to get perfect the first time.

In order to get as close as possible to the correct pocket depths, determine lifter rise at 10-degrees BTDC overlap on the exhaust side. For the intake side, you need to know lifter rise at 10-degrees ATDC overlap. This usually requires pre-assembly mock-up of the valvetrain.

Additional information you need includes the rocker arm ratio, valve angles (for both intake and exhaust), and valve free drop (how far below or above the leading edge of the valve in relationship to the deck of the head). The piston manufacturer can then calculate more accurately the required valve pocket depth for a specific combination. If you're in doubt, call the manufacturer. They can help/advise you based on their experience on what you may need and how to obtain the required data so they can process your order. Any good custom piston manufacturer is willing to work hand-in-hand with you to achieve your goals.

Cylinder Bore Length

The piston manufacturer may also ask for the cylinder bore length. This refers to the actual length of the cylinder, from bottom to top. This length is easily obtained by using a simple tape measure. Determining the cylinder bore length is important in order to make sure the piston remains properly supported at BDC.

The manufacturer uses stroke, rod length, and cylinder length to determine if the piston can be made with the proper ovality and skirt taper to operate properly, at the top *and* the bottom of the cylinder.

It also tells them if the cylinder length is too short to provide proper support for a specific combination. This can pose a problem on some LS1 combinations with custom stroker configurations, where the entire piston skirt might fall too far out of the bottom of the cylinder. Combinations with too long a stroke and too short a rod/cylinder length are the cause of this problem. This results in annoying piston slap, which, in a very short time, leads to damaged skirts and severely deformed cylinders.

CHAPTER 9

PISTON RINGS

Piston rings, in simple terms, form the dynamic seal between the piston and cylinder wall. Consider, for example, that the rings must provide a perfect seal in spite of a wide range of temperature extremes, from winter morning cold starts to prolonged high-RPM, high-load operation in hot weather. Also consider that most pistons move and grow laterally within the cylinder during operation, creating an unpredictable sealing area and contact surface.

Ring Types

Virtually all current automotive applications have three rings per piston, commonly called the top ring, the second ring, and the third ring. They each have specific features and functions.

Top Ring

The top ring is exclusively a compression ring, meaning its function is to seal the expanding combustion gases above the piston. Without an effective combustion seal, of course, these gases leak around the sides of the piston, resulting in a power-robbing process known as blow-by.

Carefully expand the ring so it expands larger than the crown of the piston. But be careful; you don't want to damage the ring by expanding the ring too wide.

During the piston's combustion cycle, cylinder pressure pushes the top ring against the cylinder wall and the bottom of the ring groove to form a seal. As pressure increases above the ring and between the ring's inside diameter and the piston groove, the ring is forced downward and outward, creating a tight seal over a wide range of engine RPM. The primary job of the top ring is to serve as a compression seal. It also helps to protect the second ring from combustion heat.

Positive-twist, barrel-faced top compression rings are the most popular designs for performance applications. The barrel face provides quick ring break-in and uninterrupted cylinder wall contact. Dykes top rings have a step cut on the top side (sort of an L-shape profile) and are designed for drag racing applications (including Top Fuel and Funny Car) to aid in gas-loading. Dykes rings require pistons designed with Dykes ring grooves.

Most modern OEM-compression rings have a premium-grade cast iron, ductile iron, or steel alloy coated with a

High-quality performance ring packages are available from the major suppliers. Some high-performance piston manufacturers supply the correct rings (per application) with their pistons.

heat-resistant graphite, plasma-molybdenum, or chromium facing material. Graphite and plasma-moly are relatively soft and porous ring facings that have excellent oil carrying capacity and comparatively fast conformability (or break-in) to the cylinder bore surface. Chrome, used today primarily by overseas OEMs, is a much harder, nonporous material.

Second Ring

The second ring, although commonly referred to as a compression ring, functions primarily as a final oil control (about 80 percent of the second ring's duty is oil control and about 20 percent for compression control). The two

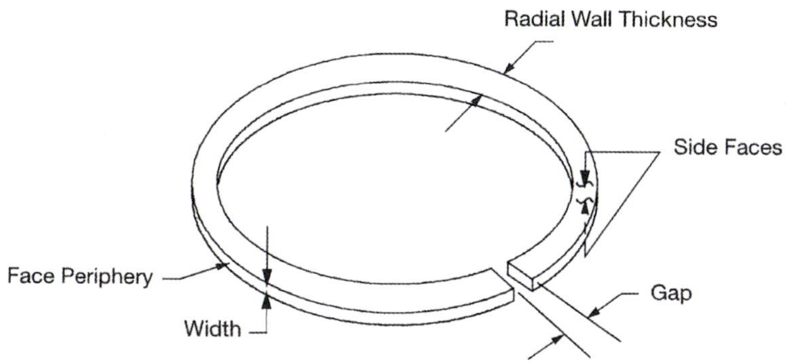

Top or second ring dimension points. (Illustration Courtesy MAHLE Clevite)

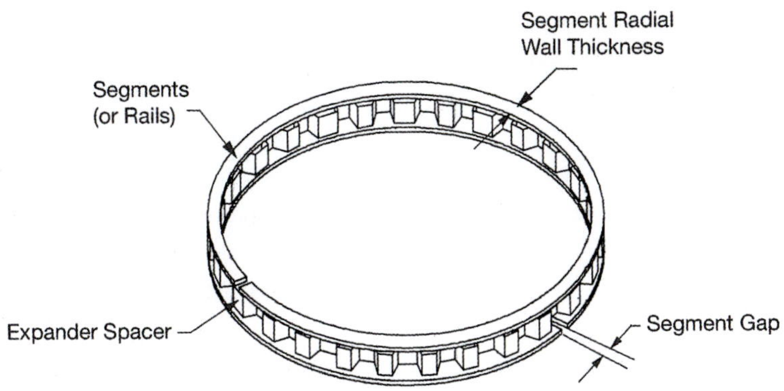

Oil ring package. (Illustration Courtesy MAHLE Clevite)

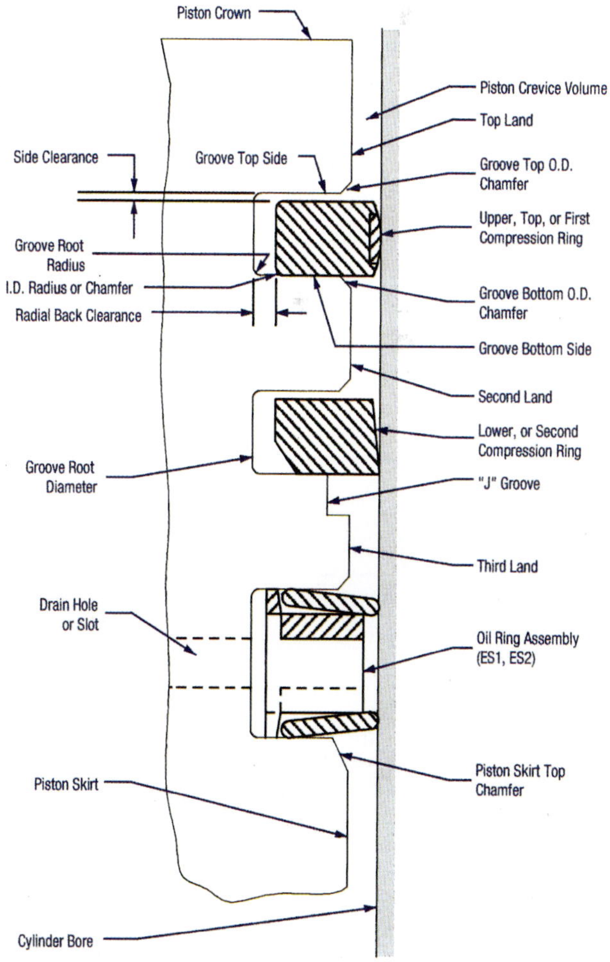

Cross-sectional view of ring-to-piston identity and dimensional reference. (Illustration Courtesy MAHLE Clevite)

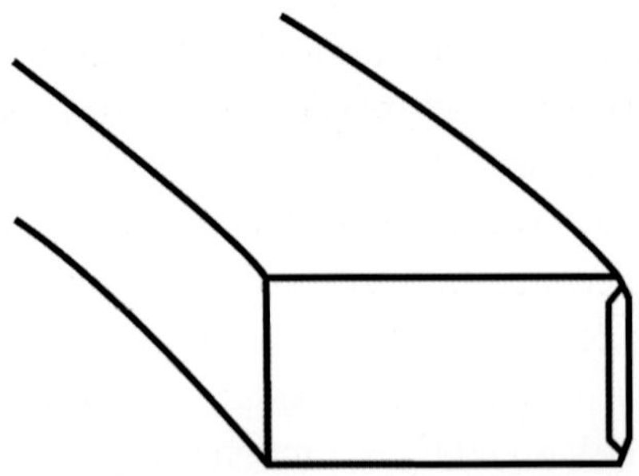

Non-twisted barrel face ring. (Illustration Courtesy MAHLE Clevite)

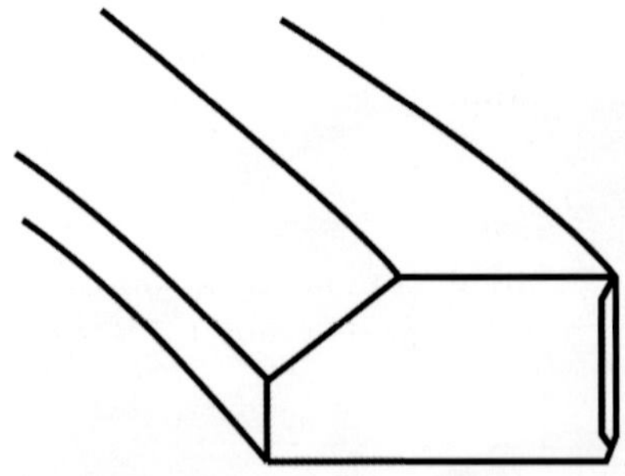

Twisted barrel face ring. (Illustration Courtesy MAHLE Clevite)

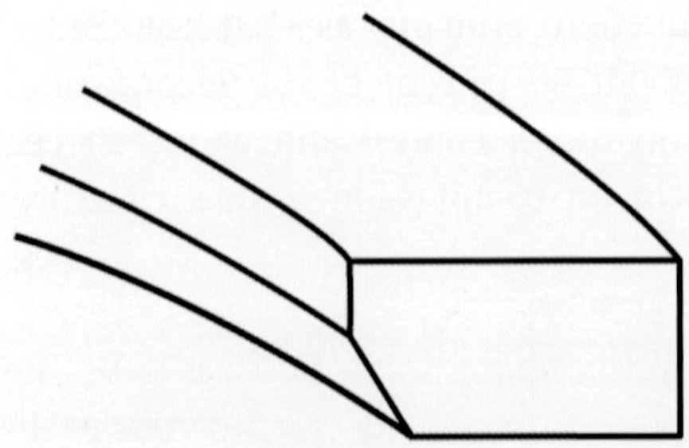

Reverse-twist taper face. (Illustration Courtesy MAHLE Clevite)

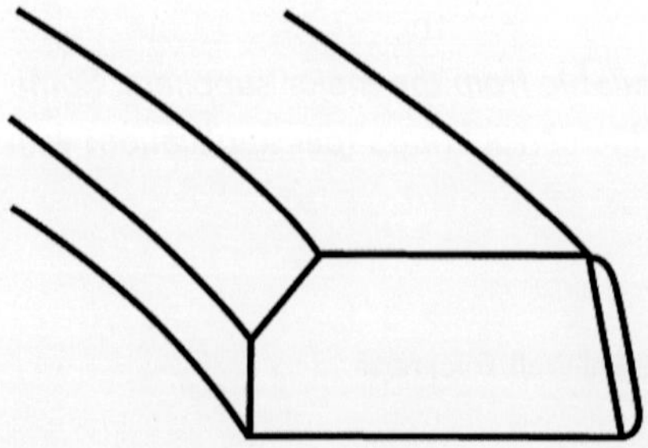

Positive-twist barrel face. (Illustration Courtesy MAHLE Clevite)

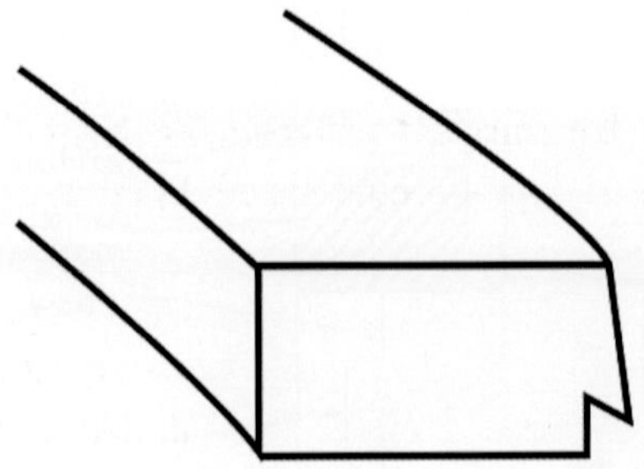

Taper hook groove. (Illustration Courtesy MAHLE Clevite)

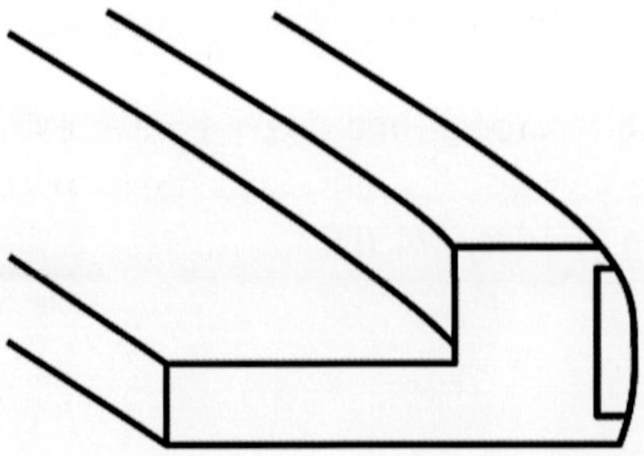

Dykes top ring (drag racing application). (Illustration Courtesy MAHLE Clevite)

common designs are RBT reverse-twist, taper face and THG hook groove (sometimes called a Napier ring).

The reverse twist, taper-face second ring is very common in performance ring sets. An inside bottom bevel causes torsional twist in the ring. The twist prevents oil migration from creeping up behind the second ring. The taper face scrapes oil from the cylinder wall for oil control. Most RBT second rings are made from gray cast iron, with no special face coating.

THG second rings also have a twist, but with no inside bevel and no taper. Instead, THGs have a hook groove on the bottom OD. THG rings offer slightly better oil control than RBTs. THGs also act as a check valve to relieve excessive combustion pressure built up under the top ring, helping to stabilize the top ring (avoiding or minimizing ring flutter).

Like the top rings, OEM second rings are typically coated with a phosphate, plasma-moly, or chrome facing to protect them from combustion-related heat. These rings usually also have a low-tension, tapered face designed to reduce drag while maintaining the ring's oil control capabilities. Most major OEM and aftermarket suppliers build a reverse torsional twist into the second ring to create a more effective seal at the piston land, preventing oil migration behind the ring.

Third Ring

The third ring serves as an oil control ring package that essentially removes oil from the cylinder wall during the power stroke. Oil control rings have multiple components, including an expander (spacer) and upper and lower scrapers (rails). The expander must maintain original tension over an extended period, provide virtually instantaneous seating, and resist corrosion. The oil rails in this classic design rest on angled expander pads that deliver lateral and vertical force to seal the top and bottom surfaces of the ring groove.

Oil rings are offered in standard-tension (for most street applications and high-output engines susceptible to bore distortion) and low-tension (for high-performance applications) designs. Standard-tension rings apply around 21 pounds of tangential pressure against the cylinder wall. Low-tension designs apply approximately 15 pounds of tension.

Tension and Gap

Ring tension is a key consideration for racing applications because the piston rings can account for nearly 15 percent of an engine's internal frictional power loss. The theory is that the lower the tension, the higher the power output. The trade-off, in many cases, is ring durability, but that's a secondary consideration in many high-output applications where rings may be changed on a routine basis.

The best way to start a debate among engine builders is to ask whether an end gap is necessary in the second compression ring. That's because there's been considerable talk about the sealing properties of "gapless" rings. The theory behind the gapless configuration is quite simple: By eliminating a secondary escape path for combustion gases that slip past the top ring, you should be able to increase compression.

Most domestic engine manufacturers have concluded that an increase in end gap in the second ring actually *improves* sealing performance due to what's known as the "Pressure Balance Theory." Without a second-ring end gap, gases become trapped between the second and top rings. As the piston moves through its power stroke, the buildup of these gases actually lifts the top ring off its land. This, of course, causes extreme blow-by and power loss. Many dynamic engine tests have proven this to be case, whether OEM or aftermarket examples. As a result, most engine manufacturers use a *larger,* not smaller, end gap in the second ring.

Piston Ring Trends

The trend in ring design today focuses on reduced friction and high ring conformability. As ring-to-cylinder-wall friction is reduced, horsepower is freed and fuel economy is increased simply due to a reduction in parasitic drag. Ring conformability refers to the rings' ability to seal within a bore that is not perfectly round (as often occurs when a bore is distorted during heat cycles and the stress of head bolt clamping). Ring sealing affects power output, emissions control, and oil consumption.

As engines continue to be smaller and lighter, block decks are becoming shorter and pistons are being downsized. The result is that rings are thinner; some are 1.2 mm, while 3-mm oil rings are common. Ring locations have gradually moved upward toward the piston dome. This trend is likely to continue for several reasons including performance and emissions concerns. For example, a smaller crevice volume (the circumferential void area between the piston and bore, above the top ring) is beneficial in reducing unburned hydrocarbons.

Also, regarding fuel economy, rings are becoming radially thinner, in an effort to reduce ring-to-wall friction. The thinner dimension also allows the rings to be more conformable (flexible) to the shape of the bore. This allows lighter ring tension, providing the bore geometry is correct.

Materials and Coatings

Although steel remains the current top choice for the top-ring position, iron (coated or uncoated) is preferred for the second ring position. The popularity of the traditional three-piece oil-ring assembly continues, using a stainless steel expander and chrome plated or gas nitrided steel rails. Although uncoated cast-iron top and second rings continue to serve well in the economy ring class, ring material is migrating toward steel for the top rings. For ring faces, manufacturers have developed superior wear coatings such as plasma-applied moly for increased durability.

Using a premium-grade material depends largely on the OEM application. For turbocharged or supercharged applications, you may choose a ductile iron with a plasma-moly coating, or steel rings with plasma-moly or chrome. Generally, most domestic applications use a moly coating, while most imports use chrome. The basic plasma-moly-coated ductile ring still does a good job on racing engines.

Molybdenum-disulfide (moly, for short) offers high scuff resistance. Used in a variety of configurations, it is also somewhat self-healing and absorbs some abrasion. Moly is longer wearing than a bare cast-iron surface, and more heat resistant than chrome. Phosphate and moly coatings enhance break-in and reduce the possibility of ring microwelding. Moly coatings are able to withstand temperatures up to 1,200 degrees F. It is applied to cast iron, ductile iron, or steel rings.

Gas nitriding provides a hardened surface finish to improve wear resistance. This coating is applied to steel or stainless steel rings.

Chrome plating continues to be one of the best wear-resistant coatings available, with a temperature resistance rating of around 800 degrees F.

On bores that have been plateau honed, the best applications are plain-faced and plasma-moly rings.

Ring Gap

For standard replacement rings, the gap is already established by the ring manufacturer. In general terms, the smaller the bore, the smaller the gap; the larger the bore, the larger the gap. Top rings generally take a .010-inch minimum gap; the second rings require a .018- to .020-inch minimum gap.

For gap tolerance, particularly on the second ring, a slightly larger gap than the top-ring gap can be used. Increasing the second-ring gap helps to balance pressure between the top and second ring. This also helps to maintain seating of the top ring, thereby eliminating blow-by pressure that might otherwise lift the top ring. Although, a small bit of blow-by past the second ring can help oil control, blowing oil back down into the sump.

If you plan to use a gapless ring anywhere on the piston, it should be used at the top ring location only. Although a gapless top ring provides great leak-down readings, the primary benefit is on alcohol engine applications, to prevent excess fuel runoff into the cylinders.

Some racing applications currently use a gapless second ring, such as those by Childs & Albert and Total Seal. Popular designs include use of an interlocking step design at the mating joint or a

Prior to measuring ring gap, it's critical that the ring must be placed squarely in the cylinder bore. A dedicated ring-squaring tool that's adjustable to bore diameter makes this task easy. Place the ring into the bore, about 1/4 inch down from the deck, and insert the squaring tool. The ring is immediately leveled in the bore.

Using a feeler gauge, measure the existing ring gap. Compare this to the piston manufacturer's specification (gaps differ between top and second rings).

Oil ring support rails (where needed to complete a floor support for oil rings where the piston pin bore intersects the oil ring groove) have a protrusion (pimple). It must face down at the center of the piston pin bore. This prevents the support rail from rotating out of position and prevents its gap from aligning with the void on the ring groove floor.

If the ring gaps are too tight, they may be file-fit by hand or by using a ring filer. Don't remove too much material. File with a few strokes, clean the ring, recheck gap in the bore, and continue the process in small increments until you achieve the desired gap. Remember to check gap (and correct where needed) at each cylinder, and keep each ring dedicated to the cylinder in which it was checked and fitted. Small variances in cylinder bores may exist, so for tailor-fit ring gaps dedicate each ring to one cylinder. For example: fit the top ring for number-1 cylinder, then for number-2, and the remaining cylinders. Next, fit the second ring for number-1 cylinder, then number-2, etc. Keep the rings organized per cylinder location.

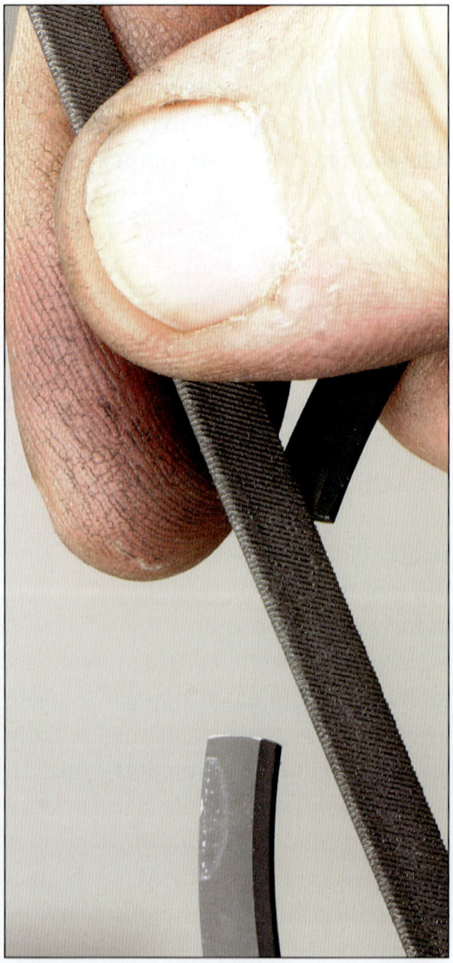

If rings are file-fit, carefully dress the filed edges to remove any burrs.

Here's an example of a piston that requires a support rail for the oil ring. Note, because of the short compression distance, that the wrist pin bore intersects with the oil ring groove. After installing the rod to the piston, install the support rail to the floor of the oil ring groove.

counterbored iron ring that sandwiches to a thin oil-ring rail, with gaps placed at opposing locations. The reason for using a gapless second ring is that it dramatically increases the anti-leakdown property.

However, a gapless second ring can create two problems a buildup of pressure that can unseat the top ring (creating ring flutter and instability) and the reduced leakdown reading provided by a gapless second ring can mask potential leakdown problems at the top ring.

Current replacement rings utilize unmodified conventional gaps on hypereutectic pistons. However, some experimentation is taking place. The Keith Black Racing hyper pistons specify a larger-than-normal gap for the top ring, with the intent of compensating for ring expansion during high-heat operation.

There is also a growing trend among OEM designs to use a slightly larger second-ring gap, simply to maintain top-ring

The support rail has a small protrusion/pimple. This must face down and be centered in the void above the pin bore.

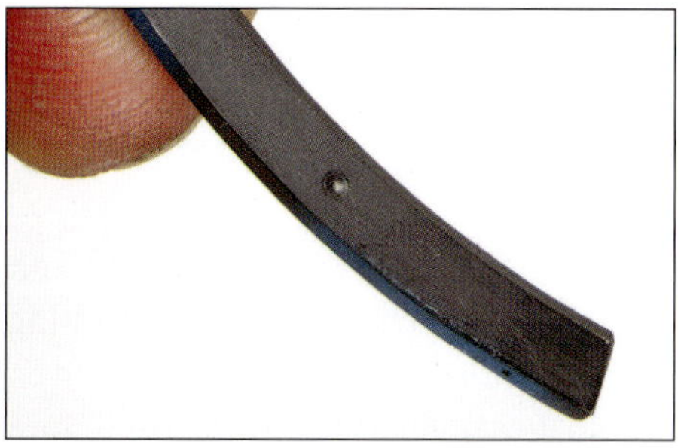

Top and second rings typically have a small dimple/dot. This is an orientation reminder. Typically a ring must be installed with the dot facing upward, but verify this in the ring manufacturer's instructions.

Once the support rail is installed, install the oil ring expander.

Install the lower oil rail onto the expander; this captures the expander. Next install the top rail.

When installing the oil ring rails, make sure that the ends of the expander butt together and don't overlap!

Instead of messing around with clumsy, antiquated and often aggravating barrel-type ring compressor tools, the slick pro-approach is to use a tapered billet compressor that has a tapered inside surface. This allows the piston and ring package to glide through the tool, compressing the rings during travel.

One-piece billet tapered ring compressors must match your cylinder bore diameter. These are available in popular increments to suit most bore diameters. If you're building multiple engines that have the same bore size, this is the way to go.

stability (for pressure balancing). This gap increase at the second ring, in some applications, may be as large as .020 to .030 inch.

Determining Ring Depth

Here is a formula to use as a guideline when selecting top and second ring depth (22 is the SAE constant):

Ring Depth = bore diameter ÷ 22

The same applies to the oil ring. Standard depth on a ring for a 4.0-inch bore previously was about .180 inch. Today, this is about .130 (for all applications, including OEM and racing).

Ring Installation Methods

Winding a ring onto a piston can easily create a helix in the ring, resulting in ring distortion that may not recover. *Always* use ring pliers to install rings onto pistons. The tapered ring tool is the best to use, especially with today's thinner, more fragile rings.

As far as synthetic oils and other ultra-slippery materials are concerned, don't use a full-synthetic assembly lube or oil at the ring-to-cylinder wall in a fresh engine. If the lube is super-slippery, the rings may take much longer to seat, or they may not seat at all. For initial break-in, it's best to use a petroleum-based engine oil.

After oiling the clean bore and oiling the ring package (and adjusting the ring end gaps to avoid gap alignment), slip the rod/piston through the compressor until the ring package is captured.

With the upper rod bearing installed and lubed, and with the piston and rod properly aligned, insert the piston skirts into the bore and allow the bottom of the compressor tool to firmly contact the block deck. Using finger pressure, push the piston/ring package into the bore until the ring package is fully inserted in the bore.

An alternative to using a dedicated cylinder-bore-diameter compressor is to use a billet tapered compressor that has a split wall and adjustment clamp. This style is available in various ranges instead of one specific bore size.

After installing clean rings to clean pistons, keep each package organized for respective cylinder locations. Don't mix and match; a small stack-up of tolerances may result in less-than-optimum ring gaps.

Camshafts

The camshaft can be viewed as the brain of the engine. Its profile dictates when and for how long the intake and exhaust valves open and close, which is a major factor in how the air/fuel package enters, remains in, and exits the cylinder. Today's cam technology has made great strides in engine performance and fuel efficiency.

Cam Types

You cannot use roller lifters with a flat-tappet cam, nor can you use flat-tappet lifters with a roller cam (unless you enjoy destroying your cam and anything else in its wake).

Rather than featuring the same ramp angle on each side of the cam lobes, asymmetrical camshafts feature different angles on each side of the cam lobes that allow cam designers to "fine tune" valve opening and closing rates.

Flat-Tappet Camshafts

A flat-tappet cam uses lifters that rub against the cam lobes (with an oil film in between). These lifters are designed to rotate in their bores in order to continually spread the load around the lifter face (if the lifter can't rotate, frictional wear concentrated across one path causes the lifter/lobe to wear prematurely). A flat-tappet cam's lobes are slightly tapered to promote lifter rotation.

While flat-tappets have a narrower radius at the peak of the lobe, roller cams require a large-radius lobe peak to maintain full lifter roller contact with the lobe. If the lobe had a narrow (small) radius, the roller jumps and chatters as it rides past the peak. Roller lifters need a nice, flat surface to continually ride on the lobe. One advantage of roller cams/lifters (in addition to reduced friction) is that rollers hold the lifter at max lift for a longer duration, which helps to load the cylinder with more air/fuel.

The 4/7 swap cam (center) and the LS cam (right) feature the second and fourth lobes from the front in an opposed position compared to the stock firing order cam.

Roller Camshafts

Roller cams utilize lifters that are equipped with a roller bearing that rolls against the cam lobe. They require a flat (non-tapered) surface on the lobes and virtually eliminate lifter-to-lobe friction and wear. Roller lifters can handle much more abrupt transitions from base circle

Roller lifters have a roller bearing and trunnion. OEM roller lifters may be adequate for most street applications. However, if you're building an engine with increased spring pressures and abrupt closing ramps, you need something beefy. Toss the OEM stuff and grab a set of heavy-duty aftermarket roller lifters that are designed for severe duty.

Bridged roller lifters install as a pair. During test fitting, make sure that the anti-rotation bar clears the lifter bosses when each lifter is on the cam base circle.

If the design uses OEM "dog bones," the OEM dog bone is dropped onto the lifter pair. This single guide prevents the pair of lifters from rotating.

Since the dog bones need to be secured to hold them in place, a spring-tempered sheet-metal securing plate is mounted to the lifter valley floor, featuring extended tension "fingers." Each finger presses down onto the center of each dog bone to prevent the dog bone from popping loose.

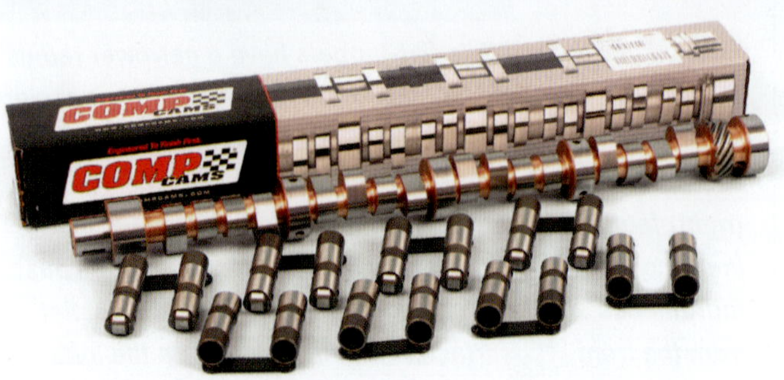

For retrofits (when installing a roller cam in an engine that originally used a flat-tappet cam) or for engines that originally used a roller cam with lifters guided by dog bones, performance aftermarket roller cam kits are available with bridged lifters.

Overhead cam engines use cam followers (no pushrods) that directly activate the valves from the cam lobes. For a high-performance build, full-roller cam followers greatly reduce friction and allow higher RPM, greater efficiency, and much longer life. These are Crane Cams intake and exhaust cams, adjustable cam timing gears, and Ferrea alloy full-roller cam followers for a Honda race engine.

to ramp to lobe peak, and can hold the lifter at maximum lift for a longer period of time. For performance and durability, roller cams offer distinct advantages over flat-tappet cams.

Another advantage offered by roller cams relates to long-term service. A flat-tappet cam establishes an individual wear pattern between each lobe and its lifter. If you plan to use the original lifters, they *must* be re-installed in the original positions. Otherwise, buy a complete set of new lifters to use with the existing cam. Since there's no wear pattern established with a roller cam, you can mix up roller lifter positions, or replace one lifter at a time, etc.

Asymmetrical Roller Camshafts

A traditional symmetrical camshaft has the same shape/angle on both sides of the lobe, which causes the valve to open and close at the same rate of speed. Asymmetrical cams have lobes with different ramp angles/shapes on each side of the lobe. Cam manufacturers can offer cams that open and close valves at different rates so they can be designed to super-tune valve operation for maximum performance. Once a secret of high-end roller-cam race engine builders, asymmetrical cam profiles have become fairly commonplace in even performance street engines.

Today this asymmetry aids in optimizing valvetrain dynamics by tailoring how fast the valve opens, while avoiding undue stress to the valvesprings, and how quickly the valve closes without banging/shocking the valve at its seat. The approach of using different profiles on the opening and closing sides of the lobe allows engineers to tailor the valvetrain to optimize specific engine requirements.

Protecting the Cam

Due environmental concerns the EPA mandated a substantial reduction of ZDDP, which is a critical protection agent in the oil mix. The EPA also eliminated ZDDP in standard engine oils that are intended for the late-model production engine with a roller-style camshaft. Your oil must contain enough ZDDP to adequately protect the flat-tappet cams during break-in and for long-term engine operation.

It's vital to protect the flat-tappet cam-lobe wear during the break-in period or your entire engine can be ruined. ZDDP is an essential compound for high-performance engines with aggressive cams and high-spring loads. Roller cams are not impacted because the rollers don't produce nearly as much friction and have no real scuff-wear issues. The EPA has directed the oil manufacturers that they can ignore older (i.e., flat-tappet cam) engines, and provide an oil that avoids converter damage so emissions are reduced. As a result, collector-car owners were left without a viable alternative.

To help ensure flat-tappet cam lobe damage does not occur, apply the specific cam lobe and lifter assembly lube that's recommended by the cam manufacturer. Many engine builders also install valvesprings with a lower rate (lighter pressure) for the break-in phase, switching to springs with a higher rate after break-in has been accomplished. You can use a lighter valvespring for break-in to help guard against lobe damage during break-in, but this isn't a definitive solution. It's imperative you use a dedicated break-in engine oil and a ZDDP additive.

In order to protect a new flat-tappet camshaft during camshaft break-in, select and use a specific break-in oil that is designated for flat-tappet camshaft break-in; these include: Joe Gibbs BR or BR30, Brad Penn Break-In Oil, and Royal Purple Engine Break-In Oil, etc.

Select and use a mineral-based engine oil (preferably a non-detergent oil in the 30W or 10W30 range), plus a 16- to 20-ounce bottle of specialty ZDDP additive.

In order to provide long-term protection for a flat-tappet camshaft: Use a specialty engine oil that contains high levels of ZDDP (around 10,000 to 12,000 ppm). These are offered by companies such as Mobil 1, Redline, Joe Gibbs, and Brad Penn.

Before attaching the timing chain (or belt), always check camshaft endplay with a dial indicator. Generally, endplay should be somewhere in the range of approximately .003 to .006 inch.

With a handle/grip attached to the cam you greatly reduce the risk of damaging cam bearings.

Use the engine oil of your choice, plus a bottle of ZDDP additive with every oil change.

For break-in of flat-tappet engines, avoid any engine oil with the API logo on the container. The small starburst indicates that the oil has been formulated for new engines, from an energy-conserving standpoint. It is for passenger car gas engines equipped with roller cams.

Cam Runout

Check any cam (new or used) for runout. Rest the camshaft on a pair of clean V-blocks; a roller-bearing cam checker stand is ideal. Position a dial indicator on a stand that is rigidly secured at the base at the camshaft's center journal. Preload the dial indicator plunger against the journal by about .050 inch, and then zero the gauge. Slowly rotate the camshaft while observing the indicator gauge. Maximum allowable runout is about .001 inch. If runout is excessive, it's best to replace the cam.

Before installing any cam, you must lube the journals and lobes (and distributor drive gear if applicable). A high-pressure lube is needed to protect lobes during break-in. While the choice of lube isn't as critical for a roller cam, when dealing with a flat-tappet cam always use the specific cam assembly lube specified by the cam manufacturer.

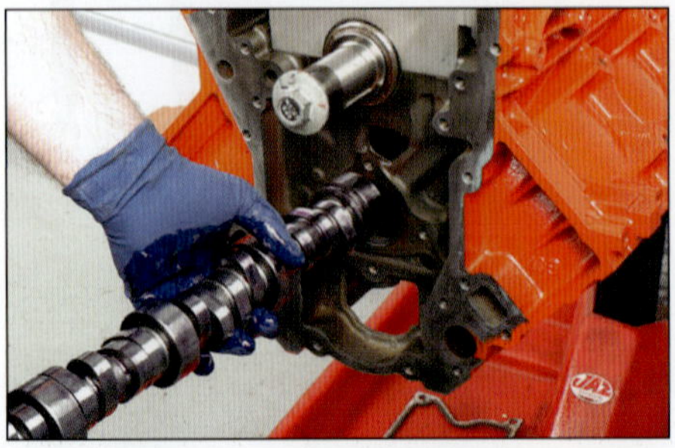

Insert the cam slowly. Each time one of the journals exits a cam bearing, the cam is no longer guided and can drop in the bore. Carefully align each journal during cam bore entry. If the cam stops, never try to force it. Gently lever it until you achieve a smooth entry into the next cam bearing.

Cam Endplay

Check camshaft endplay when the camshaft has been installed for a test fitting and before lifters are inserted. Set up a dial indicator at the cam nose if held by a retaining plate or onto the face of the cam gear. Push the cam inboard until it stops. Adjust the indicator plunger for (about .050 inch of preload), and then zero the gauge. Pull the cam forward and observe the amount of travel on the gauge. Check with your engine's specs, but in general you should have something in the range of .003 to .006 inch.

Degreeing the Cam

You can't assume the cam timing is correct, so you need to degree the cam, ensure it's machined correctly, and be sure that the timing marks on the cam and crank gears are precise. This simply means that you make precise measurements to both verify the cam profile and to properly phase the cam timing with the piston positions in order to take full advantage of the performance that the cam has to offer.

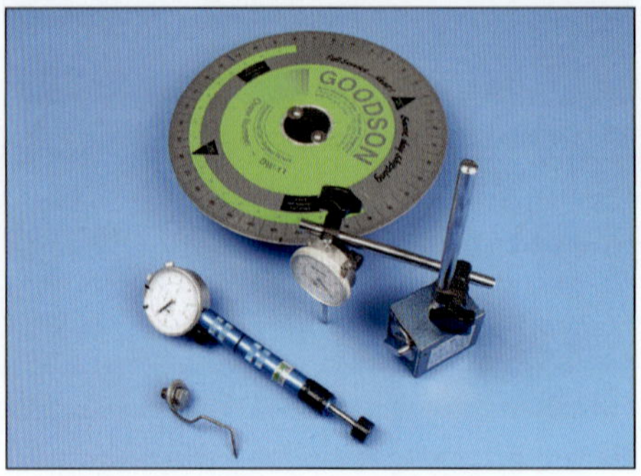

Pictured here is a Goodson degree wheel, a dial indicator with adjustable stand, a homemade wire pointer, and a Foster lifter gauge.

The quality of today's performance cams isn't in question because Crane, Comp, Bullet, Lunati, Crower, etc., utilize the latest computer imaging and precise CNC machining centers. The reason to take the time to degree a cam has more to do with crank and cam timing-gear timing-mark precision, checking for incorrectly machined cams or crank keyway locations, and stack-up of machining tolerances. If you install a new cam and simply align the timing marks on the gears, the engine likely runs just fine. But if you want to know what's going on and you want to maximize the potential of the new cam, taking the time to perform a degree check is critical.

In order to degree the cam, use a large-diameter degree wheel, a piston stop, and a dial indicator mounted to an adjustable stand. Also use a pointer that bolts to the front of the block and extends to the edge of the degree wheel (most people use a piece of coat hanger

If you rotate the crank too quickly or use a jerking motion, you can easily run past your intended point. If this happens, start over by rotating in the normal direction two more times. If you rotate in the opposite direction to try to hit your missed target, slack in the timing chain results in a false reading.

PART NO: 1449121 HYDRAULIC ROLLER SPECIAL
GRIND NUMBER HR-224/347-2S1-15 4A
ENGINE IDENT. 1997-UP CHEVROLET V-8 LS1 5.7 LITRE
VORTEC 4.8L, 5.3L, 6.0L

VALVE SETTING: INTAKE .000 EXHAUST .000 ---------> HOT

LIFT:			
	INTAKE @ CAM 347	@ VALVE 590	ROCKER ARM RATIO
	EXHAUST @ CAM 347	@ VALVE 590	1.70

ALL LIFTS ARE BASED ON ZERO LASH ND THEORETICAL ROCKER ARM RATIOS

CAM TIMING @ .004 TAPPET LIFT		OPENS	CLOSES	ADVERTISED DURATION
	INTAKE	26.0 °BTDC	74.0 °ABDC	280.0°
	EXHAUST	80.0 °BBDC	27.0 °ATDC	287.0°

SPRING REQUIREMENTS

		TRIPLE	DUAL		OUTER		INNER
PART NUMBER			99831				
LOADS:	CLOSED	130	LBS.	@	1.800	OR	1 51/64
	OPEN	305	LBS.	@	1.240		

RECOMMENDED RPM RANGE WITH MATCHING COMPONENTS

MINIMUM RPM	2400
MAXIMUM RPM	6500
VALVE FLOAT	7100

CAM TIMING @ .050 TAPPET LIFT		OPENS	CLOSES	MAX LIFT	DURATION
	INTAKE	(0.5)°ATDC	44.5 °ABDC	111°ATDC	224.0°
	EXHAUST	53.5 °BBDC	(1.5)°BTDC	119°BTDC	232.0°

REMARKS: FIRING ORDER: 1-8-7-2-6-5-4-3

A typical cam card. When degreeing the cam, use the card as your reference. Always keep a cam card in your files. This provides a handy reference in the future if you decide to swap cams.

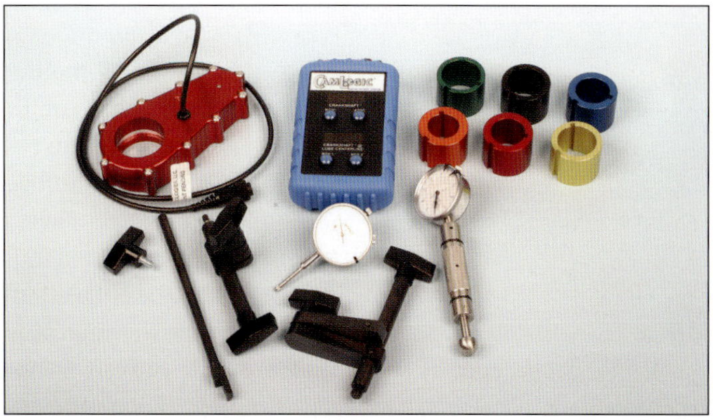

Here's a high-tech way to degree your cam, using positioning sensors and a digital control box. This is Cam Logic's system, including a crank-mounted encoder, digital display unit, cam lobe dial indicator, and a piston indicator.

wire with one end sharpened to a point and the other end looped to allow bolt-mounting).

Locating Top Dead Center

The short block must be assembled, including main bearings, crank, rods, bearings, pistons, camshaft, timing gears, and chain or belt. Depending on the design of your degree wheel, you can fasten it to the crank snout or install a crank balancer to allow wheel mounting.

With the wheel and pointer mounted, rotate the crank until the number-1 piston is at TDC. Align the pointer to the TDC mark on the degree wheel. Next, in its typical direction of rotation, rotate the crank until the piston is halfway down the bore. Install a piston stopper to the block deck at the piston bore.

You simply need something that the piston can stop against. You can purchase a piston stopper tool. It's a bridge that bolts to the block deck and has a center-mounted, adjustable bullet-nosed stopper.

Or you can fabricate a stopper by using a piece of hefty 1/4-inch-thick (or more) piece of steel or aluminum flat-stock with a tapped hole in the center (thread size isn't critical but, for the sake of example, let's use a 3/8 x 24 thread). Radius the tip (simply to avoid nicking the piston) of a 3/8-inch x 24 bolt that's about 2 inches long.

Install a jam nut onto the bolt and run the nut toward the bolt head. Thread the bolt into the hole in the flat stock. Drill two holes in the ends of the flat stock to allow mounting to the block and mount the bar to the block, positioning the bolt at the center of the bore. Thread the bolt so it protrudes into the bore to a depth where you're sure that the piston makes contact with the bolt when the piston is below TDC. Exact depth isn't critical. Tighten the jam nut on the bolt.

Carefully turn the crank in the typical direction until the piston contacts the stop. Note the reading on the degree wheel.

Next rotate the crank in the opposite direction until the piston once again contacts the stop, and note this reading. If the crank stopped at exactly the same point at the end of each rotation, you got lucky and happen to have the degree wheel adjusted exactly at TDC. In the real world, however, the two readings are likely different.

When the two readings (before and after TDC) are different, simply use the average, since true TDC is exactly halfway between the two readings. For example, if the pointer marks, say, 30 degrees after TDC at the first stop and 32 degrees at the second stop, the average of the two is 31 degrees. Simply adjust the pointer to the 31-degree mark on the wheel. Rotate the crank in the opposite direction until the piston stops, to verify your mark. The pointer should read (in this case) 31 degrees after TDC.

In order to verify the setup, remove the piston stop, and install a dial indicator to locate the indicator's plunger over the center of the bore. Or, depending on piston dome design, pick a spot where the plunger contacts a flat deck surface (quench area) of the piston.

Rotate the crank clockwise until the piston is .100 inch below block deck. Note the pointer reading on the wheel before TDC. Slowly continue to rotate the crank clockwise. The piston reaches TDC and begins to run downward. Stop when the piston is again .100 inch below deck, and note the pointer reading on the wheel. The readings for BTDC and ATDC should be the same.

With the degree wheel adjusted properly, you're ready to degree the camshaft.

Locating Cam Lobe Centerline

The procedure for finding lobe centerline is different for symmetric and asymmetric cams.

A dial lifter gauge, such as the one made by Foster Tools, allows you to directly read the camshaft lobe.

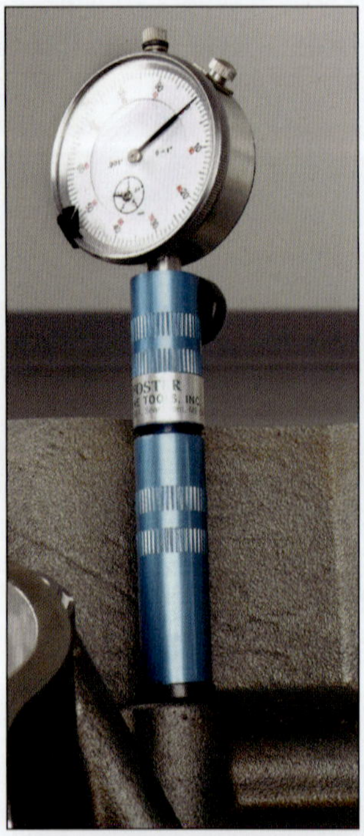

Installed lifter gauge. This acts as a temporary lifter that directly follows the cam lobe profile.

Symmetric Cams: Although there are several methods that can be used to degree the cam, the centerline method is considered the easiest and most reliable. With the number-1 piston at TDC and the degree wheel's pointer at 0 degrees, position a dial indicator with the plunger on the intake valve retainer. Make sure that the indicator plunger is parallel to the valvestem (not cocked at an angle).

Rotate the crank clockwise until the camshaft's number-1 intake lobe reaches maximum lift. Zero the dial indicator gauge (don't disturb the plunger; simply rotate the gauge to zero). Rotate the crank counterclockwise until the dial indicator gauge reads .100 inch. Next, rotate the crank clockwise until the gauge reads .050 inch before maximum lobe lift. Observe and record the degree wheel pointer reading.

Continue to rotate the crankshaft clockwise until the indicator gauge moves past zero to .050 inch on the closing side of maximum lobe lift. Record this reading on the degree wheel.

Rotate the crankshaft smoothly to avoid jerking and running past your desired marks. If you run past the target mark, start over. Don't reverse direction to compensate, since slack in the timing chain affects your accuracy.

Add the two numbers from the degree wheel readings and divide by 2. The result represents the location of maximum lift of the intake lobe in relation to the crank and piston, which is the intake centerline.

For instance, if the first degree wheel reading was 100 degrees, and the second degree wheel reading was 120 degrees (100 + 120 = 220). Half of that number equals 110, indicating the cam's intake centerline is 110 degrees. Compare this to the intake centerline listed on your cam card, which should be 110.

Once you've determined what your cam's intake centerline actually is, the next step is to find out how your cam is currently timed (advanced or retarded). If the intake centerline is greater than the lobe center angle, the camshaft is advanced. If less, the cam is retarded.

For example, if your cam has a 110-degree intake centerline and the lobe centerline is, say, 108 degrees ATDC, your cam is currently timed with 2 degrees of advance. If the intake centerline is 112 degrees ATDC, the cam is 2 degrees retarded.

Advancing the camshaft opens and closes the valves sooner, which increases low-end torque. Retarding the cam sacrifices some low-end power but provides

Crank position is constantly displayed, as is cam lobe position.

The crank position encoder slips onto the crank snout (using one of their crank-specific keyed aluminum adapters). The encoder is then secured to the block face to prevent encoder body rotation.

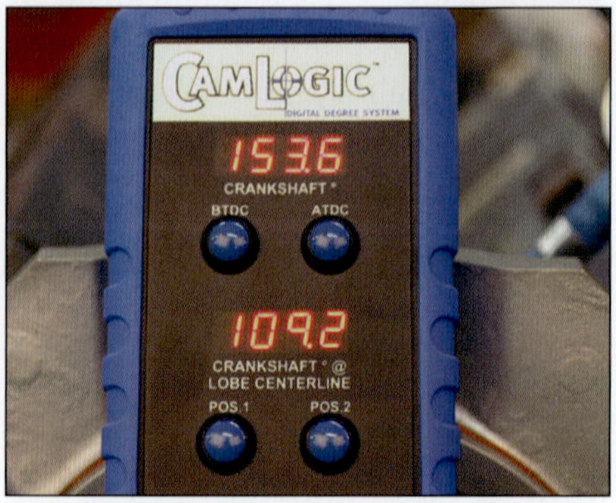

The digital display unit constantly monitors crank angle (top display) and camshaft lobe position (lower display). Using this system is surprisingly simple and quicker than using a degree wheel.

more top-end power. The only way to tune the cam timing to your preference is to test it. Dyno testing may get you into the ballpark, but the best test is always in the car, on the track (quite often what looks ideal on a dyno monitor isn't what the vehicle needs for top performance on the track).

Altering cam timing is relatively easy. Crank gears are available with multiple keyways (giving a choice of zero or advance/retard positions), and adjustable cam gears are available with increment references for adjustment.

Asymmetric Cams: On an asymmetric cam the valve opening and closing rates differ on each side of the lobe, When degreeing this type of cam, take the lift reading directly at the lifter or cam lobe. Either set up a dial indicator onto the lifter (on the edge of the lifter for a hydraulic roller lifter or in the pushrod pocket for a solid roller lifter), or use a cam lift gauge (such as from Foster Tools).

This is a straight-bodied housing with a dial indicator gauge at the top and a spring-loaded plunger at the bottom. With the lifter removed, insert this gauge into the lifter bore, with the plunger contacting the base circle of the lobe (the shortest lift of the lobe). Finger tighten the gauge body base and zero the gauge. The gauge expands when turned and locks into the bore. It allows you to monitor cam lobe movement directly against the cam lobe instead of reading at the valve.

Slowly rotate the crank clockwise, until the lobe rise indicates .050 inch on its opening ramp. Record the degree wheel reading. Continue to rotate the crank clockwise until the gauge reads .050 inch on the lobe's closing ramp. Record the degree wheel reading. Compare your measured opening and closing wheel readings with the cam card. You can then determine if you want to advance or retard the cam.

Adjusting Cam Timing

Advancing the cam helps to increase low-end torque by starting the intake opening sooner. Advancing also increases the clearance from the exhaust valve to the piston but decreases the clearance from the intake valve to the piston. Retarding the cam increases high-end power by keeping the intake valve open longer. Retarding the cam increases the clearance from the intake valve to the piston but decreases the clearance from the exhaust valve to the piston.

Lobe Separation Angles

The lobe separation angle (LSA) can have a dramatic impact on performance characteristics.

A tighter LSA decreases piston-to-valve clearance, places the torque curve on a lower RPM range, and increases maximum torque, but results in a narrow power band. A tighter LSA also helps to increase cylinder pressure because the transition time between valve opening and closing is shorter. It also increases the chance of detonation due to higher cylinder pressure and increased compression. A tighter LSA also reduces engine vacuum and idle quality.

A wider LSA results in the opposite characteristics, including increased valve-to-piston clearance, higher vacuum, less chance of detonation, etc. Performance-wise, a wider LSA broadens the powerband, reduces maximum torque, and raises torque to a higher RPM range. In a nutshell, tight LSA is better for bottom-end torque and wide LSA is better for high-RPM running. A nice compromise is an LSA of 110 to 108 degrees.

GM LS Cam Lifters

All LS engines are designed to use roller lifters. OEM lifters are hydraulic, but the performance aftermarket offers both hydraulic and solid roller cam lifter kits.

Although the small-block Chevy had lifter bores located in a valley under the intake manifold, the LS lifter bores are located in the block decks, so there's no need to disturb the intake manifold when servicing the lifters. Instead of installing lifters one by one or in bridged pairs, a bank of four lifters are guided in a plastic composite "lifter tray."

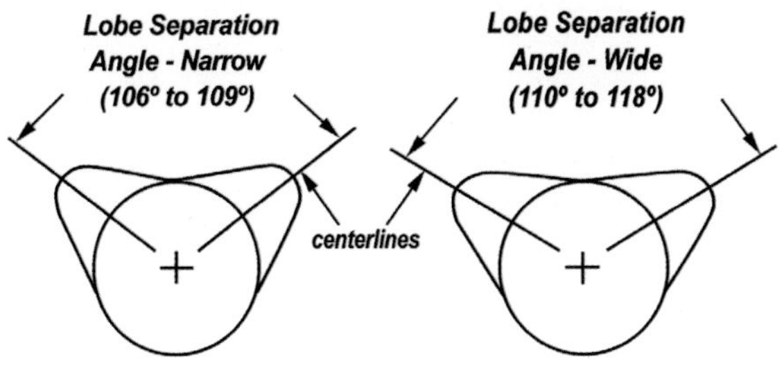

Lobe separation angle (LSA) refers to the number of degrees that separate the centerline of the intake lobe from the centerline of exhaust lobe. (Illustration Courtesy Lunati)

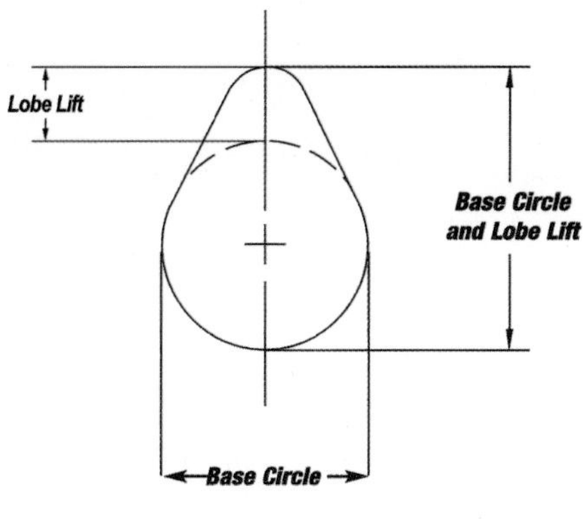

The camshaft base circle is the concentric base from which the lobe extends. The base circle position is where valve lash adjustment takes place. (Illustration Courtesy Lunati)

The tray has four lifter bore cavities. Each lifter has two opposing flats at the top of the lifter body. These flats engage into flats in the plastic bucket assembly. These flats keep the roller lifters in plane with their cam lobes, preventing the lifters from rotating (same principle as using dog bones or bridged-pair lifters).

Insert the bank of four lifters into each tray and lightly snap in place. Install the entire lifter tray assembly through a recess in the upper section of the block, and align the lifters into their bores. Secure the tray to the block with a center-located bolt. Install the lifters to the block before installing the head gasket because the head gasket and head traps the lifter assemblies.

After the camshaft is installed, install the assembled lifter trays and bolt them to the block. Because the lifters were pushed up into the tray cavities, they're held away from the cam. Simply use a pushrod to manually pop each lifter down to meet its cam lobe. The upper flat section of the lifters remain guided in the trays.

A nice aspect of this lifter tray design has to do with future camshaft service. When you want to remove and re-install or swap cams, there's no need to remove the intake manifold. Loosen or remove all rockers to remove all valvespring tension. Remove the pushrods. Rotate the crank two full revolutions. This causes the cam lobes to push the lifters far enough up into the buckets to hold the lifters away from the cam. Then remove the cam, insert the new cam, and pop each lifter down onto the lobes with a pushrod.

Although this method usually works as designed, there is always the possibility that one or more lifters might fall out of the bucket retainer. If that happens, you are forced to remove the cylinder head(s) in order to remove the buckets, retrieve the lifter, and re-install.

As a safety measure to avoid this concern: Once all lifters have been popped up into the buckets and are away from the cam lobes with the camshaft retainer plate removed and before removing the camshaft, insert a pair of 1/4- or 5/16-inch-diameter metal rods at lengths of about 24 inches into the two oil galley holes located just above and at each side of the front camshaft bore. Insert these rods all the way through the block until they *gently* bottom-out. Now you can remove the cam without the worry of lifters dropping into the cam bore. Once the cam has been replaced, remove the rods and use a pushrod to pop each lifter down. Inserting these temporary rods avoids the potential for any of the lifters to accidentally drop out of their trays.

Special Firing Order Camshafts

Every engine is designed with a specific cylinder firing order. However, special camshafts are available to alter the original firing order. Primarily intended for specific racing applications, a special

Once the cam is installed, rotate it to verify smooth rotation. If the cam feels tight you may need to remove the cam, clean the bearings of lube, and relieve any high spots on one or more cam bearings with a bearing scraper. If you're not experienced at this, it's best to let a skilled shop handle this for you. Once cam installation is finalized, install the cam retaining plate (designs vary depending on engine design).

In order to degree the cam, install the timing setup. With number-1 piston at TDC, align the timing marks on the crank and cam gears according to the OEM locations (usually the cam gear dot is at 6 o'clock and the crank gear is at 12 o'clock, but this can vary depending on engine design). Snug the cam gear bolts. Just remember to apply thread locker and to torque these bolts to spec once you've set your cam timing.

When installing the LS lifters to the buckets, insert the lifters deep enough to snap them in place. In this up-position, the lifters are held away from the cam lobes.

The lifter bucket holds a bank of four roller lifters with the roller tips in plane with the cam.

Note the series of small tapered walls in the top of the lifter bucket; they are designed to help drain oil to the lifters. However, this OEM design still tends to hold too much oil and drains slowly. A simple mod involves drilling a couple of 5/16-inch holes around the base of the cavity for faster oil drain.

Flat-tappet cams have lobes that are slightly tapered, to force the flat-tappet lifters to rotate. This lobe taper also causes the cam to walk fore/aft slightly. Since a roller cam doesn't need (and can't use) this taper, a roller cam needs a method to prevent this walk. Many roller cams have a stepped nose, designed to be stopped from walking forward by a retainer plate. Otherwise, a cam button is needed; this drops into the center of the cam nose and takes up the space between the cam nose and the rear of the timing cover.

firing order (SFO) camshaft is sometimes preferred, which allows the race engine builder to address certain combustion heat and crankshaft disturbance (harmonic) issues. The reason to alter the firing order is to obtain a smoother running engine with enhanced (even) fuel distribution, while increasing crankshaft and main bearing durability. Although slight horsepower gains may or may not be obtained in the process, the goal is to decrease engine harmonics

Although SFO camshafts were born from racing development, car manufacturers have adopted this concept. For example, General Motors has taken advantage of this for Gen-3 and Gen-4 LS engines, in order to further reduce harmonics and increase engine durability. The LS camshafts have a 4/7 and 2/3 swap, from earlier-generation engine firing orders.

In racing applications, the benefit of using a special firing order addresses the needs of engines that tend to operate at peak engine speeds for extended periods of time (such as endurance racing), and for engines that are called upon to produce peak power very quickly (such as in drag racing). Crankshaft harmonics are often caused when two adjacent cylinders are firing in close succession. By relocating the firing order, it's possible to create a smoother operation and enhanced acceleration curve.

For comparison, let's compare the firing order between early-generation Chevrolet V8 engines and the later LS platform. Early small-block and big-block engines had a firing order of 1-8-4-3-6-5-7-2. With this firing order, a companion cylinder reaches TDC at the same time as its counterpart, one on the power stroke and one on the exhaust stroke. The LS engine, as an example, swaps cylinders 4 and 7, thereby creating a new firing order of 1-8-7-3-6-5-4-2. In certain drag race engine applications, a 4/7 swap helps to reduce fuel distribution and heat issues that may have previously been caused by cylinders number-5 and -7 firing in succession.

Changing the firing order is not a matter of simply repositioning the spark plug wires, however. In order to alter the firing order, a special camshaft is required. With the special firing order (SFO) camshaft in place, you must then remember to arrange the spark plug wires in the corresponding order. Most leading race camshaft manufacturers offer SFO camshafts for those who wish to perform this change.

Camshaft Needle Bearings

You have a choice between conventional babbit cam bearings and needle-type roller cam bearings. As with most mods, each has its pros and cons. The advantages of moving to needle bearings in the cam bores include less friction at the cam journals and a more precise maintaining of the camshaft centerline during engine operation.

A conventional babbit cam bearing has an oil hole, allowing oil to lubricate the cam journal. The oil clearance between the cam journal and the bearing is filled with oil so that, during camshaft rotation, the cam journals ride on a film of oil (just like rod and main bearings do).

A camshaft needle roller bearing allows the cam to ride directly on the bearings. Oil splash is sufficient for lubrication, and the needle bearing housing blocks off the cam tunnel oil passages, so there is a slight reduction of oil temperature.

In order to install camshaft bore needle bearings, the cam bore diameter must be increased to about .040 inch. The result is about a .002-inch interference fit for the bearings. This must be done on an alignment boring fixture (or in a CNC machine).

Needle bearings require a camshaft with large-diameter journals (large-diameter core). This results in a beefier cam core that is less prone to flexing, which is great when using ultra-high-pressure valvesprings.

The downside to needle bearings is that the labor and machining costs to modify a block to accept needle roller bearings is roughly $600 to $700. Aftermarket race blocks are available already machined for needle bearings but you must also run a steel-billet camshaft with needle bearings, which raises cost a bit more. A potential glitch is that needle bearings *may* promote more valvetrain harmonics than babbit cam bearings, where the cam journals ride on an oil film that helps to reduce harmonics.

If you want the ultimate and don't mind paying the price, needle cam bearings are the way to go. If you're simply building a strong street engine where continuous high engine speed isn't a factor, it's a waste of money.

Camshaft needle roller bearings are commonly available in sizes ranging from 50-mm ID with 58-mm OD to 55-mm ID with 63-mm OD. If you move to needle bearings, you must order a camshaft that has the appropriate-diameter journals.

Distributor Gears

Matching the distributor's driven gear (if your engine has a distributor) to the camshaft gear is important. Depending on gear metallurgy and compatibility, the gear on the distributor needs to be compatible with the type of metal on the camshaft's distributor drive gear. If they're incompatible, you can easily and quickly strip the teeth off the distributor, sending debris throughout the engine's oiling system.

Today's steel roller cams are heat-treat hardened to resist wear. As a result, you can't use just any distributor driven gear. A sintered-iron distributor gear gets chewed up, sometimes even during engine break-in. Most manufacturers of steel cams recommend using a bronze or high-silicon copper alloy gear. Lunati also sells its Everwear distributor gear, which is a specially treated metal composite that is compatible with all steel cams. There is also a fairly new generation of "poly" gears that are compatible with any cam, regardless of the cam's gear material.

Composite distributor gears (made of carbon ultra-poly) offer strength and wear resistance. Bronze gears are great (and safe for the cam), but they eventually wear out, resulting in ignition spark scatter. Bronze gears are fine for racing where the engine (and distributor) are serviced routinely. For street or racing, a poly gear is a safe choice that provides longer life. A common bronze gear costs around $45, while a poly gear can run a a little more than $100.

If you have a cast-iron camshaft, you can use a cast-iron gear or a composite poly gear. If you have a ductile iron camshaft, you can use a hardened steel or composite distributor gear. If you have a billet-steel camshaft, your best choice is a bronze gear or a composite gear.

A hardened steel cam gear can quickly wipe out an iron distributor gear, sending plenty of debris throughout the engine. Don't play guessing games; make sure that your distributor gear is compatible with the cam.

If in doubt, a bronze gear is a safe choice, but remember that a bronze gear eventually wears and may need to be replaced once in a while. A better choice is one of the new poly gears that are compatible with all camshaft gear materials and last longer. This makes poly gears a good choice for the street.

Cylinder Heads

The cylinder head largely controls the all-important process of burning the air/fuel mixture. If your engine is to operate at the optimum, the heads must be properly inspected, measured, and corrected if necessary. Blueprinting heads involves attention to several areas, including deck surface, intake and exhaust port shape, valve selection, valve seating depth, and combustion chamber volume.

Many of today's OEM heads offer impressive performance and have the potential to be modified to further enhance performance. The GM LS engine family is a good example. LS heads flow some pretty impressive numbers straight out of Detroit. The L92/LS3 head, for example, can be further enhanced via CNC porting to increase intake and exhaust runner volume, multi-angle valve seat machining for superior valve seating and flow, and lightweight valvetrain components to enhance engine speed/acceleration.

During guide honing, the gauge is used to check/verify guide ID through the entire length of the guide.

Cast Iron versus Aluminum

In an apples-to-apples comparison, aluminum heads are lighter than cast-iron heads. So, if your goal is to reduce weight, aluminum is a no-brainer choice. Another variable relates to potential detonation. Since aluminum releases heat faster than cast iron, the theory is that aluminum heads are less likely to promote detonation. As a result, you can run a slightly higher compression ratio on the same fuel. Generally speaking, you can run upwards of 11:1 compression ratio with aluminum heads on premium 92-octane pump gas, whereas you may be limited to around 10.5:1 with iron heads.

Bronze valve guides can be trimmed to provide needed retainer clearance, providing enough height remains for seal installation.

In the old days, when OEM aluminum heads were installed, there were common problems with head warping. However, with the higher quality of metallurgy in today's aftermarket aluminum heads, this problem has been greatly minimized.

Another thing to consider (and it's only a consideration, not a problem or negative) is that aluminum is a softer material than iron. So more attention must be paid to areas such as female threads and specific pressure points such as valvespring seat areas. With a quality aluminum casting, as long as you don't cross-thread or overtorque fasteners, you shouldn't have any problems with thread stripping. And the same holds true for cast-iron heads.

However, some builders feel more confident by installing harder stainless steel thread inserts in areas such as spark plug holes, intake manifold bolt holes, valve cover bolt holes, and accessory-mounting bolt holes (for alternators, power steering brackets, etc.). I've built many engines with aftermarket aluminum heads and have never stripped-out a single bolt-hole thread. I see no reason to install thread inserts unless it's necessary to perform a repair.

Symmetrical exhaust ports (big-block Chevy pattern on a Dart Big Chief II head).

Symmetrical intake "cathedral" ports on an LS head.

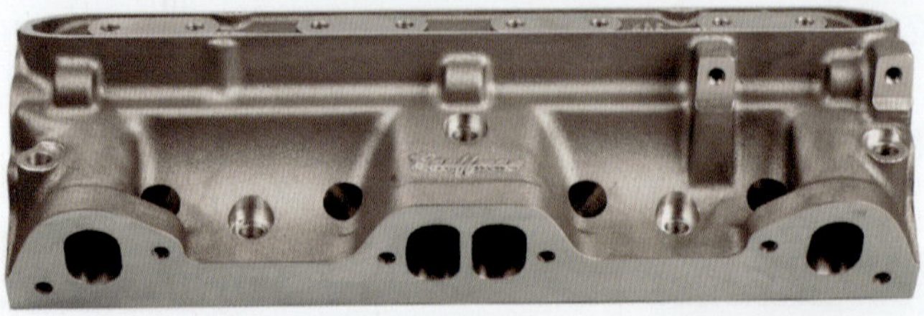

Asymmetrical exhaust ports on a big-block Pontiac head (oval ports at ends and D-ports at center).

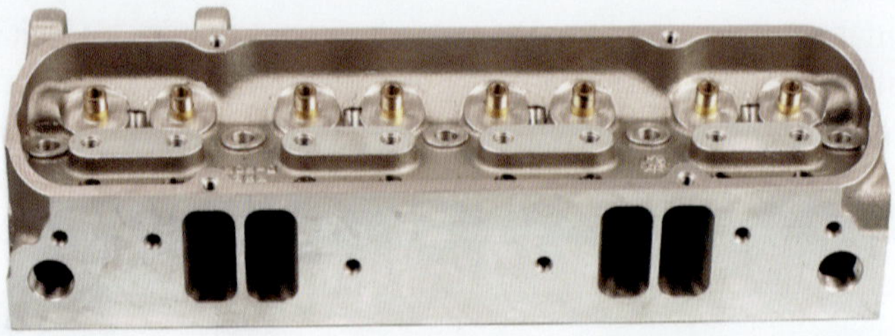

Small-block Ford intake ports. Matching ports to gaskets and heads is critical because of the close proximity of the pairs of ports.

Special spring compressors are required when removing/installing valves and springs on many OHC cylinder heads because of the tight confines.

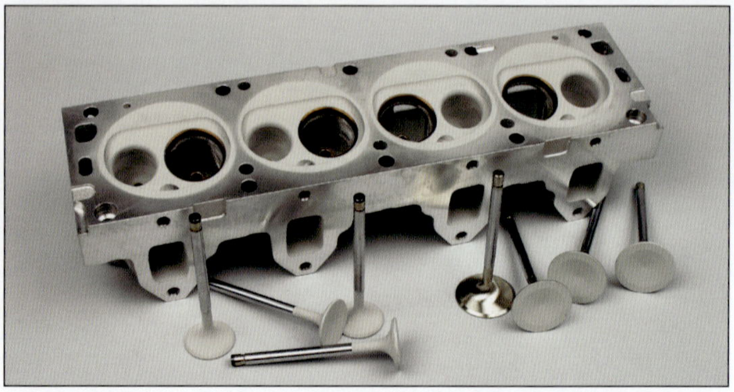

Thermal barrier (ceramic) coatings applied to combustion chambers, exhaust ports and exhaust valve throats can provide increased combustion efficiency and reduce heat soak.

Valvesprings are a different story. A cast-iron head may offer enough hardness in the parent casting to allow valvesprings to mate directly to the spring pocket. But with aluminum heads, always install hardened-steel spring seats or cups between the aluminum and the spring. This prevents the springs from digging into the aluminum. This not only prevents damage to the aluminum surface but prevents the spring installed height from changing.

Also, when installing any aluminum cylinder head, you must use hardened washers and lubrication under all bolt heads or nuts (if using head studs) because of the softer material. The importance of torquing the cylinder head fasteners is just as important with aluminum as with cast iron. Always follow the head manufacturer's specs for torque value, tightening steps, and tightening sequence.

Other advantages of aluminum heads include easier repairs. It's much easier to TIG-weld aluminum than to weld repair cast iron. Aluminum heads also can be heat-straightened if warped, with much less concern about stress cracking.

Today's cylinder head technology has been elevated by quantum leaps compared to OEM and even aftermarket heads from only a decade ago. Outstanding performance heads are available from industry leaders such as Dart, Trick Flow, Brodix, AFR, Edelbrock, RHS, Kaufman Racing (Pontiac), Indy, Koffel (Mopar), etc. Cylinder heads are available for most applications, in a wide variety of options, including bare/unfinished, finished but bare, CNC ported, as-cast with CNC-profiled chambers, and fully loaded bolt-on and go versions.

Deck Surface Finish

The cylinder-head deck surface finish is measured in Ra (roughness average). The lower the number, the finer the finish. Generally speaking, for cast-iron heads, the accepted range is around 30 to 110 Ra. With aluminum heads, the finish needs to be finer, ranging from 30 to 60 Ra, and with some heads even finer at 20 to 30 Ra. A precision measuring instrument, referred to as a profil-o-meter, is used to measure surface finish. Today's

Deck surface finish on aluminum heads, especially when used with MLS head gaskets, should be in the 30- to 60-Ra range (with an even smoother 20- to 30-Ra specified for some applications).

When measuring a cylinder head deck for flatness, use a precision machinists' straightedge and a feeler gauge. Here front-to-rear length is checked.

Also measure the head deck diagonally from corner to corner, at all four corners. Generally speaking, maximum allowable deviation is about .002 inch.

precision resurfacing machines can be calibrated for the desired finish.

Deck Flatness

Check every cylinder head, new or used, for deck flatness. This is particularly important for heads that have been previously heat cycled on an engine. With the deck surface clean, use a precision machinist's straightedge (a hardware store ruler won't do) and a feeler gauge. Always check with the manufacturer's specifications for maximum allowable deviation, but generally shoot for a maximum deviation of .002 inch.

Flatness must be checked along the length of the head from front to rear, on intake and exhaust sides of the main deck, and diagonally from corner to corner in an X pattern. MLS (multi-layer steel) head gaskets are more sensitive to deck flatness, in some cases requiring even a tighter maximum allowable flatness. Although some OEM specs may suggest a slightly more acceptable out-of-flat condition on V-6 or inline-6 heads, for performance use, it's best to use about .002 as the maximum.

When a cylinder head deck (or block deck) is milled, the intake deck drops, which can create problems with port alignment and manifold bolt hole alignment. A rule for 90-degree Chevys (small- and big-block) is that for every .005 inch removed from the block or head deck, remove about .006 inch from the cylinder head's intake deck, and about .008 inch from the manifold's end rails.

As you can imagine, trigonometry formulas are involved here, and I simply don't have the space to go into detail on this calculation. The easy way is to take advantage of specialty machining fixtures available from sources such as BHJ.

Valves

Intake valves are normally larger in diameter than the exhaust valves for a reason. As the piston moves away from the deck, this creates a vacuum pull for the intake air/fuel. The larger the intake valve, the more air/fuel can be pulled into the combustion area. As the piston moves up on the exhaust stroke it pushes exhaust out, so the exhaust valve doesn't need to be as large (pushing air is more efficient than pulling air). The specific size of the valves is dictated by displacement and (with regard to the intakes) valve lift. Of course, the real estate available in the combustion chamber limits how large the valve heads can be. (For more details about valves, see Chapter 12.)

Valve Angles and Canted Valves

Intake and exhaust valves are placed at angles relative to the piston centerline for reasons of flow and fit. If the valves are "straight up" (in-line with the piston centerline), the flow area is limited and the head needs to be much taller to accommodate the valves.

When I refer to valve angle, I'm referring to how the valve is angled toward the intake side (with the valve tip angled closer to the intake side and farther away from the exhaust side). This angle is called the valve's "roll" angle. Commonly used valve roll angles range from 9 to 23 degrees.

One of the factors considered when designing a head (and its valve angles) is port cross-section. The shallower the valve angle, the larger the cross-section can be. For instance, with a shallower valve angle, the cylinder head can have a taller intake port cross-section.

An example is the LS cathedral-port head (LS1, LS6, LS2, LS3, L92), which has a 15-degree valve angle. Typical early-generation small-block Chevy heads had a 23-degree valve angle. The closer you bring the valve to vertical (zero degree reference), the more you can increase intake port height. Shallow

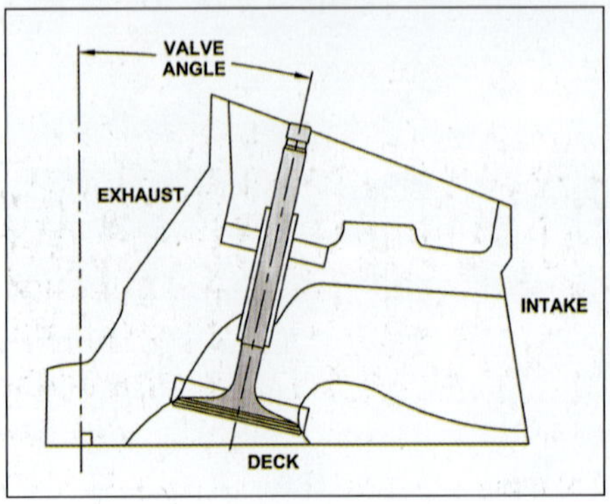

Valve roll angle refers to the valve's inboard/outboard angle relative to 90 degrees from the block deck.

In some cases where large-diameter pushrods are required (especially with some multi-angle valve heads where valve canting is featured), it may be necessary to grind the pushrod passages for clearance. Here a Big Chief II head is clearanced for 1/2-inch-diameter pushrods. Don't get too carried away. As little as .010 inch is sufficient clearance. If you grind too far, you can easily open up an intake port, resulting in a vacuum leak.

valve angles also allow small-volume combustion chambers.

Some cylinder head designs, such as Trick Flow's twisted-wedge Ford heads, utilize different angles for intake (15 degrees) and exhaust valves (17 degrees), which differs from the original Ford 20-degree in-line valve angles. These modified valve angles allow the head designer to place the valve in a better position for flow and to move the spark plug location for improved flame travel.

Canted valves have a layout where, in addition to the roll angle, the intake and exhaust valves for the same cylinder are angled relative to each other (viewing the head from overhead, the valve angles differ from left to right), resulting in compound angles. Canting is basically used in order to accommodate valve fit in the head when needed.

In-line valve layouts have all valves at the same angle (identical roll angle). On a twisted-wedge head, for example, roll angles differ between intake and exhaust. Canted valves are common on big-block Chevy designs. Hemi applications (with hemispherical piston domes) feature opposing valve angles.

When deciding among valve angles and airflow volumes, it's really best to consult the cylinder head manufacturer to select the head design that best suits your application. Greater port volume usually accommodates engines with larger displacements and smaller port volume is suited for smaller displacements. Depending on port cross-section, other engine variables, and the intended use of the engine, it's difficult to make blanket statements.

Valve Seats

Valve seats can be cut (or recut) in cast-iron heads, providing the seat area is not damaged. Otherwise, the seat area can be countersunk and a seat insert can be interference fit and installed, which can then be cut with the appropriate angles. Aluminum heads always require hardened seats (seat materials can include ductile iron, silica bronze, beryllium copper, powdered metal, and tungsten carbide).

Typical seat angles in a common three-angle seat has a bottom cut of 60 degrees, a seat cut of 45 degrees, and a top cut of 30 degrees. Top cut refers to the angle location closest to the flow exit (closest to the combustion chamber). Seat cut is between the top and cuts. Bottom cut refers to the angle cut that is positioned deeper, closer to the airflow inlet.

The reason for these three angles is to promote smoothness of the airflow without obstructions (picture a trumpet or tulip shape). The bottom cut (depending on how much material is available) generally has a width of about .100 inch. Seat width (the 45-degree angle) is critical for valve sealing in the closed position and heat transfer (the greater the contact width, the more the potential heat transfer).

For most street-performance, flat-tappet cam applications, the seat width is cut at the 45-degree angle of about .040 inch for intake valves and about .055 inch for exhaust. As valvespring rates increase, seat widths increase to around .060 inch for intakes and .080 for exhaust valves. With high-pressure springs, as found in all roller cam applications, a narrower seat flows better than a wide seat, but a narrow seat isn't as durable. Wider seats offer more heat transfer (necessary to pull heat away from the valves), which is why exhaust seat widths need to be greater than intake seat widths.

In order to promote airflow even more, seat transitions can be slightly radiused to allow a smoother path for airflow. This should only be done by a skilled machinist and isn't necessary at all for a street engine.

Valveguides

Performance valveguides are typically made of cast iron, brass, bronze, aluminum bronze, or powdered metal. Cast-iron heads require no separate guides; the cast-iron material itself is

Valve seats for today's aluminum performance heads are, depending on the application, made of ductile iron, silica-bronze, beryllium-copper, powdered metal, or tungsten-carbide. These are silica-bronze seats installed in a Trick Flow cylinder head. Seat material must be compatible with the type of valve material being used.

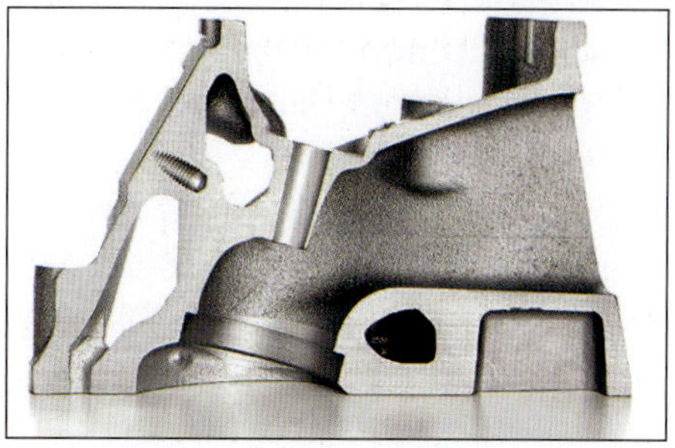

With shallower valve roll angle, the cylinder head can have intake port runners with increased cross section. This is a 6.2-liter LS3 head with 15-degree valve angle. (Photo Courtesy General Motors)

adequate, providing the guide hole is appropriately sized and is concentric with no wear. They can also be restored by oversizing the hole and installing a guide insert that is then reamed to size. However, aluminum heads must use guide inserts.

Valveguides wear (become oval shaped) due to the lateral operating pressures (side loads) of the valvestems against the guides (because of the force exerted by the rockers as they sweep across the valvestem tips). The higher the cam lift, the more angular force is applied.

Valveguide height needs to be checked in relation to spring retainer clearance when the valve is fully open (this is when the bottom of the spring retainer is closest to the top of the guide and valve seal). During test fitting, if the bottom of the retainer contacts the top of the guide or seal, clearance must be made by using a different seal or reducing the height of the valveguide. Ideally, you need a minimum of about .060-inch clearance between the retainer and valveguide when the valve is fully open.

Valvesprings

A set of quality, high-performance valvesprings should have springs closely matched for height and pressure rating. However, each spring should be checked for open and closed pressure at specified heights, using a valvespring tester.

Quality control of today's premium springs is excellent, but checking each spring in a set is always a good idea. A spring that falls out of the specified range can be replaced in order to achieve a closely matched set, and greater efficiency.

Ports and Runners

When I refer to ports and runners, the term "port" refers to the hole at the deck of the intake or exhaust path. The "runner" is the "tunnel" path through which the flow occurs. The size and shape of the cylinder head's intake port directly affects the power curve and torque. It's important to match the shape and size of the intake manifold port to the cylinder head's intake port. This promotes a full flow and avoids interruptions and/or restrictions to the flow as it moves from the manifold runner into the head. This is referred to as port matching, where any deviation in the transition from where the intake manifold ports meet the cylinder head intake ports is corrected.

As a result of the casting process, slight shifts can occur where the intake manifold's ports don't align perfectly with the ports in the heads. The basic goal of port matching is to eliminate any overhang that interrupts airflow. Along with eliminating any obstructions in the airflow, you want to maximize the flow. For instance, if the intake manifold ports are slightly smaller than the ports in the cylinder head, excess material at the manifold ports can be removed. (See Chapter 15 for more information on port matching.)

Common intake ports are somewhat oval while others are somewhat rectangular (oval ports are usually not a perfect oval and rectangular ports are not perfect rectangles; these are simply terms to describe the type of port shape). Bigger numbers for the intake runner volume may sound cool but, as with other aspects of an engine, bigger isn't always better. As runner volume increases, so does the amount of air/fuel that can pass through.

Sounds good, but as runner volume increases, this larger area can also slow the air/fuel velocity, which reduces throttle response. Smaller runners promote more velocity and crisper throttle response. The velocity at which the air/fuel mixture can pass through is critical. The mixture must travel fast enough to keep the mixture in suspension. Otherwise, fuel droplets can fall out and not burn as part of the rich mixture. In other words, you want the fuel/air mixture to travel through as a fog or mist instead of fat raindrops. In a fuel-injection setup, the flow is mostly a "dry flow" of air until it reaches the back of the intake valve, when fuel is introduced.

Add to this equation the engine displacement. As displacement increases, the engine needs more runner volume. But there's a limit of practicality. If the runners are too large for the engine, you may make more power, but the power is only at higher RPM. Smaller runner volume is generally better for the street; larger volume is generally better for top-end speed at the track. You also have other variables, depending on the specific engine: bore/stroke, cam profile, etc.

There's no such thing as one size fitting all when talking about intake runner volume. As with other aspects of the engine, runner volume needs to be matched to the specific application. That's where you rely on the recommendations of cylinder head manufacturers.

The exhaust port shape and size is less critical. Basically, as long as the exhaust gas can move out of the head with no restriction at the entrance of the exhaust manifold or primary header tubes, you're good to go. Keep in mind that the exhaust primary tubes should not be notably larger than the cylinder head's exhaust ports, though, because this can have a detrimental affect on exhaust scavenging.

Many buyers tend to focus strictly on the port/runner volume numbers (again assuming that more is better). In reality, port length and cross-sectional area tells a more accurate story. A short runner length with a large cross-sectional area can actually flow better than a long runner. Again, this depends on the specific head design and other variables in the engine combination.

Porting and Runner Reshaping

One of the biggest mistakes a builder can make is to go hog-wild by regrinding the inside of the intake runners. Although smoothing the runners and eliminating all rough casting surfaces, lumps, and bumps inside the runners may look really cool, it isn't necessarily going to make more power. In fact, this approach can kill power.

I can't make blanket statements here because different engine designs and different head designs are affected, well... *differently*. In some cases, the head manufacturer intentionally leaves some surfaces rough in order to promote breakup of the air/fuel mixture (to keep it atomized). In other cases, a head can benefit from a smooth, uninterrupted finish. The head design may have included a small hump with a drop-off inside the runner in order to promote swirl. Before you go nuts with a die grinder, first understand what runner shape and surface finish is most beneficial.

Proceed with Caution: Modifying the ports and runners is a critical procedure if you expect to get the most out of your heads. This is a job best left to a skilled professional. Each runner (and port) volume also needs to be equal at each location in the heads. If you don't have access to a flow bench to accurately measure and monitor runner volume and velocity, don't touch them. Chances are you'll do more harm than good.

With that said, there are a couple of mods that are always good to make, including port matching and volume equalization. Always match the intake manifold outlet port to the intake port on the head. This holds true for the intake manifold gasket as well. Make sure that the gasket has enough of a footprint to seal, without overhanging into the intake path, which creates a restriction. If the cylinder head manufacturer recommends a specific intake gasket, it's best to follow suit.

CNC-Ported Cylinder Heads

No longer are CNC-ported heads reserved only for high-end race programs. In the past several years, and with enhancements in manufacturing technology, CNC-ported cylinder heads have been made a viable option for popular engine applications. CNC porting is more exact than hand porting cylinder heads, with an unlimited ability to exactly duplicate an original shape many times over.

Here are a few of the common advantages of CNC porting:

- Port volume, cross-sectional area, combustion chamber volume, and shape are consistent.
- Valve seats and ports can be blended seamlessly.
- Surface finish can be manipulated by machining.
- Cylinder head performance and flow characteristics can be duplicated from engine to engine.
- There is no casting core shift; all features are precisely machined.

Finishing Bare Cylinder Heads

If you buy a set of bare/unfinished heads, size the valveguides and finish-cut the valve seats. Bronze valveguides are honed (using a dedicated valveguide hone) to achieve approximately .0016-inch oil clearance with 11/32-inch valvestems (check for recommended stem oil clearance with your specific valves). Intake and exhaust seats are cut to angles of 15, 45, and 60 degrees, using a contour cutter.

The goal is to establish exactly the same depth (valve to seat) for each valve, to eliminate creating different closed-valve chamber volumes. Each valve seat is first rough-cut to accept each valve. The distance from the head deck to the face of the valve is then measured, and each valve position in the head is marked for plus/minus depth difference. By using a single valve location as the reference, each seat is then finish-cut to sink the valves to match the reference valve depth. Again, the goal is to achieve equal chamber volumes.

The seats and bowls are then blended to eliminate any sharp edges or overhangs that disrupt airflow. For this example, teflon valve seals were installed to the .560-inch guides. These seals allow enough room for up to .580-inch lift before coil bind or hitting retainers. Whenever you're dealing with an aluminum cylinder head, it's necessary to install a hardened seat at the base of the valvespring, to prevent the

Bare cylinder heads usually have raw, unfinished valve seats that allow the machinist to create the desired valve depth.

Prior to honing new valve guides, the guide ID gauge is adjusted for valve stem diameter plus specified oil clearance.

The initial cut is checked for uniformity and concentricity. Note the witness cut through the dye.

Once initial cuts have been made, valves are installed to check for valve depth. The goal is to establish equal depth of all valves.

After measuring (and recording) each valve's depth location, finish cuts are made to equalize all valve depths.

Measure the distance between the top of the valveguide seals to the valve locking grooves as a reference prior to installing the springs. This helps when checking retainer-to-seal clearance at full-lift/valve open.

A burette makes it easy to measure combustion chamber volume. This is done by installing the valves (with a light smear of lithium grease on the valve faces) and a spark plug to provide a liquid seal. A clear plexiglass plate with a drilled hole is adhered to the deck with grease. Colored liquid is introduced from the burette, while monitoring the volume of liquid (in cubic inches) required to fill the chamber. The volume of each chamber is recorded and compared. The chamber with the largest volume can be used as the zero point of reference. Material may then be removed from other chambers or by sinking the valves deeper to create a matched equal-volume set of chambers.

spring from digging into the softer aluminum. You also need a *locating* design to prevent the spring from walking around on the head.

If your heads don't have machined reliefs at the spring base areas, you can use spring cups (featuring an OD lip to capture the spring's outer diameter) or spring locators (featuring ID lips to register the spring ID). For example, if your valvesprings have an OD of 1.437 inch and an inner-spring ID of .640 inch, you can use a set of spring locators that fit over .560-inch guides with a .010-inch clearance (the inner locating lip of the spring locators provide an ideal fit to the inner spring ID). Specs for these inside locating shoulder types are .690-inch ID of inner spring, .060-inch locator thickness, 1.550-inch OD, .570-inch ID.

Inspection and Reconditioning

If you're starting with an OEM engine core, never plan to reuse the stock cylinder head bolts. Most of today's OEM head bolts are the torque-to-yield type, which can easily lose their elasticity if the heads have ever been serviced. Don't take a chance. Toss them and replace them with new OEM bolts, high-performance aftermarket head bolts, or high-performance aftermarket head studs.

Because the GM LS engine is extremely popular today for performance builds, include a few LS-specific tips where appropriate. For example, when removing LS heads, be aware of the four inboard 8-mm "pinch" bolts. On a head that is dirty and covered with oil sludge, these bolt heads may be difficult to see. If you're not aware of them, you can easily damage the head if you attempt to pry it off with force

Remove Springs

Before you disassemble the head, use a mallet to apply a single hit to the tip of each valve. This helps to break the valve locks free from their embedded positions. With the heads removed and placed on a workbench, use a valvespring compressor tool to compress each spring. With the spring compressed about 1/4 inch, remove the pair of valvestem locks using a pick or a pencil magnet. Carefully relax the spring and remove the retainer and spring. Don't be concerned with keeping the springs in order if you plan to replace them. Even if you plan to reuse the springs, be sure to check them for rate and height anyway; specific location at this point doesn't matter.

Remove Valves

Before removing the valves, perform a vacuum test on each intake port and each exhaust port in order to inspect for

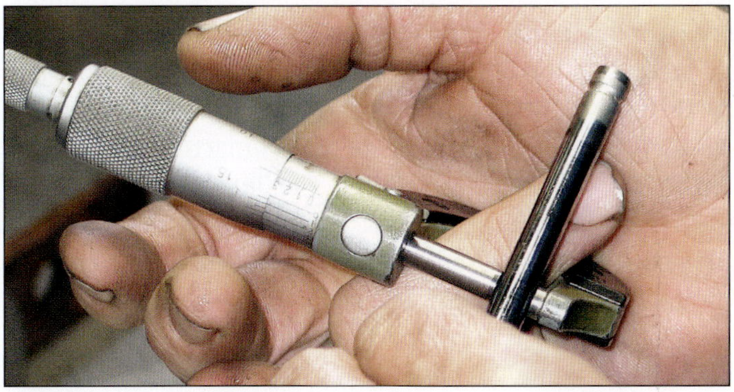

When measuring valve stems, measure at a minimum of two locations, including the center and just under the retainer lock grooves.

A runout gauge is used to check valve seat runout. A centering mandrel is inserted into the guide and the gauge is rotated along the valve seat. Any runout is unacceptable. This usually requires replacing the seat and cutting the new seat to establish zero runout.

existing valve seat sealing. With a valve closed, against its seat, a vacuum pull should hold steady. If vacuum leaks, the valve is not sealing and must be corrected by recutting or replacing the seat and/or the valve. Note the vacuum reading for each seat location as a reference.

Even if you plan to perform a complete seat and valve service, the initial data provides a reference point. If you plan to reuse the original valves, organize them for location reference. A simple method is to drill a series of holes in a wooden yardstick to accept the valves. Place a dividing mark at the center of the stick and label one side for the left bank and the other side for the right bank. Label the holes according to valve location on the heads.

Remove the valveguide seals and discard them. Use a pair of pliers or specialty seal pullers and a twisting motion during removal. If you have aluminum cylinder heads, a steel spring seat is located on each valvespring boss. The steel seats prevent the springs from gouging into the soft aluminum surface. Remove the spring seats. If you plan to use new springs of the same diameter, you may wish to reuse the original seats.

If you plan to reuse the original valves, use a micrometer to measure the stem diameter. Take the measurements immediately below the keeper groove, at the center of the stem and above the fillet. If any deviation exists, this indicates stem wear, so discard the valve. Also inspect each valve for runout by resting the stem on V blocks and checking with a dial indicator as the valve is slowly rotated. Any runout means that the valve is bent and must be discarded.

Check Head for Damage

After complete disassembly of the heads to a bare state (remember to remove any plugs, such as threaded plugs or expansion cap plugs), the heads must be thoroughly cleaned in a jet wash or hot tank to remove all traces of contaminants. If media tumbling is performed, the heads must then be rewashed to remove any particles.

Before performing cylinder head assembly, inspect the head for cracks and signs of coolant leakage (which could be the result of a crack or casting porosity). Depending on the fault (crack or porosity), the head may be rescued by a skilled cylinder head shop using tig welding and/or vacuum resin impregnation. Depending on the extent of the damage and the estimated cost of repair, make a judgment call to decide if the head should be repaired or replaced.

Check Head for Straightness

Place the head on a clean workbench and carefully check the deck surfaces for warpage. This includes the block deck, intake deck, and exhaust deck. Use a precision machinist's steel straightedge and a feeler gauge. Only use a precision-ground straightedge. Avoid using a scrap piece of metal that may not be straight and flat.

Position the straightedge lengthwise on the deck from front to rear at the center of the combustion deck and insert a feeler gauge between the straightedge and the deck. OEM specs usually call for an allowable maximum of .003 inch along a 6.00-inch distance, or .004 inch along the entire length. Although this tolerance range may be acceptable for a routine rebuild, ideally there is no gap whatsoever. If any deviation is found, the deck should be resurfaced.

Perform the same flatness check with the straightedge positioned from each corner to the opposing corner in a diagonal manner. The same tolerance specifications apply. If any warpage is found on the combustion deck, resurfacing is required. If more than .005 inch must be removed in order to achieve flatness, a thicker head gasket may be required in order to retain desired compression ratio and alignment from head to intake manifold.

Deck Milling

Restoring surface finish and/or deck flatness is one reason to mill the deck. Another reason is to reduce combustion chamber volume when fine-tuning to achieve a target compression ratio.

Although (in general) a .005-inch clean-up cut shouldn't affect intake manifold mating angle, a more severe material removal can result in a mismatch between the head and the intake manifold. Depending on head design, this can also bring the valves (especially with oversized valves) dangerously close to the edges of the cylinder bores or pistons.

Instead of milling the head decks parallel to the existing deck plane, the decks can be angle milled. This involves positioning the head at an angle relative to the cutter to remove more material from one side of the head (inboard or outboard) than the other side. For instance, .100 inch might be removed from the exhaust side of the deck, to meet zero at the intake side (removing material from the exhaust side while leaving the edge of the intake side alone). If the heads are angle milled, the top of the head bolt-hole locations of the head must then be spot-face milled to allow the underside of the bolt heads to retain full contact with the heads. Otherwise, the bolt heads meet the head bosses at an angle, eliminating full contact/support for the head bolts. The intake manifold deck surface of the heads must also be angle milled to retain full and even contact with the manifold. As part of your head inspection, perform the same flatness check at the cylinder head's intake deck. Typical specification calls for a maximum allowable deviation of .0031 inch. The exhaust deck specification is generally an allowable maximum of .005 inch. Some builders of older engines perform angle milling (taking more material from the exhaust side) in order to reduce the valve angle and promote a more direct airflow. With today's readily available high-performance heads and advanced designs, this is rarely needed.

Other Inspections

Whether new or used, inspect all valvesprings. First measure each spring's pressure at the installed height, then at the maximum-lift specification, and check for coil-bind height. Compare your findings with the manufacturers' spring specification.

When checking dual springs, perform this measurement check with the spring retainer in place because of the locating step in the retainer.

With dual springs, you probably have about .100-inch difference between inner and outer spring height, so the inner spring can enter coil bind before the outer spring.

Each valvespring must be checked on a spring tester. The spring is compressed to its specified installed height while recording spring pressure.

Closely examine combustion chambers for burrs or other sharp spots. These can lead to hot spots during engine operation, resulting in detonation (pre-ignition). If any sharp edges are found, lightly smooth them out.

If valve seats require refacing, the valve seat angles of 30, 45, and 60 degrees must be maintained. Each valve seat must be inspected for runout, using a valve-seat runout gauge. This check allows you to verify valve seat concentricity, relative to the centerline of the valveguide. An OEM specification may provide a maximum-allowable valve-seat runout of as much as .002 inch; for a blueprinted precision build, a preferred maximum runout should be less than .001 inch.

Inspect each valveguide's inside diameter using a dedicated valveguide bore gauge. When measuring the inside of the guide, check diameter from the top to the bottom of the guide, inspecting for taper wear. If you find any taper, replace the guide. Based on the valvestem diameter of the valves to be used, the guide diameter must allow the proper amount of oil clearance. Refer to the cylinder head manufacturer's spec for recommended valveguide clearance. Depending on the application, this clearance can run from as little as .0012 inch for intake locations and .0016 inch for exhaust locations to as much as .0015 for intake and .0025 for exhaust. Worn guides can be replaced or drilled out, reamed, and fitted with bronze guide liners that are then honed to size.

Flow Bench Testing

The only way to verify airflow of a cylinder head (aside from taking the head manufacturer's word for it) is to run and monitor airflow at a constant pressure. The pressure drop is measured across the port/chamber area, determining the ratio of the pressure drop across the calibration. Readings are then monitored through the range of valve lift (from the moment of opening to maximum lift, and especially just before the valve closes. Pressure is readjusted to calibrate test pressure before each check (at 10 percent of valve movement, 20 percent, etc.). Controls on the flow bench allow the operator to switch from sucking air through intakes to blowing air through exhausts.

A flowbench measures this airflow in relation to an applied pressure differential, in inches of a water column. Wet flow testing uses a solvent similar to fuel in specific gravity. Wet flow testing allows you to not only measure airflow, but how fuel/air is distributed within the port as the valve operates.

A special pressure differential (PD) valve, such as the one made by RTS Tools, uses a special valve with a hollow stem where the stem passage connects with orifices in the PD valve's face. This allows you to determine not only how much volume (in cubic inches) exists in

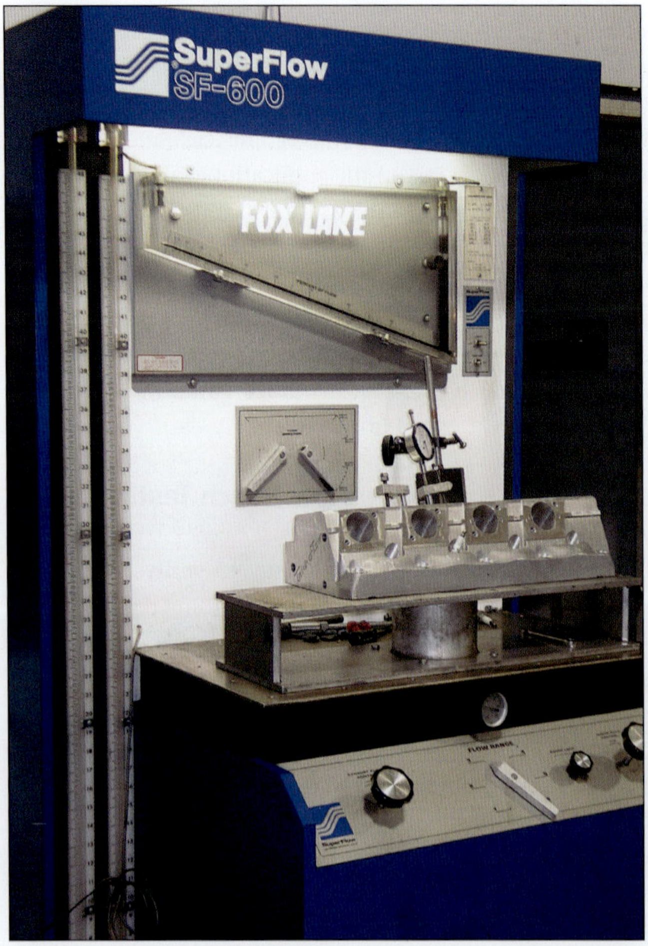

Flow benches measure intake and exhaust port airflow at various stages of valve lift (including max lift). This allows the head specialist to verify and record original flow and to verify the results of porting modifications. Without a flow bench, modifying ports is a guessing game. This is a SuperFlow SF-600 flow bench at Fox Lake Performance's cylinder-head shop. The two levers on the lower wall of the unit allow pressure to be diverted to the intake or exhaust paths.

the port, but where the fuel/air is being distributed and in relation to valve movement (based on valve lift). There is not space here to delve into flowbench testing, but this is how advanced race engine builders develop power-boosting cylinder head performance (by obtaining base data and being able to monitor port volume and flow as they modify their ports). Having this data removes the guesswork when choosing cylinder heads.

Cylinder Head Installation

You must first ensure the block decks and cylinder head decks are clean and dry. Then carefully wipe the decks with a fast-drying solvent and a lint-free rag. Examine the deck surfaces closely to verify that nothing is trapped between the decks such as oil, particles, lint, etc. Then you can begin installing your heads.

Position the head gasket (using the deck dowels to locate the gasket). Pay strict attention to head gasket orientation to make sure that alignment is correct and that critical coolant passages are not blocked. (Always refer to the head gasket manufacturer's instructions for gasket location.) Your hands must be clean and dry while handling the head gaskets. Some MLS gaskets have a seal coating that may be sensitive to finger acids. Make sure that the head is fully seated on the dowels by gently taping the head with a rubber hammer.

If you're using head bolts, always measure each bolt for length and compare to installed depth (bottom of female threaded holes to top of cylinder head bolt locations) to verify that bolts don't bottom-out. If they do, you don't achieve clamping load.

If you're using head studs, clean them with a fast-drying solvent to remove any buildup of preservative grease that may be present on the threads and to remove any particles that may have accumulated during handling. Apply a bit of oil to the bottom threads (or use a locking compound if you prefer a more permanent stud mount). Install the studs only hand tight. Tightening them completely can result in slight splaying of the stud angle. The clamping load is achieved when the nuts are tightened.

Locking the studs in place is really only needed if you plan to service the engine frequently (as with a race engine). However, if you decide to lock the studs in place, *do not* use an anaerobic compound (Loctite, etc.) because it expands when cured. Depending on cylinder wall thickness, this can result in excess pressure against the cylinder walls, which can lead to cylinder wall cracking. Instead, use JB Weld or a similar compound. Again, unless you plan to routinely service the heads, simply lube the lower stud threads before installation.

Apply a film of moly lube (such as ARP Ultra Torque) to the upper stud threads and to the bottom of the nuts. Install the flat washers, install the nuts, and tighten. Begin by tightening all nuts to 15 ft-lbs. Continue to tighten in steps until you reach the final recommended torque value for your application. For example, if a head calls for a final value of 110 ft-lbs, tighten at 15, 30, 55, 85, and then 110 ft-lbs. Always follow the specific tightening pattern recommended by the head manufacturer or by the aftermarket stud or bolt manufacturer. This is generally a spiral pattern, starting at the center and working outward in a clockwise spiral pattern. Follow the same tightening pattern during each step.

CHAPTER 12

VALVES, SPRINGS, RETAINERS AND LOCKS

To realize optimal performance, maximize airflow around the valve so air/fuel mixture quickly enters the combustion chamber, efficiently burns, and then expeditiously exits the cylinder. This chapter delves into the intricacies of optimizing the valve and valve parts for your particular setup. The valves and related valvetrain hardware need to precisely match one another from cylinder to cylinder, but these parts must also meet the performance standards you're targeting. The valves required for a race engine are far different than valves required for a street engine.

Valve Inspection

When inspecting a used valve (assuming you're considering placing it back into service), check valvestem diameter in three locations: about 1 inch below the tip, at the center of the stem length, and at the lower area of the stem (about .500 inch above the throat and/or undercut). If you note any difference in stem diameter among these three measurements, discard the valve.

Check each valve for runout or bending. If a valve has more than .001 inch

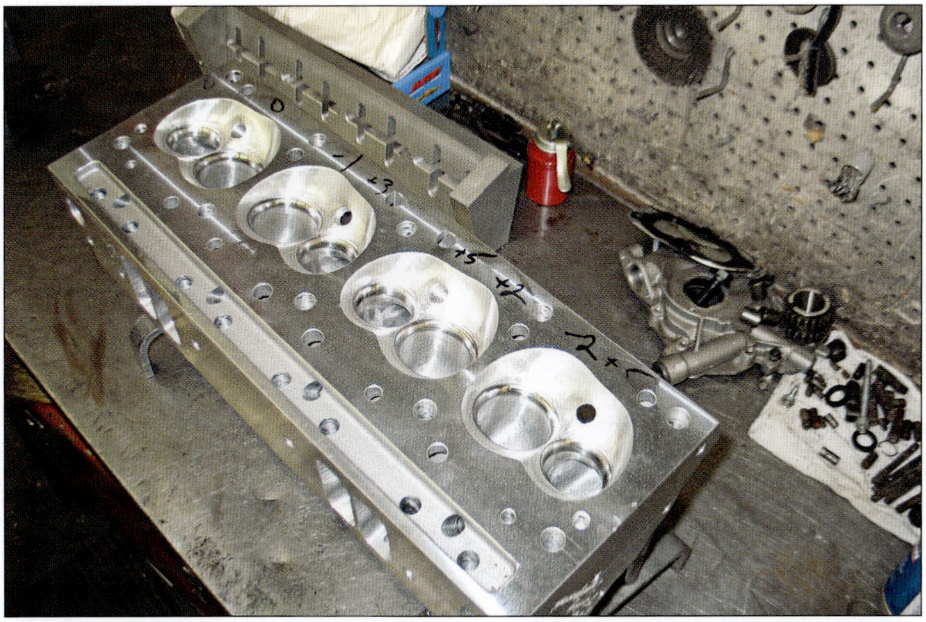

This Pontiac big-block head has been checked for initial valve depth. Note the depth markings adjacent to each chamber. In this example, initial depths ranged from -0.001 to +0.005 inch.

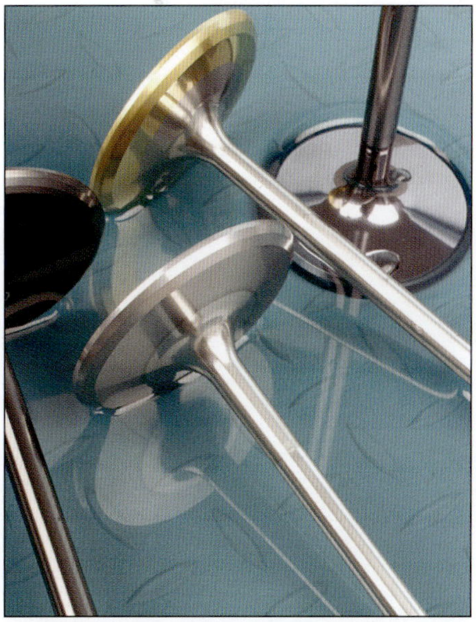

Today's high-performance aftermarket intake and exhaust valves are available in a variety of materials and with specialized coatings and surface treatments.

runout, don't use it. Don't even think about trying to straighten it out. Runout can be checked by placing the valve on a small pair of V-blocks or by fixturing it onto a dedicated valvestem runout tool.

Monitor runout with a dial indicator while slowly rotating the valve. Also check the valve for concentricity, which is the valve head in relation to the centerline of the stem. Stack-ups in runout and concentricity variation prevent optimum sealing at the seat, and under high-RPM use, this can also lead to eventual valve failure (as in the head snapping off the stem).

Also inspect all valves for cracks/faults. A visual inspection may enable you to find surface cracks (at the valve face areas and the entire valve head as it tapers to the stem). Surface-flaw checks can be performed with magnetic particle inspection or with a dye penetrant system. In order to find cracks that are below the surface, an ultrasonic inspection station is useful. Inspection of exhaust valves is particularly important due to the metallurgical stresses experienced by exhaust heat.

Depending on the type of valve, it may or may not be reconditioned by regrinding the faces. Some valves originally had a specialized treatment, such as diamond-like coating (DLC), that provides a harder surface. You aren't able to regrind without ruining this coating, but you can always send the valves back to the manufacturer for retreatment. However, if the valves are far enough gone to require refacing, you may as well replace them (if the application involves racing.

Valve refacing (seating face) can be done by a refacing machine with a collet or chuck or by centerless grinding. A facing machine has a collet or chuck that secures the stem and the valve is spun against the grinding wheel to create fresh face angles. This secures the valve within a tolerance range that should be .0005 inch or less.

Centerless grinding eliminates the collet or chuck. Instead, the stem is secured along most of its length, rotating the valve on its own centerline (this eliminates any variables of runout on a chuck and spindle). The valvestem is secured in a pneumatic support. Canted rollers feed the valve against a precision stop. The valve spins against the spinning grinding wheel. Theoretically, centerless grinding is much more accurate, although advances have been made with collet machines that now rival the precision of centerless grinding.

However it's achieved, the goal is to grind the valve face while eliminating runout. I do not delve too deeply into reconditioning valves here, because the discussion throughout this book focuses more closely on high-performance and racing applications, rather than individual component reconditioning.

Valve Seats

Valve seats (in the finished form) are responsible for three basic functions: to provide a seal at the closed-valve position, to promote optimum airflow when the valve opens, and to help transfer heat away from the valve head when the valve is closed. The seats need to be hard enough to last a reasonably long period, soft enough to prevent damaging the valve, and offer enough heat transfer to prevent burning the valve.

The location height of the valve seats in the cylinder head is critical to achieve consistent valve-to-piston clearances and uniform combustion chamber volumes. When your goal is to blueprint an engine, this is part of the task. With the valves installed to the head (minus springs) and the head upside-down on a bench, a

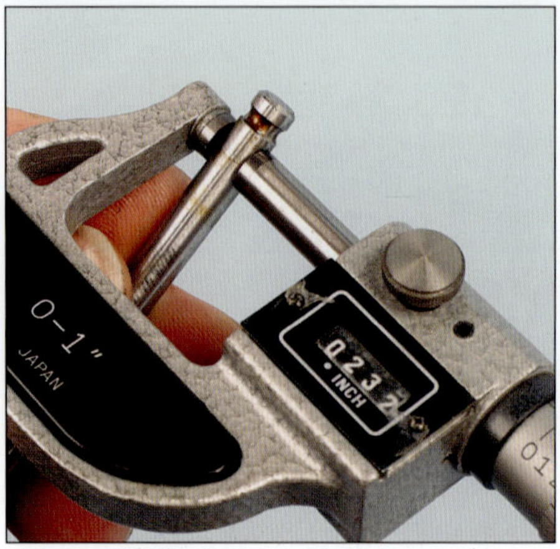

Measure stem diameter just below the lock area.

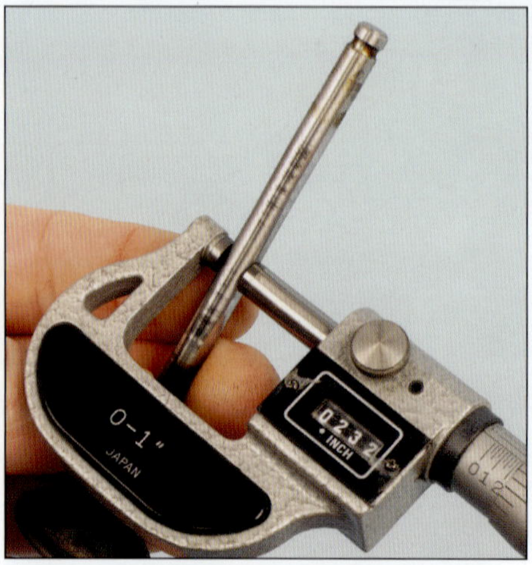

Use a micrometer (not a caliper) to measure valvestem diameter. Take three measurements. Here a stem is measured at its center length area.

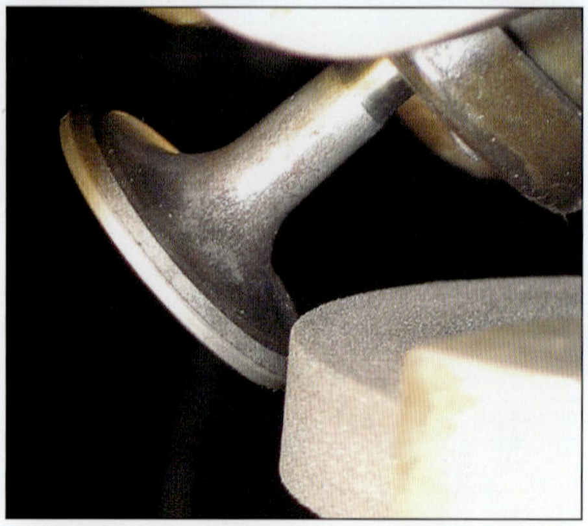

Valves can be refaced (on a collet machine or a centerless grinding machine). Remember that regrinding the sealing face of the valve sinks the valve farther into the seat, increasing chamber volume and decreasing compression ratio.

valve depth checking gauge can be used to determine the distance between each valve face and the head deck.

Uneven installed valve depths can be corrected by cutting the problem seats a bit deeper or replacing the seat(s) and recutting to move the valve closer to the deck. If you don't pay attention to this, you may think (on paper) that your compression ratio is, say, 11.25:1. In fact you may have chambers that range from 11.15:1 to 11.37:1, for example.

Every time the cylinder heads are freshened up (reconditioned), the seats are recut and the valves sink deeper into the port areas. By the same token, every time the deck surfaces are "kissed," the valves move closer to the combustion chambers. Recutting the seats results in sharpening the short-side radius of the port, slightly hindering flow. Moving to large-diameter valves, and/or replacing seats with new seats tends to improve flow because you're moving the valve closer to the combustion chamber and allowing a softer radius. A variety of seat materials are available, including beryllium copper, ductile iron, stellite, chromium, nickel alloy, cobalt, and powdered metal.

TECH TIP

Insert Fit

Valve seat inserts are interference-fit to the heads. For installation into an aluminum cylinder head, the interference fit should be in the following range:

Seat Material	*Fit (inch)*
Cast-iron	.003 to .005
PM	.005 to .007
Beryllium copper	.004 to .0045

These are generalized specifications. Always follow the interference fit recommended by the seat manufacturer. ■

Beryllium-copper seats are commonly recommended with titanium valve locations. This alloy seat material has about 98-percent copper and provides decent heat transfer and holds up to severe valve closing pressures.

Beryllium copper offers better heat-transfer performance than, say, bronze or iron. Recently, copper-nickel alloy seats have been developed that cool as well as, or sometimes better than, beryllium copper. This has been done in large part due to health concerns about machining beryllium, which is considered a toxic material.

Some titanium valves have hardened faces to accommodate sintered-iron or bronze seats.

Titanium valves, though lighter in weight than stainless steel, tend to run hotter, so they need a seat that is able to pull heat away quicker.

Although PM seats are certainly acceptable for a variety of applications, they may not be the best choice for ultra-high-performance/racing use. Even though they allow the manufacturer to tailor the material for hardness and heat transfer capabilities, in some cases this type of valve seat tends to work-harden when cut, which sometimes makes recutting during servicing difficult.

If you're rebuilding cylinder heads for high-performance or racing use, you're better off by choosing ductile iron or nickel-copper, which is easier to machine and compatible with performance valve materials.

Seat Angles

Multiple angles are common and a three-angle valve job is the most common for performance applications. The sealing angle refers to the contact between the valve head and the valve seat. This is usually a 45-degree angle. The top angle is located at the outer perimeter of the seat (the largest-diameter surface, closest to the piston). The bottom angle is the smallest-diameter surface, closest to the valvestem. It's easy to get them mixed up. With the cylinder head in an upright (as installed) position, you might think that the "bottom" cut is closest to the head deck and piston.

The bottom cut serves two purposes: to provide less turbulent airflow transition and to reduce the width of the 45-degree sealing angle. The reduced width of the sealing angle promotes better airflow. (The bottom angle is often referred to as the throat angle.) The intake valve's location and width of the 45-degree sealing angle is commonly moved, or biased, closer to the combustion chamber. Most exhaust sealing contact areas are generally placed in the middle of the 45-degree angle to aid in durability and to keep the contact a bit farther away from sharp edges, which otherwise could promote valve burning.

The two extra valve and seat angles have nothing to do with sealing the valve when closed. These angles are provided to promote improved airflow as the air/fuel

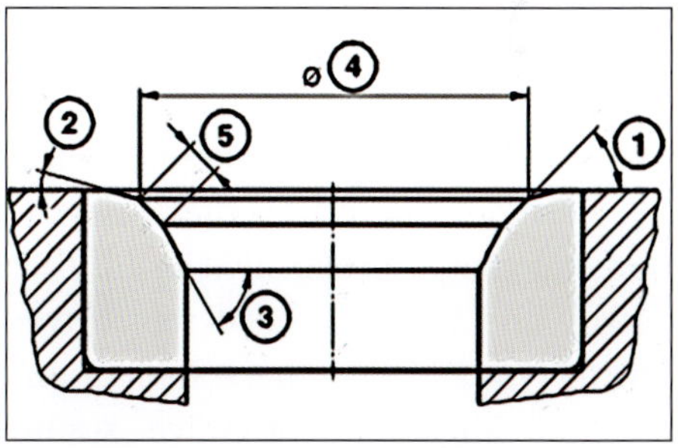

Illustration of valve seat angles.

1. ***Seat (contact with valve) angle is generally 45 degrees***
2. ***The top angle blends the top of the seat into the combustion chamber***
3. ***The bottom angle blends the seat ID into the port***
4. ***Top of the 45-gauge diameter, used to locate the valve at its mating***
5. ***Width of the 45-degree cut (varies depending on application)***

mixture exits the valve pocket and transitions into the combustion area. Making this flow path smoother theoretically improves efficiency and minimizes air/fuel flow turbulence. OEM valves and seats are usually cut with valve sealing as the priority, while aftermarket performance angles are considered to improve airflow in the pursuit of higher performance.

For racing applications, serious research has been invested to determine optimum specific angles and angle-cut widths. In many cases, this is closely held, top-secret data and you can understand why. An engine builder is reluctant or unwilling to share this info after investing hundreds of hours in research and development on the computer, in the lab, on a flowbench, and on the dyno.

As a result of trial and error and the availability of today's computer-aided software and CNC machining capabilities, many race engine developers have taken advantage of as many as five (or more) angles in order to optimize airflow. This might involve a more drastic (steeper) top cut in the 35-degree range and a 50- to 70-degree undercut just below the primary seat. The best setup for a specific engine and specific application results from research and testing. It's all about optimizing the airflow to best suit the particular RPM band, peak power, and torque needed for a given engine type and given track application.

Seat angle (where valve sealing occurs) is usually 45 degrees. A typical bottom cut (where the flow begins its path from the intake runner) may be at 60 degrees. A typical final angle may be 30 degrees. In many cases, the same angles are used at intake and exhaust locations, but the width of the valve-to-seat contact (at the 45-degree angle) usually varies. Exhaust valves are usually fitted with a slightly wider contact area (say, .060 to .080 inch) than the typical .040-inch-wide contact at the intake valves in flat-tappet cam engines. Roller cams with higher spring pressures often require wider seat contact areas in the .060-inch intake applications and .080-inch contact widths in exhaust locations, to help prevent seat damage or "pounding out."

This book focuses on performance and race spark-fired engines, but you should be aware that some Diesel engines have seat contact angles other than 45 degrees. Don't assume that every engine uses a 45-degree seat. Although a narrower seat contact area generally flows better due to the heat transfer between the valve head and the valve seat), exhaust seat contact areas are usually wider to allow more efficient heat transfer, which aids in cooling the valves. An exception is the professional drag race engine application, for which the builder may take advantage of narrower seat contact widths to optimize airflow in these short-run applications, with durability taking a lower priority because these engines typically experience more frequent teardowns and freshening.

Performance Valve Technology

A variety of valve materials and designs are available for performance applications. This information is provided to help you make an informed decision.

Stellite is a hard coating applied to valvestem tips and faces to increase surface hardness and increase resistance to wear. This treatment is commonly used on stainless steel valves. In an effort to increase heat dissipation in exhaust valves, some valvestems are hollow and are partially filled with sodium. The sodium liquifies at about 200 degrees F, allowing it to travel up and down inside the valvestem, aiding in heat transfer.

Some valves have hollow, empty stems simply as a way to reduce weight. Hollow-stem valves are usually finely micro-finished on the inside of the stem to remove potential stress risers that might otherwise exist as a result of the drilling process.

Stainless Steel

Most high-performance stainless steel valves are manufactured as a one-piece forging, most commonly using a stainless steel material called EV-8. High-quality stainless-steel valves are also equipped with a welded-on hardened tip.

Titanium

Titanium valves offer substantial weight savings and are best suited to engines that require quick acceleration and experience extended high RPM. Reducing valve weight reduces rocker arm wear and allows the use of lighter springs, which in turn reduces frictional loads at contact areas between the lifter and the cam lobe.

Titanium valves, while substantially more expensive than stainless steel valves, are typically about 45 percent lighter than steel or stainless steel valves of the same size. Titanium valves are a good choice where valvetrain weight needs to be reduced. Titanium valves are not pure titanium, but are a blend of copper and other

Notice the visible coloration difference at the valve throat. This Del West titanium intake valve has been treated with a chromium nitride (CrN) coating on the valve head and face.

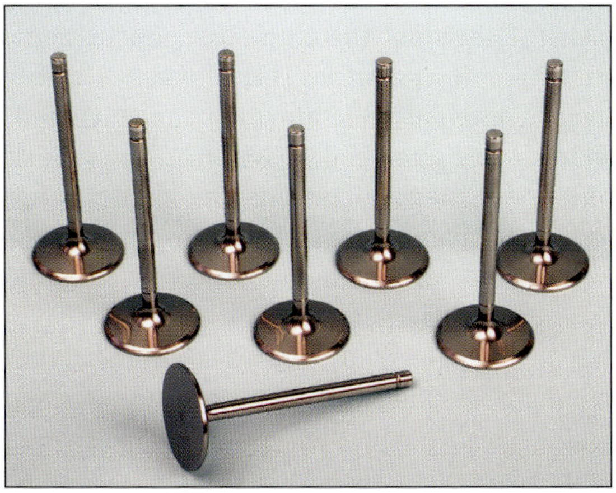

When handling any valves, especially titanium, it's best to prevent acids from your skin contaminating the surfaces. Wear latex gloves (or something similar) or keep the valves protected by a thin application of oil. The point is to keep the valves clean. Treat them with care.

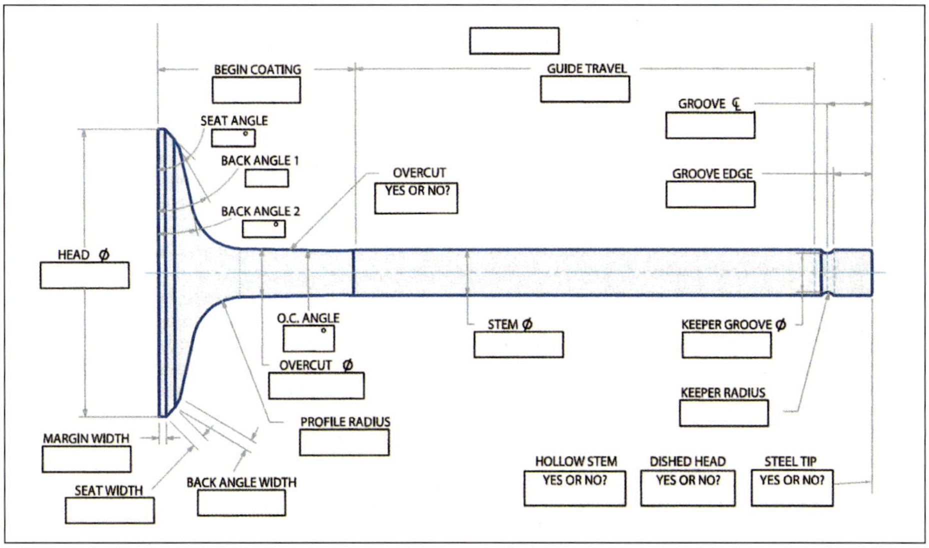

Here's an example of a performance valve nomenclature and custom valve worksheet. (Illustration Courtesy Del West Engineering)

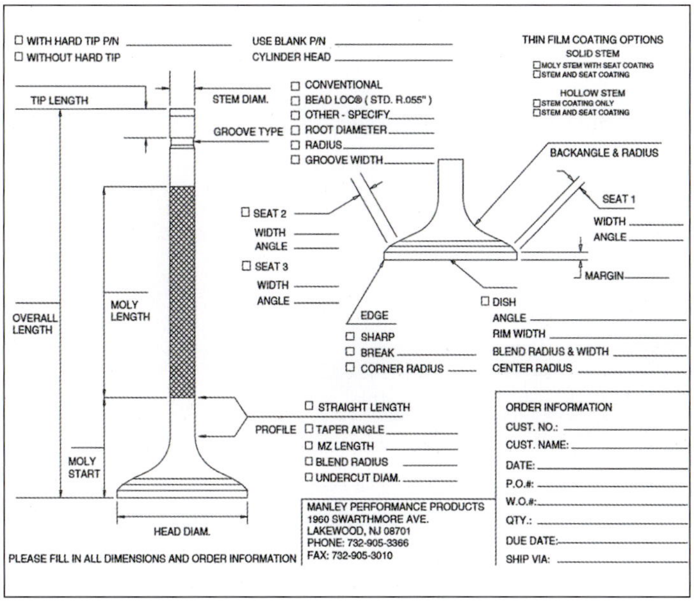

This custom-order chart is an example of the data required when ordering custom valves. Variables include stem diameter, overall valve length, lock groove type, head diameter, type of head edge, width and angles of mating face area, and seat margin. (Illustration Courtesy Manley Performance Products)

alloys. The advantage lies in its excellent weight-to-strength ratio. The disadvantage involves higher cost and increased difficulty in remachining during valve service due to its tendency to gall when cut with incorrect cutting bits.

Although not considered a disadvantage, it's important to know that titanium valves are relatively softer than other materials. As a result, the valve tip must be protected from rocker arm pressure. This is handled by adding a hardened-steel lash cap or by the application of a specialized hard coating at manufacturing. If you purchase titanium valves, the manufacturer should provide information relative to the need for lash caps.

If a titanium valve requires a hardened lash cap, the valve tip is usually coated with ceramic or other hard-surface coating in order to protect the titanium from the friction generated between the lash cap and the valve tip that could otherwise result in galling. Some titanium valves have a deep, diamond-like treatment that does not require a lash cap. Again, when using titanium valves, it's critical to verify if lash caps are required.

The retainer lock groove is another area of concern with titanium valves. In order to address potential galling of the titanium, the groove area is treated with a specialized positive vapor deposition (PVD) coating. Some manufacturers warn against handling titanium valves with bare hands, cautioning that skin acids may affect the valve's coating. A common recommendation is to wear latex gloves or to apply a light coat of oil to the valves before handling. If the titanium valve face is treated with a PVD coating, never use an abrasive lapping compound in an effort to enhance valve seating.

Additional cautions about titanium valves involve the valve seat material. Most titanium valves require a relatively soft seat material such as nodular iron or nickel-bronze; a too-hard seat material

This is an example of a Del West titanium valve that is treated with chromium nitride. This coating offers the additional benefit of allowing valve compatibility with ductile iron or beryllium-copper seats.

can quickly beat a groove into the valve face. However, some titanium valves are treated with chromium nitride to provide a hard protective surface, making them compatible with ductile-iron seats.

None of these issues should be regarded as titanium shortcomings, but you must be aware of the concerns if you plan to lighten valve weight with this material.

Nimonic 90 and Inconel

Other exotic valve materials include Nimonic 90 and Inconel. Nimonic 90 is a nickel-chromium alloy. Its claim to fame is its ability to withstand extreme temperatures and is sometimes preferred for nitro-methane and extremely high-boost-pressure turbocharged applications.

Inconel is a nickel-based alloy also designed for extremely high-heat conditions. It is suited for extreme-pressure nitrous and forced-induction race engine applications.

Valve Coatings

Valve coatings provide enhanced performance. The information that follows was provided by Del West Engineering and Xceldyne.

PVD occurs because of a physical reaction. Inside a vacuum-chamber plasma environment, metals are deposited via evaporation, sputtering, or arcing fragments of the metals that are physically moved onto the substrate. In other words, there is no chemical reaction that forms the coating onto the substrate.

Chemical vapor deposition (CVD) occurs because of a chemical reaction. The process exploits the creation of solid materials directly from chemical reactions in gas and/or liquid compositions or with the substrate material. The product of that reaction is a coating material that condenses on all surfaces of the part and inside the vacuum chamber plasma environment.

Diamond-like carbon (DLC) coating is a thin-film coating applied via a plasma-assisted chemical vapor deposition (PaCVD) process. This coating combines very low frictional resistance and extreme hardness. These coatings are used to reduce wear and friction for rapidly reciprocating components, where friction reduction is a primary goal. Common applications include finger followers, tappets, and piston pins.

Chromium nitride is a thin-film coating also applied using a PVD process. A cathodic arc is discharged at the target to evaporate the chromium into a highly ionized vapor, which is done in a partial pressure of nitrogen inside a nitrogen tank. This method deposits chrome onto the stem. This provides a higher level of adhesion than the PVD sputtering method in which a glow-plasma discharge bombards the material and converts some material into a vapor. This process is commonly used for titanium, steel, and nickel valves.

An example of a titanium nitride coating on the valve face area.

Thermally sprayed coatings can provide thick coatings over a large area at a higher deposition rate than other coating processes such as PVD or PaCVD. These are coatings that include plasma spraying and high-velocity oxygen fuel (HVOF) spraying; they are widely used to protect valvestems and tips.

Thin-film coating options such as chrome nitride, titanium aluminum chrome nitride, diamond-like carbon, and amorphous silicon carbide are used during the valve design process based on

A hard coating applied to the stem tip and lock groove of a titanium valve avoids potential galling where contact exists from a lash cap and the valve lock.

A specialized titanium nitride coating on the valve face provides superior protection from heat and carbon deposit buildup.

the suitability of the coating properties for the specific engine application and historical post-engine teardown feedback and analysis.

In certain applications, a combination of coatings may be used on an individual valve.

For example, the "ductile" properties of a chrome nitride coating (hardness 1,600 HV) is selected for application to the valve tip, while the low-friction attributes of a diamond-like carbon or amorphous silicon carbide coating (friction coefficients .1 or less) is chosen for application to the critical valve-seat head region.

Low-friction and inert thin-film coatings are compatible for dry fuels, which include low-sulphur content or alcohol-based fuels. The application of a coating on the valve head and valvestem can be considered as a solid lubricant, minimizing adhesive wear between the valve-seat or valve-guide interface. Adhesive wear, also known as scoring, galling, or (worse case) seizing, results when two solid surfaces slide over each other under pressure. Surface projections, or asperities, plastically deform and eventually weld together under the high localized pressure. As sliding continues, these bonds break. This creates cavities on one surface and projections on the other. Tiny abrasive particles can also form, causing additional wear.

Specific to applications associated with excessive exhaust gas temperature, hybrid coatings (platinum-, palladium-, or niobium-based) have been examined as a means to retard embrittlement of the base titanium material. The idea is to minimize the ingress of oxygen through the coating and represent novel strategies to yield robust coatings for ultra-high temperature environment applications.

Valvesprings

Valvesprings are critical engine components that must be matched to the rest of the valvetrain components. If one fails, there is potential to destroy an entire engine by "dropping a valve." There is a wide variety of valvesprings available today to suit most applications; single, dual, triple, and beehive. They are defined by ID/OD, with or without dampener, installed height, seat load, open load, coil bind, natural frequency, and rate.

Engine valvetrains must be viewed as a system. Valvesprings are the control mechanism for the system, and are measured in linear force (in-lbs) or load, not pressure (lb/in^2) as sometimes referred to.

In a typical pushrod V-8 engine, the valvetrain system for one valve consists of the cam profile, lifter, pushrod, guide plate, rocker arm, rocker arm stud, valve, valvespring, valvespring retainer, valvestem locks, valveguide, valvestem seal, and valve seat. All of these components must work together to balance the forces in the valvetrain system.

Valvesprings must balance or counteract the forces in the valvetrain system, and keep all system components in contact with one another. These forces are created by the masses of the moving components of the system. Too little valvespring control force and the components of the system become unstable and out of control (valve float). In contrast, too much valvespring control force wastes mechanical energy and could overstress the components in the system, causing them to fail. The goal is to minimize component mass, yet maintain enough strength for the applied loads, and to use valvesprings with the appropriate control force.

Most performance aftermarket camshaft manufacturers recommend spring specifications for a particular cam profile. In most cases these recommended spring specifications were found by taking an average of generally accepted component masses and testing them in different load

Typical dual-spring with dampener spring. The bottom of the retainer must fit snugly into the ID of the inner spring.

Using steel retainers on beehive springs allows you to save about the same amount of weight as using titanium retainers on full-size non-tapered springs.

The smaller OD retainers used on beehives also provide additional rocker arm clearance.

scenarios and RPM ranges. This ensures the natural frequency of the valvetrain system is tuned for the application.

Spring Designs

Spring designs have evolved over the years. Traditionally, springs were made using round wire. In order to handle higher loads with more radical cams, double springs (an inner spring supplementing the outer spring) and triple springs (inner and outer along with an additional dampener spring) were developed. Although able to handle higher loads and speeds, this added more mass. Ovate wire, although more expensive, provide increased spring wire cross-section (allowing the creation of a stiffer spring in the same available space). Ovate wire isn't round. It's oval shaped with a bit of a teardrop, similar to the shape of a camshaft lobe.

Another method of providing necessary spring pressure while saving weight and space is to use "beehive" springs, which taper in diameter from bottom to top. This allows the use of small-OD retainers (which, in steel, may be as light as a large-diameter titanium retainer), which improves rocker clearance.

One advantage of the beehive design, aside from size and weight economy, is the tapered spring's ability to better absorb harmonics, for increased stability. There is debate about the advantages of this design; some builders embrace the beehive design and others prefer constant-diameter dual springs. Although beehive springs may accommodate certain applications using solid or hydraulic flat-tappet cams or hydraulic roller cams, solid roller cams require heftier springs, with conventional double or triple springs preferred.

When using aluminum heads, you need hardened steel seats to prevent the springs from digging into the aluminum. The seat also locates the spring to keep it from walking eccentrically. The choice is spring cups or spring locators. Spring cups have an OD raised lip to capture the OD of the outer spring. Spring locators have a raised shoulder at the center to register to the inner spring ID. This spring seat/locator has a raised inboard shoulder that captures the seat onto the guide boss and provides a centering reference for the inner valvespring. Measure the diameter of the raised shoulder and make sure that it clears the inner diameter of the inner spring. The inner spring ID should be a close fit to the outer diameter of the locator's shoulder. Consider a maximum clearance of .050 inch (where the inner spring's ID is .050 inch larger than the shoulder OD). Excessive spring wander eventually wears out the guide and guide seal.

Whenever you're using aluminum cylinder heads, hard-steel spring seats are a must. They prevent the valvesprings from digging into the softer aluminum. These spring seats are available in various styles, including those that center onto the guide boss. The outer edge of the seats may or may not have raised lips. In order to prevent the springs from walking eccentrically, the base needs to capture the springs. If the cylinder head has a flat area with no spring cup recess, the spring seats must have raised lips to capture the springs. Steel spring seats are available in various thicknesses to suit installed spring height requirements.

Choosing Valvesprings

Valvesprings control the energy transmitted from the camshaft to the valves. It's critical to understand the springs' role and dimensional/clearance concerns.

Valvespring manufacturers apply an antirust coating, which should not be removed. Do not clean new springs with any type of solvent. Handle springs carefully to avoid causing nicks or burrs that could lead to stress riser failure.

When measuring spring load, place the retainer on the base of the spring and note the thickness of the retainer where it makes contact with the outer spring. When measuring installed height, deduct the thickness of the retainer.

Before installing springs, install one valve at a time along with its retainers and locks. Fully pull up the retainer to seat the valve in the closed position. Using calipers, measure the distance from the outside step of the retainer and the spring seat. After measuring and recording each valve's location, examine the data. The shortest height recorded is the installed height for all springs (spring shims may be used to adjust the remaining valvespring installed heights). Final installed height should be within .020 inch of the specification provided by the cylinder head or camshaft manufacturer.

Before removing the retainers, also measure the distance between the bottom of the retainer to the top of the valveguide seal to verify that the retainer doesn't contact the top of the seal or guide. Note the maximum valve lift of the camshaft with rocker arms. The distance between the retainer and guide/seal should be at least .090 inch greater than the maximum valve lift. If this clearance is too tight, you may be able to machine the top of the guide to achieve enough clearance.

Inspect for spring coil bind with the springs installed. With the valve in its fully opened position at maximum lift, maintain at least .060 inch. However, depending on the specific type of spring, the recommended minimum clearance may be tighter. Always refer to the spring manufacturer's specifications. If clearance between coils is too tight, corrective choices include using springs that are rated for greater valve lift, obtaining longer valves, machining spring pockets deeper, or choosing a different spring retainer design.

Be sure to closely inspect for possible contact between the rocker arm and the spring retainer when the valve is in its closed position. If there is a clearance issue, it may be corrected by using a different rocker arm design, or by a combination of different pushrod and valve lengths (shorter valve and longer pushrod).

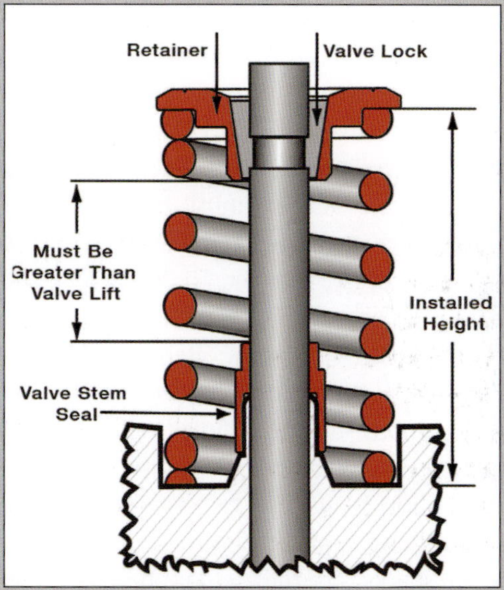

With dual springs, the inner spring is typically .100 inch shorter than the outer spring. The retainer must be designed with steps to accommodate the inner spring(s). (Photo Courtesy Comp Cams)

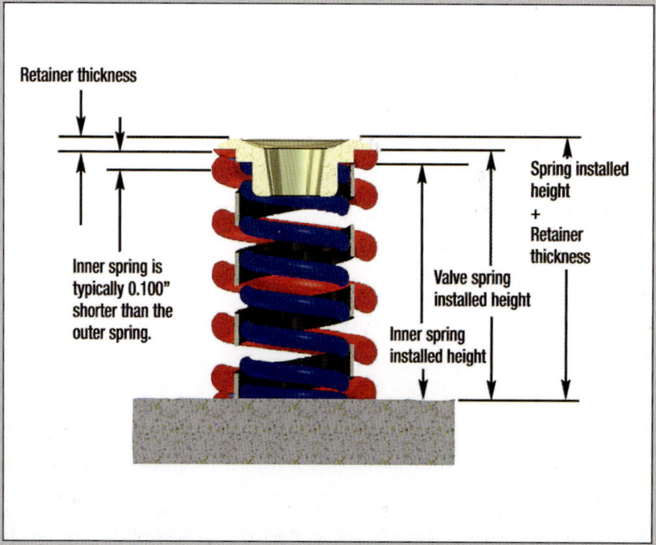

Installed height is measured from the bottom of the retainer, where the spring makes contact, to the bottom of the spring where it contacts its seat. In the valve-closed position the distance between the bottom of the retainer boss to the top of the valve guide seal must be greater than the max valve lift in order to prevent the retainer from hitting the seal. (Photo Courtesy Comp Cams)

Altering rocker arm ratio also affects spring requirements. As you increase rocker arm ratio, a correspondingly stronger spring is needed. This is fairly easy to determine. The rule is to increase open spring pressure by the same percentage of rocker arm ratio increase. You can use this formula:

Percent of Change =
(new ratio – old ratio) ÷ new ratio

For example, if you have 1.6:1 rockers, but you decide to swap to 1.7:1 rockers, the formula works like this:

1.7 – 1.6 = .1
.1 ÷ 1.6
.0625 x 100
6.25 percent

So, in this case moving from 1.6:1 rockers to 1.7:1 rockers represents an increase of 6.25 percent. As a result, you increase the valvespring's open pressure by the same amount. If the original springs had, say, 500-pounds open pressure, you need springs with an open pressure of 6.25 percent more, or 531 pounds.

It is important to check the coil bind of the valvesprings when increasing rocker arm ratios to determine if there is a clearance issue with any of the components. To calculate, use this formula:

Lift Increase =
(cam lift ÷ old rocker arm ratio) x
new rocker arm ratio

For example, a .550-inch-lift small-block Chevy cam with 1.5 rockers changing to 1.6 rockers:

.525 ÷ 1.5
.350 x 1.6
.560 inch

An increase of .010-inch valve lift

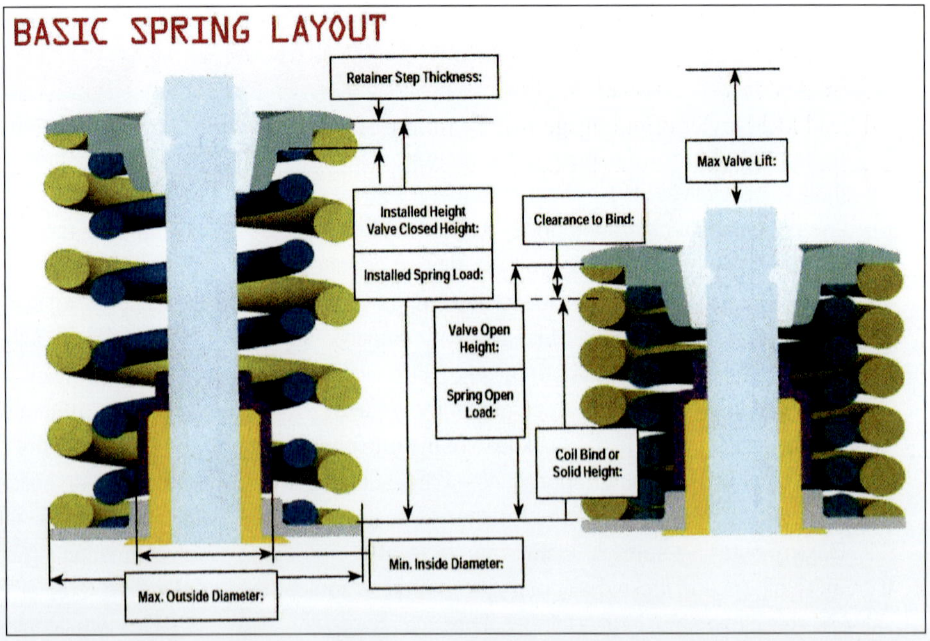

Example of basic valvespring layout and applicable dimensions. (Illustration Courtesy PAC Racing Springs)

Spring Break-In

Similar to seating piston rings or breaking-in a flat-tappet cam, new valvesprings should be conditioned (broken-in) to aid in long-term spring life. During initial engine start-up, run the engine at 1,500 to 2,000 rpm until it reaches normal operating temperature. Shut the engine off and allow the springs to cool to ambient temperature.

Spring Pockets

Pocket clearance refers to the clearance between the ID of the spring pocket (where the spring rests on the cylinder head) and the spring's OD.

Excess clearance allows the spring to move (dance) on the pocket, which leads to spring instability and causes wear at the bottom of the springs and the spring seat. A spring cup may be installed to register the spring OD. A spring locator may be used to register the spring ID. If there's no clearance, the spring binds in the pocket, which places stress on the bottom coil, limiting the bottom coil's ability to grow (via heat and compression).

Retainer Fit

The bottom of the retainer must fit the spring with a minimal clearance (generally .005- to .010-inch clearance to .001-inch interference. If it's too tight, you apply too much stress to the spring. If too loose, you increase wear. If the springs have duals or triples, this applies to each spring as it mates to the retainer.

Along with spring fit, consider the connection between the retainer and the valvestem. This involves the valve locks (keepers). The installed outer angle of the keepers must match the inner taper of the retainer (7-degree lock with a 7-degree retainer, 10-degree lock with a 10-degree retainer, etc.). The traditional 7-degree design is often changed to the 10-degree style, which provides additional surface area to optimize load spreading.

Also, the ID of the keeper pair must match the valvestem diameter and the style of lock groove on the valve. The shape of the retention band inside the keepers must be compatible with the shape of the locking groove on the valve.

The majority of valves have a square-cut groove, with the keepers featuring a square-cut male rib. However, radiused (called bead lock) styles offer engagement with less-isolated stress points (radiused instead of sharp edges). Many titanium valves have a bead-lock design, due to the nature of titanium, which is very sensitive to stresses (nicks, sharp edges, etc.). The point is to be aware of the need to match the retainer and keeper sizes and designs to accommodate your valves and springs.

Valvespring Applications

Many of the following tips were provided by Crane Cams. This information will help you select the correct valvespring for a particular application.

Flat-Tappet Cam and Lifters

Small-block engines generally require 105 to 125 pounds of seat pressure. Big-block engines (due to heavier valves) usually require 115 to 130 pounds of seat pressure.

Flat-tappet open pressures should not exceed 330 pounds. Open pressure should be at least 220 pounds for engines that rev to 4,000 rpm. For higher engine speeds, open pressures should be at least 260 pounds with stock-weight valves (lighter valves require less open pressure). Be aware that open pressures of 280 pounds or more can cause press-in rocker studs to pull loose, so screw-in studs are needed when open spring pressures exceed about 280 pounds.

Hydraulic Roller Cams and Lifters

These need high spring seat pressures in order to control the heavier roller lifters as well as to control the more aggressive opening and closing rates common to roller cams. Small-block engines generally require seat pressures in the 120- to 145-pound range. Big-block engines usually need seat pressures in the range of 130 to 165 pounds.

General-purpose street small-block engines with hydraulic roller cams require at least 260-pound open pressures for applications up to about 4,000 rpm. Healthy (moderate performance) hydraulic roller small-blocks prefer open pressures in the 300- to 360-pound range. Serious-performance small-blocks can use up to about 400 to 435 pounds open pressures for reasonable valvetrain life.

Big-block general-purpose street-roller-cam engines require at least 280 pounds open pressure for engine speeds up to about 4,000 rpm. Moderate performance big-block roller setups generally require open pressures in the 325- to 375-pound range. Superior big-block roller cam setups can use springs in the 450-pound range (again, I'm referring to also obtaining decent valvetrain life/durability).

Open spring pressures that exceed 360 pounds of open pressure should really be mated with billet-steel roller lifter bodies, not OEM cast-iron roller lifter bodies.

Solid Lifter Roller Cams and Lifters

Solid lifter roller cams are designed with more aggressive opening and closing rates, requiring high seat pressures to avoid valve bouncing. High-strength, one-piece valves are needed for durability.

Seat (closed) pressures generally are required in the 180- to 200-pound range for moderate performance engines. High-end race applications such as Pro Stock and blown alcohol applications commonly use around 340 to 370 pounds of seat pressure.

Open pressures for street/strip applications are usually in the 350- to 450-pound range. Circle track and bracket racing applications generally run in the 450- to 600-pound range. Really serious, extreme-output drag engines and short-track circle race engines may require 600 or more pounds (open pressures, depending on the application, can run as high as 900 pounds).

Always pay attention to the spring specs recommended by the cam manufacturer. If in doubt, go a bit higher instead of a bit lower. In maximum-output engines, it may be beneficial to run different spring rates on intakes and exhausts (because intake valves are larger and may provide heavier mass, and because the intake valve is opening against higher cylinder pressure, a high-rate spring may be helpful on the intake valves).

After selecting and measuring your new springs, but prior to installing, it's a good idea to compress the springs (up to the coil bind height) about three times. This relieves any stored stress/energy and helps them to stabilize for more accurate measurements. This should be done on a quality valvespring compressor. Be sure to wear safety glasses whenever compressing valvesprings on any spring compressor tool.

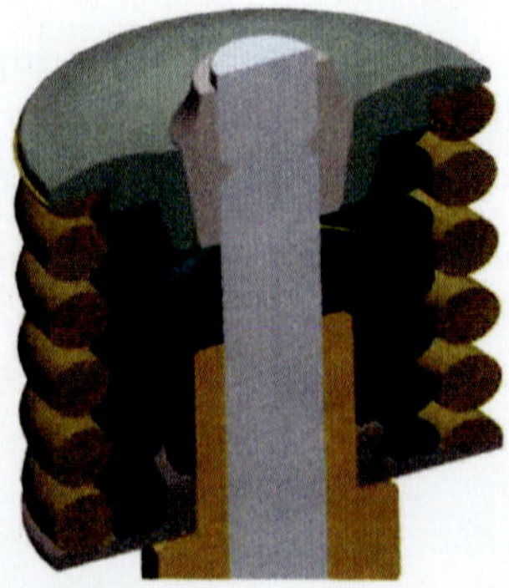

COIL BIND CALCULATION EXAMPLE	
Installed height	1.950
Valve Lift (minus lash)	-0.785
Spring Coil Bind Height	-1.105
CLEARANCE	0.060

Example of coil bind determination. Installed height of this spring is 1.950 inches. Subtract combined valve lift (minus lash) of .785 inch for a coil bind height of 1.105 inches. This provides a safety clearance of .060 inch to avoid coil bind. (Illustration Courtesy PAC Racing Springs)

Spring Installed Height

Installed height refers to the dimension measured from the bottom of the spring retainer to the point where the top of the outer spring contacts the retainer to the spring pocket in the cylinder head when the valve is closed. Installed height affects spring tension, and is the determining factor for spring closed tension. Check your camshaft specification card. It lists the suggested spring installed height for that cam with the cam manufacturer's recommended springs.

For example, if the card notes that a specific valvespring part number should be installed at 105 pounds at 1.700 inches, that means that if the spring is installed at a height of 1.700 inches, this should place 105 pounds of tension with the valve closed.

Altering the spring installed height directly affects tension. If spring installed height needs to be reduced, a shim may be placed under the spring. An alternative is to select a retainer with a deeper dish, or a different-style valve lock that changes the retainer location on the valve. Spring tension increases as installed height is decreased, and spring tension decreases if the installed height increases.

Using a valvespring tester, compress the valvespring to its recommended installed height and record the seat pressure at that dimension.

Spring Open Pressure

Valvespring open pressure represents the pressure (in psi) that's placed against the spring retainer at maximum valve lift. The spring must provide sufficient pressure in order to control the lifters against the cam as the valve changes direction from opening to closing. Insufficient spring pressure causes the lifters to bounce over the cam lobe, which causes valve float. This can also lead to premature camshaft wear.

If valvespring pressure is excessive, the pushrods experience greater stress and are more likely to flex, which also leads to valve bounce. Heavier valves can handle higher spring open pressures,

A valve height checker (barrel micrometer) allows you to determine the exact installed valve height and compare this to required installed spring height. With the steel spring seat in place, the mic is dropped over the valve stem and the retainer and locks are installed to secure the valve. With the checker mic adjusted to full height (eliminating freeplay), the height reading is recorded. Additional shims may be added under the springs to adjust spring height.

A valve height checker (basically a barrel-style micrometer) provides an easy and accurate method to measure installed valve height. You can also use a telescoping gauge (contacting one end to the spring seat surface and the other end to the spring contact surface of the retainer), but an incriminated height gauge provides the best and most reliable accuracy.

Light checking springs are useful for measuring purposes, to avoid unnecessarily fighting valvespring pressures during height checking, cam degreeing, etc. Note that this cylinder head (a Dart Big Chief II race head) has machined valvespring pockets. A steel cup-style spring seat is used here, since the machined recess in the head captures the raised outer lip of the spring cup.

while lighter valves are better suited with lower spring pressures.

Spring Coil Bind

Valvespring coil bind occurs when the spring is compressed to a point where the coils touch each other during maximum lift. As mentioned earlier, you must check for coil bind. Using a valvespring checker press, coil bind can be measured by placing a retainer on top of the spring and carefully compressing the spring until the coils touch each other. The distance from the bottom of the retainer to the bottom of the spring is the coil bind height.

To calculate maximum spring travel, subtract the coil bind height from the installed height. Remember that spring travel must be at least .060 inch greater than the full lift of the valve. This safety margin is needed to prevent coil bind and overstressing the spring. If the coils bind, this creates a dead-stop during valvetrain operation, so something's gotta give: pushrod bending, rocker arm breakage, lifter and/or camshaft damage, etc.

Coil Clearance

The easiest way to measure coil clearance (prior to assembly, that is) is with a bench-mounted spring tester. Install the valve into the cylinder head (remember to dedicate each valve to each valve location in each head. That way, once measured and clearanced, you know that the particular valve fits properly in a specific location. Don't assume that you can mix up the valves). With the valve fully inserted into the (intake or exhaust) location, with the valve fully seated, install locks and the retainer. Pull up on the retainer and measure the distance between the underside of the retainer (where the spring contacts the retainer) and the spring seat. If you're using hardened spring seats, be sure to install these seats for this check.

Place the spring on the tester and compress to the spring's installed height. The pressure tester displays the seat pressure at that installed height.

Then determine the maximum (gross) lift of the valve, based on your cam lift and rocker arm ratio. The difference between the installed height and (gross) lift indicates how much additional travel is available to the spring. For example, if the installed height of the valvespring is 2.000 inches, and the maximum lift of the valve is .5000 inch, the open spring height should be 1.500 inches.

Sufficient seat pressure is needed to keep the valves from bouncing when they return to their seats in the closed position. When valves are allowed to bounce on their seats, cylinder pressure is reduced and it can lead to tulip deformation of the valve head. If this deformation occurs, it is possible for the valve head to break off the stem.

If the engine is equipped with hydraulic lifters, the springs must exert enough pressure at the valve lifter to keep the lifter plunger centered in its travel. Otherwise, the lifter can "pump up," causing the valve to be held slightly off its seat. This can easily be mistaken for an ignition or fuel system related misfire.

The choice of engine oil can also affect hydraulic lifter operation. If oil pressure is too high and/or the oil is too

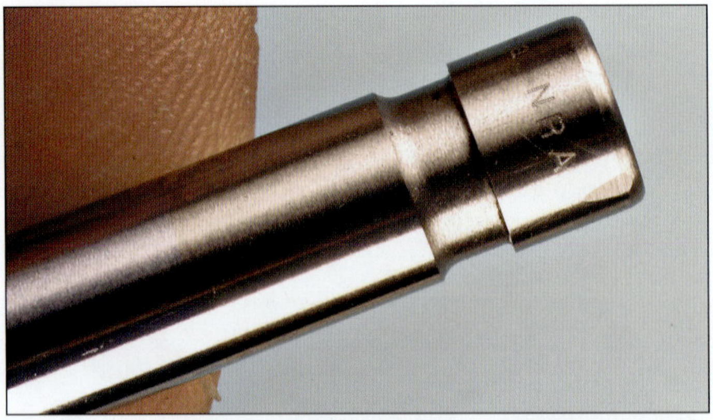

Here's an example of a valve with a traditional square-cut groove. Always inspect the groove style in order to verify the type of locks required. Valve disengagement and failure will result if the wrong style lock is installed.

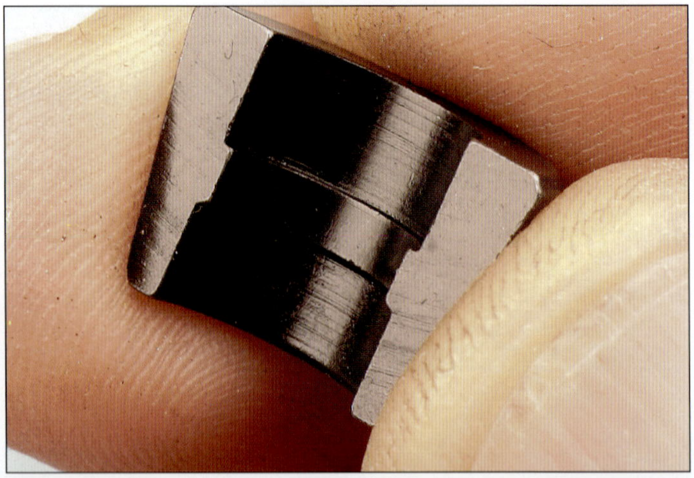

A square-cut lock is shown here. Mismatching lock styles leads to a catastrophic failure.

high in viscosity, this can also lead to lifter pump-up.

Valvespring Seats

If you're running aluminum cylinder heads, you must install hardened steel seats under the springs. This prevents the springs from digging into the aluminum. The steel spring seat also serves another function: to keep the spring centered and to prevent spring "walk," which can lead to excessive valvestem side deflection. Excessive spring wander eventually wears out the guide and guide seal.

Spring seats are available in two basic styles: spring cups and spring locators. Spring cups have an OD raised lip to capture the OD of the outer spring. Spring locators have a raised shoulder at the center to register to the inner spring ID. If using spring cups, measure the diameter of the raised lip and make sure that it clears the OD of the outer spring. If using spring locators, measure the OD of the raised shoulder and measure the ID of the inner spring. In either case, you should have a fairly close (snug) fit. Consider a maximum clearance of .050 inch (for example, with a spring locator, where the inner spring's inside diameter is .050 inch larger than the shoulder OD).

Retainers and Locks

Always check for potential interference between the retainer and valveguide, especially if you've moved to a high-lift camshaft. When the valve is fully opened, it's critical that the bottom of the retainer doesn't contact the top of the guide or guide seal. This is most easily done by installing light checking springs (to avoid unnecessarily fighting the compression force of the intended springs). During test fitting (with the cam, valves, pushrods, and rockers installed), rotate the cam and observe the clearance between the retainer and guide seal at each valve location. At maximum lift, maintain at least .050- to.060-inch clearance.

If you're planning to use hydraulic lifters, unpressurized lifters will give you a false reading. It's best to substitute solid lifters of the same length as your hydraulics for checking purposes. The plunger in a hydraulic lifter depresses, and this prevents you from reading the cam's true maximum lift.

Pay attention to the design of the valve's locking groove. Most valves have a "square cut" groove that requires locks (keepers) with square-cut keys on the ID of the locks. However, some performance valves have a bead-lock design with a radiused bead-type groove on the valvestem that requires a bead-style lock. If you mix them up (square-cut locks on a bead-grooved valve or bead locks on a square-cut groove), the retainers pop off and you drop the valves (not a good thing).

Also, retainers and locks are machined with a specific degree match-up: the angle of the outer surface of the locks matches the inner wall of the retainer bore. These angles must match (7-degree locks with 7-degree retainers, 10-degree locks with 10-degree retainers, etc.). We can debate the advantages of different degree designs all day long but, basically, remember that you must match the assemblies.

If a valve is designed with a radius lock groove, you must use a bead lock.

This is a bead-lock valve lock. Remember, this style is only to be used with a bead-lock valve.

When purchasing retainers, make sure that the inside diameter matches the angle of the outside diameter of the valve locks. If you have 7-degree valve locks, you must use 7-degree retainers. If the locks are 10-degree, use 10-degree retainers, etc. If you mismatch the lock/retainer angle, you'll have problems.

Always buy the retainers and locks from the same manufacturer. Even though retainers and locks may be listed at 7 or 10 degrees for example, there may be slight differences in angles between manufacturers. If you buy Crane retainers, buy Crane locks. If you buy Comp Cams retainers, but Comp Cams locks, etc. Sometimes parts from different manufacturers match up, but sometimes they don't.

Lock/retainer packages are available in three basic styles: rotating, clenching, and semi-clenching. OEMs tend to use rotating styles that allow the valves to rotate during engine operation. This ever-changing, valve clock position places the seat contact in random locations to extend the service life of the valves and seats. However, this isn't what you want for any high-performance or racing setup, since valve rotation diminishes the optimum sealing (as set during the build).

Hardened steel lash caps are designed to protect the relatively soft tip of a titanium valve although titanium valves are now available with built-in hard tips. Lash caps are also useful in fine-tuning rocker arm geometry and valve lash range and are available in various thicknesses. The lash cap should fit snugly onto the valve tip.

Clenching styles lock the valve in place (via a slight interference fit), preventing rotation. Semi-clenching styles holds the valve tightly in its clock position, while allowing a small bit of rotation. For all-out racing or any high-RPM use, it's best to lock the valve with a clenching-style lock/retainer setup.

Valve Lash Caps

Lash caps (made of hardened steel) are individual caps that are installed onto the valvestem tip. These are only required in certain applications, either for achieving desired valve lash or to protect a relatively soft valvestem tip material. Lash caps are commonly used with titanium valves (since titanium is a relatively soft material) in order to prevent the valve tips from deforming. However, some titanium valves have hardened tips (or thermally sprayed on diamond-like hardness) that eliminate the need for additional lash caps.

Lash caps are offered in various thicknesses in order to accommodate lash requirements. Lash caps should fit snugly onto the valve; loose enough to be finger installed and removed but tight enough to eliminate excessive bouncing, rotation, and relative motion. If you're using lash caps, you must keep close watch on fit (especially for valve lash), and keep them organized with the individual valves.

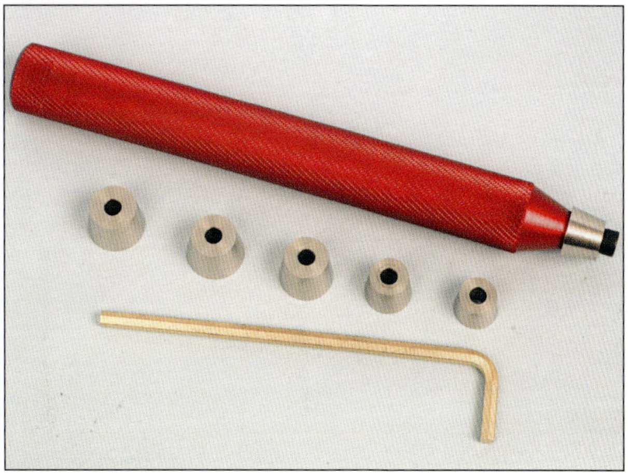

Since valve retainers and locks are available in 7-, 8-, and 10-degree mating angles, it's easy to make a mistake by mixing these up if you're dealing with various angle designs. This tool from PAC Racing has an assortment of cone adapters to easily allow you to identify the lock and retainer angles. (Photo Courtesy PAC Racing Springs)

CHAPTER 13

Rocker Arms

Cam timing events are transmitted to the lifter, pushrod, rocker arm, and then the valve. You need strong rocker arms for particular applications, so the valves are precisely actuated. Rocker arms are installed on all overhead-valve engines and transmit lifter movement to the valves. Because the rocker arm is essentially a lever, the length of the rocker (its pushrod contact point to pivot), increases valve lift relative to cam lobe lift. In other words, the rocker's length provides a ratio-multiplier that magnifies the lift of the cam lobe to create increased valve lift.

Rocker Arm Ratio

Determining gross valve lift involves considering both the camshaft lobe lift and the rocker arm ratio. The ratio of the rocker arm (length of fulcrum) increases the effective valve lift beyond just the camshaft's lobe lift. Here's the formula:

Gross Valve Lift =
cam lobe lift x rocker arm ratio

For example, if a camshaft's intake lobe lift is .365 inch and a rocker arm ratio of 1.6:1 is used, the formula works out like this:

Gross Valve Lift =
.365 x 1.6
.584-inch gross lift

If you're considering changing the rocker arm ratio and you know what the gross lift is and what your current rocker-arm ratio is, here's a simple formula to calculate current lobe lift:

Current Lobe Lift =
current gross valve lift ÷
current rocker arm ratio

The rocker arm lock nut is properly tightened down and then valve lash will be measured.

During test fitting, verify that the rocker tip is centered to the valvestem tip. If you're using pushrod guideplates, the guideplates may require adjustment for positioning.

For example, if your current gross valve lift is .600 inch and you're currently using 1.5:1 rockers:

Current Lobe Lift =
.600 ÷ 1.5
.400-inch lobe lift

Now that you know what the cam lobe lift is, multiply that by the rocker arm ratio that you plan to use. That gives you the new gross lift.

New Gross Valve Lift =
lobe lift x new ratio

For example, let's say that you're considering moving from a 1.5:1 rocker to a 1.6:1 rocker, using the same camshaft. The formula works out like this:

New Gross Valve Lift =
.400 x 1.6
.640-inch new gross lift

As a rule, each ratio increase (in whole numbers) adds about .035 inch of increased valve lift. Changing from 1.5:1 rockers to 1.6:1 rockers increases valve lift by about .035 inch. This provides increased valve lift and quicker valve opening and closing rates. Although this sounds great, before you rush to a higher rocker arm ratio, be aware of the following:

- Increasing rocker arm ratio decreases piston-to-valve clearance, so you need to check this.
- Valvespring retainer-to-valveguide clearance must be checked, as this clearance decreases.
- Check for valvespring bind at full lift.
- Longer pushrods are likely required with high-ratio arms, so make a careful pushrod length determination. You are measuring for precise pushrod length anyway.

As you know, rocker arm ratio refers to the distance between the centerline of the rocker arm pivot to the center of the pushrod cup. The higher the ratio, the shorter this distance becomes. Unfortunately, most published OEM rocker arm ratios cannot be trusted. Due to mass-production tolerances, the listed ratios are theoretical.

For instance, a set of OEM rockers that are listed as having a 1.5:1 ratio might actually involve a range of, say, 1.49 to 1.54:1. For high-precision fulcrum lengths and superior consistency, the performance aftermarket is the way to go. High-quality aftermarket performance rockers are produced with a much higher level of precision and consistency. Increasing rocker arm ratio is a neat way to increase power. You get the effect of a high-lift cam without the need to increase duration. The valve is opened faster, lifted farther off its seat, and closed later.

Guideplates

Stud-mounted rocker arms allow the rockers to pivot around the axis of the stud; shaft-mounted rockers maintain a constant rocker alignment. Stud-mounted rockers require some type of guiding or alignment to prevent the pushrods from walking around. Depending on the cylinder head design, closer tolerance pushrod slots in the head or steel guideplates take care of this.

Guide plates have pushrod slots. The slots must be checked for edge burrs (deburred if necessary). During test fitting, slowly rotate the crank and check each pushrod for depth clearance at its guideplate slot (primarily at max lift). Especially when using a high-lift cam and/or high-ratio rocker arms, you may need to deepen the slots, but do not widen the slots.

If your head requires guideplates, you must use hardened pushrods to prevent pushrod galling. Most guideplates are stamped from steel and may have sharp burrs around the pushrod slots. Be sure to inspect each guideplate and deburr any sharp edges before installing them.

Rocker arms with higher than stock ratios locate the pushrod cup closer to the rocker arm's central pivot point. During test fitting, closely check for pushrod-to-guideplate-slot clearance at maximum lift. Although aftermarket performance guideplates usually have longer slots for this reason, but you still need to check this clearance. If you lengthen (deepen) the pushrod slots, be careful to only make the slots deeper. Don't oversize their width.

Also check for pushrod clearance in the heads; a high-ratio arm moves the pushrods from the original intended location in the head. If you need additional clearance at the cylinder head's pushrod slots, you can do this with a hand-held grinder (with the head off the block). Don't get too carried away. A clearance of around .010 inch is adequate. Removing too much material can be dangerous and may open-up a hole in an intake runner.

When installing guideplates, with the rocker studs finger tightened in place and with rockers and pushrods installed, you should be able to wiggle the guideplate from front to rear (in relation to the engine) in order to center the pushrods in the width of the guideplate slots (guideplate mounting holes usually allow a bit of movement and adjustment). Once the slots are centered to their pushrods, fully tighten the plates (using the rocker studs).

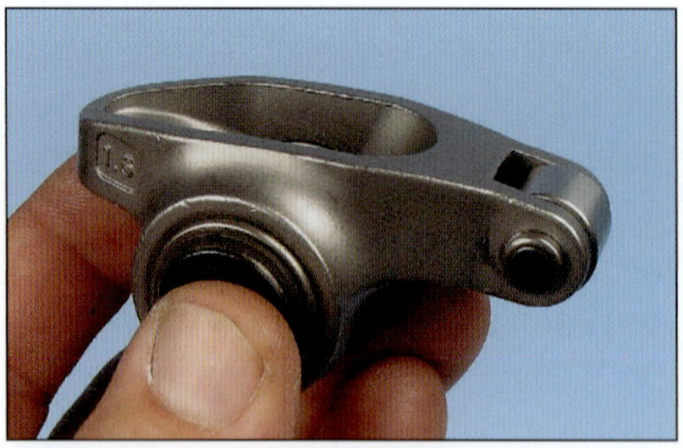

A full-roller rocker has a needle-bearing trunnion for the pivot point and a roller bearing for valve tip contact.

Full-roller rockers made from forged aluminum billets that are CNC machined provide lighter weight and high strength. Pictured here is a set from Harland-Sharp for the GM LS engine.

Performance rockers made from steel forgings offer superior full-roller performance and strength that far surpasses OEM designs.

Roller Rockers

A stock OEM rocker arm (in most cases) rubs on the valve tip and pivots on a ball fulcrum, causing friction. A full-roller rocker arm has a needle bearing trunnion (the pivot) and a roller bearing at the rocker nose that contacts the valve tip. This virtually eliminates frictional losses. Less friction results in quicker response, freer revving, and potentially more horsepower.

Aftermarket roller rockers are also available with a roller bearing at the nose but an OEM-type pivot, and some OEM rockers are available with a needle bearing trunnion pivot but a non-roller tip. A full-roller rocker is a good choice for any application. However, using full-roller rockers becomes more important when dealing with small-diameter valvestems and/or high lift and spring pressures.

If you run 5/16-inch-diameter or smaller valvestems with high lift and high spring pressures, having a roller bearing at the nose dramatically reduces the chance of valve deflection. Also, because of the larger arc imposed on the rocker due to a high-lift cam, roller tips greatly reduce the chance for valve tip damage. If cost isn't a factor, go to full-roller rockers.

Shaft-Mounted Rockers

Shaft-mounted rocker designs are available for most applications (depending on cylinder head design). They pivot on a common shaft instead of individually on rocker studs or bolts. The advantage of a shaft mount involves improved valvetrain stability. When shaft mounted, the rockers can't individually move around the axis of the valve, so rocker-to-valve and rocker-to-pushrod stability and consistent contact is improved.

A shaft-mount design may have one common shaft that provides the pivot axis for the entire bank of rockers on a cylinder head. Other setups have each cylinder's pair of rockers (intake and

Full-roller rocker designs that have common-shaft mounting are available for specific engine applications in which all rockers pivot on one full-length common shaft.

Racing applications commonly require high strength and reduced weight. These steel Mohawk roller rockers from Jesel are a good example. Notice that weight has been shaved where possible, while retaining strength for use with high valvespring pressures. (Photo Courtesy Jesel)

Depending on the cylinder head application, shaft-mounted roller rockers are available in which each pair of a cylinder's rockers ride on their own common shaft, rather than a full-length common shaft.

Bridged roller rockers (as on this Pontiac 455, equipped with Harland Sharp rockers) feature each adjacent pair of rockers pivoting on a common shaft. This keeps each pair of rockers in alignment, preventing them from moving out of plane relative to the valves and pushrods.

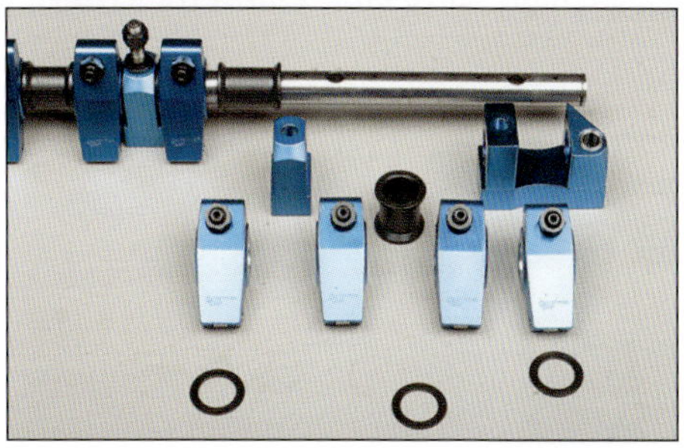

Rockers that feature a full-length common shaft mount are often adjustable, allowing you to fine-tune each rocker's location for proper valve alignment.

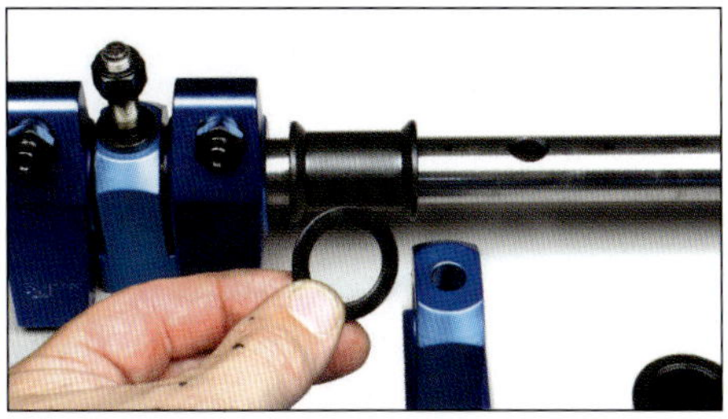

Spacer shims are placed between rocker-to-stand locations and between rocker-to-barrel spacers. Barrel-type spacers provide spacing between rockers.

Adjusting shaft-mounted rockers (where a common full length shaft is featured) can be a time-consuming process in order to position each rocker's roller tip centered to each valve. Chances are you won't achieve alignment on the first try. To re-adjust, use a caliper to measure the required change, disassemble the shaft, reinstall the shims, and try again. It's not that difficult, but it takes patience.

This Jesel shaft-mounted roller rocker system (for use with Dart Big Chief II big-block Chevy heads) includes rockers, rocker stands, and a stand-height checker. Due to the offset valve design of this Dart head, Jesel developed a unique rocker system that requires a special setup.

With valvespring removed, the stand-height checker drops over the valve stem, while the opposite end rides on a checker shaft that rests in the rocker stand shaft groove. The checker allows you to dial in correct rocker arm geometry. The valve stem tip should be flush with the upper surface of the checker. If the valve tip is lower (below flush with the checker), correction means moving to longer valves or milling the head's shaft stand bosses. A quicker, easier, and cheaper fix is to use the proper-thickness valve lash cap. If you're using titanium valves that don't already have hard tips, you need lash caps anyway. If the valve tip sits too high (above flush with the checker), shimming the rocker shaft stands is required.

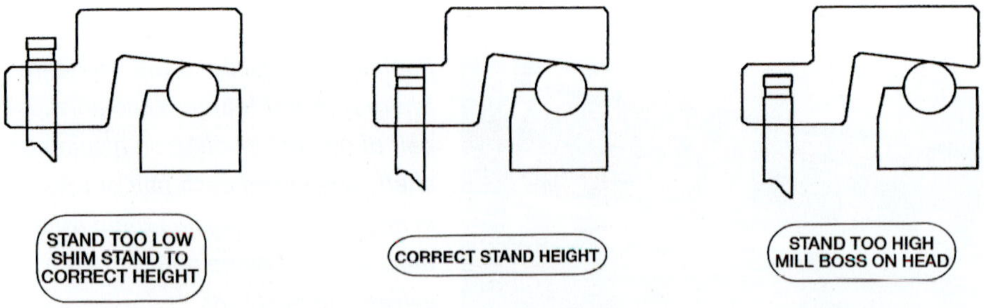

Left-to-right while using the Jesel checker: valve too high, valve flush and correct, valve too low. (Illustration Courtesy Jesel)

The Jesel rocker system for the Dart Big Chief II race head is a good example of how a unique design is sometimes required. Occasionally different shapes or offset rockers are needed to accommodate a unique valve layout or angle design. Because of the requirements of this particular head, the Jesel system has three different rocker arm designs: a straight arm with a 90-degree shaft position for two of each head's intakes, a straight arm with an off-angle shaft position for all exhaust locations, and an offset intake arm (two per head).

Initially tighten rocker arm adjustment to zero lash with the cam on its base circle.

For hydraulic lifter applications, with the cam on its base circle, and after initially setting to zero lash, tighten the adjuster by no more than 1/2 turn, then lock the adjuster with its jam nut.

exhaust) pivot on a shaft that holds one pair of rockers (so each pair of cylinder rockers have their own shaft). Especially for high-RPM and heavy-loading applications, shaft mounts offer increased stability because the rockers can only pivot on their shafts and can't move laterally across the valve tips.

Break-In and Adjustment for Roller Rockers

Granted, there's no frictional issues with a full-roller rocker at the valve and at the pivot/trunnion, but make sure you apply moly lube to the pushrod-to-rocker contact (to each upper end of the pushrod and to the contact, cup, or ball at the rocker). Rocker manufacturers include a high-pressure lube in the kit or they recommend a specific lube.

It's highly advisable to soak all roller rockers in a tub of clean, 30-weight engine oil for at least 1/2 hour before installing (I usually soak mine overnight). This allows oil to penetrate into the needle bearings and bearing axles.

Adjust valve lash according to the cam manufacturer's instructions where solid lifters are used (whether roller or flat-tappet). For hydraulic lifter setups with roller rockers, place the cam on its

With pushrods and rockers installed, hand-rotate the crank through two full revolutions, then remove the rocker.

base circle (where the lifter is at its lowest point) and gently adjust the rocker to achieve zero lash, then tighten the adjuster by 1/2 turn (this provides a small amount of preload to the hydraulic lifter).

Even if you've already measured and checked for proper pushrod length, it's a good idea to perform a final check. With the rocker removed, paint the valvestem tip with a black marker. Install the rocker, adjust lash, and rotate the crank 2 full turns. Remove the rocker and check the witness mark on the valve tip. It should be at the center of the valve. If the mark is biased toward the intake side, the pushrod is too short. If the mark is biased toward the exhaust side, the pushrod is too long.

OHC setups use cam followers that directly transmit valve action from the cam lobes.

Rocker Arm Materials

Rocker arms may be stamped steel, ductile iron, investment cast steel, aluminum forgings, extrusions, machined billet aluminum, or powdered metal. These material choices are for OEM as well as performance aftermarket arms.

If a 100-point restoration is the goal, you likely want to use whatever type was originally installed (likely stamped steel, ductile iron, cast aluminum, or steel forgings). If you have a late-model LS engine, you may opt to stick with PM rockers. However, for upgrades with power and durability in mind, you want to reduce friction and increase rocker strength, and you want rockers with correct and consistent arm ratio. This means upgrading to high-quality aftermarket rockers that are made of investment-cast steel, forged steel, forged aluminum, extruded aluminum, or billet aluminum.

Because steel is generally stronger than aluminum, and depending on how radical the build is and the anticipated spring loads and high-RPM use, steel may be a better choice in certain applications. But there are various opinions. The best thing to do is to talk to the rocker arm manufacturer for a recommendation.

Both types of materials are used in high-output racing engines. Sometimes it boils down to race sanctioning body rules or simply on how much you want to spend. From an expense standpoint, from less money to more, ductile iron, cast aluminum, forged aluminum, and steel rockers are your choices.

PM rockers are in common use by many OEMs (the GM LS engines are an example). Although the OEM trunnion/bearing design is poor (see page 142 for more details), the PM construction is surprisingly good. Generally, PM rockers are okay up to about 400 hp. Beyond that, I don't even consider them.

OHC Followers

Overhead cam (OHC) engines don't have pushrods that connect lifter movement to rocker arms. Instead, cam "followers" provide a direct leverage connection from the valve tips to the lifters or cam lobes to the valve tips (depending on design). For a high-performance or race application, aftermarket upgraded followers, which have roller contact (instead of friction/sliding contact) at both the valves and lifters, are available. Just as in roller rocker upgrades on overhead valve (OHV) applications, roller cam followers greatly reduce frictional losses and allow for higher and more consistent engine revs.

Valve Cover Clearance

Nice big, beefy aluminum roller rockers may provide the performance and durability that you desire, but remember that dimensionally larger rockers may

This Ferrea aluminum full-roller cam follower is being fitted to a Honda race head. The large needle-bearing roller rides under the cam lobe, and the adjusters ride directly on the valves. A follower pivot shaft is inserted through the head to provide a pivot point.

not clear OEM valve covers. Since most custom engine builds are likely fitted with aftermarket valve covers (which are readily available in taller heights), this really isn't an obstacle, but consider this clearance in order to decide if you need taller covers.

A simple way to determine this is to place a cover onto the head, with no bolts. While a buddy slowly rotates the crank by hand, place your hand on top of the valve cover. If rockers begin to contact the cover, you can feel the cover bump up. If so, you need taller valve covers. Before you invest in a pair of spiffy valve covers, test rocker clearance using a spare cover with the same height of the ones that you're considering.

Upgrading OEM LS Rocker Arm Trunnions

The GM Generation-3 and -4 LS engine platform provides a quantum leap beyond the earlier small-block Chevy engine. Overall, it's a great design with plenty of horsepower potential and good durability. However, a notable weakness involves the rocker arm. The majority of OEM LS-platform engines have PM rocker arms, which are surprisingly strong and well suited for street operation. The downside involves the trunnion and needle bearing system.

Under high loads and high engine speeds, the cageless bearing outer housings can work their way out of their bores, allowing the needle bearings to work their way out of the bore and scatter into the engine. There's no guarantee that this failure will occur, but especially with an upgrade to a high-lift cam and stiffer springs, the potential for the failure increases. If you find one or more mystery needle bearing during a routine oil change, it is clear evidence that at least one rocker arm is losing its trunnion bearings.

For performance use, get rid of the OEM powdered-metal trunnion and the cheap uncaged bearings in favor of a durable aftermarket trunnion kit upgrade. This is an OEM setup. Notice the small bearing needles? How'd you like to have a bunch of these floating through your engine? Seriously, if you plan to stand on an LS engine with OEM rockers once in a while, you need to upgrade the trunnions and bearings!

If you plan to retain the factory PM rockers, a simple and inexpensive fix is to perform a trunnion bearing upgrade using a readily available kit.

Comp Cams, as an example, offers a high-quality trunnion bearing upgrade kit (PN 13702-KIT). This kit, applicable to LS1/LS2/LS3/LS6/LS7 OEM rockers (as well as iron-block versions such as LQ9, etc.), includes new trunnions, new caged and prelubed bearing assemblies, circlips, and rocker arm mounting socket-head cap screws (along with two thick, installation-assist washers). The new trunnions have extended tips with circlip grooves. Instead of the bearing housings relying on a press fit, the additional circlips serve to prevent possible bearing assembly walk-out.

These bearings aren't going anywhere, so the rocker bearing issue is no longer a concern. The new trunnions and bearings are also made of high-quality materials and are designed to withstand high RPM and high spring pressures. An upgrade kit replaces the cageless OEM needle bearing assemblies and PM trunnions with very high strength, premium 8620 steel-alloy trunnions and fully caged roller bearings for vastly improved durability.

Comp Cams offers part number 13702-KIT, a must-have upgrade for OEM rocker durability.

The trunnion-bearing swap procedure can be handled using a bench vise or hydraulic press with appropriate-size inserts or with a specialty tool designed for use with a bench vise. An example is Summit Racing's tool (P/N SME-906011), which makes this job easy.

Trunnion Upgrade Procedure

Before beginning, clean the rockers with solvent or in a hot tank. After removing the original bearings and trunnions, you have the option of additional surface finishing, by having the rockers tumbled in a media tumbler or by a light blasting with glass bead, soda, walnut shell, etc. Avoid dwelling the blasting pressure in the trunnion bores, valve contact tips, or pushrod cups, and do not run a reamer through the trunnion bores. Do not sand blast; it is far too aggressive.

Place the tool vise fixture onto one jaw of a bench vise. Place the OEM rocker into the fixture's V-saddle.

Align the (black) presser to the OEM bearing, with the counterbored end of the presser facing the bearing. Tighten the bench vise to press-out both OEM bearings and the trunnion (needles fall out). Inspect and clean the rockers as needed.

Place the presser onto the short end of the Summit kit's mandrel. Slide a new bearing onto the long end of the mandrel (make sure that the stamped

Summit Racing's new trunnion vise tool (PN SME-906011) allows you to easily and safely remove the original bearings and trunnions and install the upgraded Comp Cams trunnion kit using a common bench vise instead of a hydraulic press. Removing old trunnions and bearings and installing the new Comp Cams upgrade kit in all 16 rockers can be done in as little as 30 to 45 minutes without breaking a sweat.

symbols on the bearing faces the mandrel's flange).

Slip the aluminum alignment sleeve onto the long end of the mandrel, seated against the bearing.

Slide the mandrel assembly into the rocker arm bearing bore (the aluminum alignment sleeve enters the bore first to align to the rocker body).

Using a bench vise, press the bearing into the rocker body until the mandrel's flange bottoms-out onto the rocker body. Perform this single-bearing installation to all 16 rockers before proceeding with trunnion and opposite-side bearing installation.

Place the rocker arm body onto the Summit vise fixture, with the previously installed bearing facing the fixture.

Slide the new trunnion halfway into the bearing bore (leaving the outside end of the trunnion exposed to align the remaining bearing). Apply engine oil to the outside of the bearing. Slip the remaining bearing onto the exposed trunnion (make sure that the stamped symbols on the bearing cage face outward).

The Comp upgrade kit has high-strength steel trunnions, caged bearings, and peace-of-mind circlips. This setup will never allow bearing needles to fall out. The Comp upgrade kit also includes a pair of thick steel washers that are used during the press-fit of the new bearings.

Place the kit's thick steel washer onto the end of the trunnion (against the bearing face).

Tighten the bench vise until the washer bottoms-out against the rocker body.

Verify that the trunnion rotates freely, and that a slight amount of endplay (side to side) exists, in the range of .004 to .008 inch. If you don't have any endplay, adjust the bearing depth accordingly.

Using snap-ring pliers, install one circlip to each end of the trunnion. Verify that the clip is fully seated in the trunnion clip groove.

Install the upgraded rocker arms to the cylinder heads using the new socket head cap screws supplied in the Comp Cams kit. Be sure to rotate the trunnion so that the flat recess side of the trunnion's bolt hole faces upward (the socket head cap screw must seat against this recessed flat side of the hole). Tighten the cap screw to 22 ft-lbs, following the OEM installation procedure.

Install the new upgrade kit using a press or rocker arm vise tool. It's critical to smoothly draw the OEM bearings out and draw the new bearings in. Do not try to bang the old trunnions out or attempt to install the new bearings with a hammer.

During assembly, the kit's washers provide positive stops during the press-in of the bearings.

Using the Summit kit's presser (black piece) push the bearings and trunnion out of the rocker. The presser has a shallow counterbore that aligns with the stock bearing. Summit's specialty fixture is shown on a bench vise. Push the old trunnion completely out of the rocker along with both OEM bearings.

Using the Summit kit's mandrel, place the new bearing onto the mandrel and then the aluminum alignment sleeve. This aligns the new bearing with the rocker arm body. Push a new caged bearing into one side of the rocker.

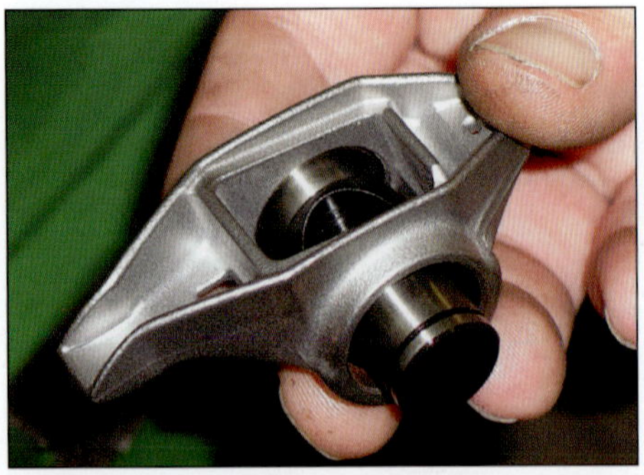

Insert the new trunnion so that one end engages the newly installed bearing. With the new trunnion in place, insert the next bearing onto the trunnion (it slips on). Make sure that all bearings are installed with the stamped lettering side facing outward!

Using a bench vise with Summit's specialty fixture to secure the opposite side, press the remaining bearing into the rocker, so that the outer face of the bearing is flush with the outer surface of the rocker body. Apply a bit of engine oil to the outside of the new bearings to ease press-in.

Here the second bearing is in place on the trunnion.

Before installing the circlips, verify that the trunnion rotates freely and that a small amount of endplay is present (.004- to .008-inch endplay). Using snap-ring pliers, install one circlip onto each end of the trunnion. Rotate the trunnion and inspect to verify that each circlip is fully seated.

Place one of the kit's washers over the exposed end of the trunnion, against the outer face of the new bearing.

This overhead view shows a fully assembled rocker with Comp's trunnion upgrade. Note that one side of the trunnion-mounting hole is recessed, that is, countersunk with a flat surface. This recessed side of the hole must face upward during rocker installation; the opposite side of the trunnion is round, to seat onto the rocker stanchion.

Depending on the version of your LS cylinder heads, some have integral rocker pedestals while others require separate pedestal rails. This 5.3L LS has rocker rails. Hand-snug a couple of rocker bolts to center the rail before fully tightening. Also, be sure to apply high-pressure assembly lube to all valve tips, rocker pushrod cups, and rocker valve tips.

If using aftermarket ported cylinder heads, check to see if the intake rocker boltholes are open to the intake runners. If so, apply thread sealant to the intake rocker bolt threads. OEM head intake rocker boltholes are generally not open to the runners.

Pushrods

Pushrods are often taken for granted by novice engine builders; they need to meet the loads and performance demands of a particular engine. In many cases, if someone is building, say, a small-block Ford engine, they tend to simply order the replacement size (diameter and length) listed for that particular engine, without any regard to valvetrain geometry. Many folks assume that "if it's listed for my engine, it must be the right one."

Pushrods are exposed to a number of extremes in terms of shock, vibration, and deflection. Choosing the correct pushrod in terms of material, diameter, and length is critical in order to optimize valvetrain efficiency.

Pushrods transmit the reciprocating motion of the valve lifters to the rocker arms. It may sound simple, but there's more to the story. Although transferring motion that's created by a high-performance camshaft profile, pushrods are exposed to extreme shocks and vibrations as the valves are forced to open and slam shut (in high-performance engines, via high or extreme valvespring pressures). The more aggressive the cam and the higher the spring pressure (and the higher the engine speed), the greater the stress experienced by the pushrods.

Check for Runout

Before installing any pushrod (new or used), check it for straightness. This can be done in several ways. You can slowly roll the pushrod across a piece of flat glass and look for a gap between the pushrod and glass surface. If the pushrod doesn't roll smoothly across the glass (wobbles or bumps), it's bent and must be discarded.

You can chuck the pushrod in a drill press and spin it. If it wobbles, it's bent. Be sure to eliminate any variable, such as a drill press with spindle or chuck runout or a pushrod that isn't chucked squarely. Be careful not to overtighten the chuck, which damages the pushrod wall surface.

You can use a pushrod checker. This is a bench-top stand that positions the pushrod horizontally. As you slowly rotate the pushrod, a dial indicator monitors runout at the pushrod's center. Pushrods should have zero runout. If bent, it's scrap.

Pushrod Length

Even if the valvetrain theoretically requires standard production length pushrods, don't assume they are correct. Test install the heads that are fully torqued to spec and have exactly the type of head gaskets used in final assembly. Using a pushrod checker, individually measure for optimum pushrod length at every intake and exhaust location. Record the results and, if necessary, order custom-length pushrods. It is important

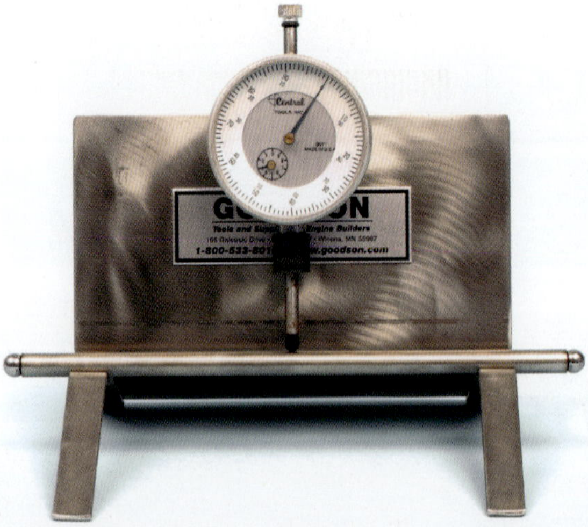

An easy and accurate way to inspect for pushrod runout is with a pushrod-checking stand with a dial indicator. Here we're using a checking stand from Goodson Tools.

to greatly minimize or eliminate a tolerance range. Will this make a difference in the final result? Maybe or maybe not. The point is to eliminate as many variables (however small) as possible.

Each pushrod length is critical in obtaining optimum valvetrain geometry. The goal is to allow the rocker arm to contact the tip of the valvestem at the center of the valvestem tip during rocker arm operation. Whether you're rebuilding an engine to OEM stock specs or building a custom race engine, never assume that off-the-shelf pushrods are the correct length.

Stock replacement pushrods may function in a stock engine, but that doesn't necessarily mean that the center of the valve tip is in the optimum position. If the pushrod is too short, the rocker arm may concentrate its contact at the outboard area of the valvestem. If it's too long, contact may be concentrated too far inboard on the stem tip. If the rocker-to-valve is off center, it can result in side loading the valve, inducing an off-center path of the valvestem within the guide. Although this may not be critical in a low-RPM grocery-getter, it can eventually cause premature guide wear or other problems under high-load, high-RPM use.

Basically, when the cam lobe is at its peak (maximum lift), the rocker arm tip should be centered on the valvestem tip. I'm focusing on optimized high-performance engines here, not by-the-numbers rebuilding a stock low-performance engine, so you must be aware of the importance of optimizing valvetrain geometry.

A number of engine modifications can alter this geometry. One is running high-lift camshafts, having a smaller base circle, which lengthens the distance from lifter to rocker, which relocates the rocker-tip-to-valve contact path). Another is milling the cylinder heads and/or block decks, which shortens the distance, requiring shorter pushrods. Running longer valvestems is another mod. A number of variables can throw the rocker arm contact path at the valve out of whack.

In order to measure and check for pushrod length, the camshaft and lifters must be installed in the block (solid lifters should be used for checking, even if you intend to run hydraulics). Although you can use your real-world springs, this job is much easier if you use light checking springs (available from any cam and spring manufacturer, such as Summit or Jegs. With the valves and springs installed, install the cylinder head using the exact type and thickness of head gasket that you intend to use during final assembly. Torque the head to spec.

Checking Pushrod

Use a special pushrod to check length. This is an adjustable-length pushrod specifically designed for determining pushrod length. These are available in a variety of lengths and come in two- or three-piece construction with threaded

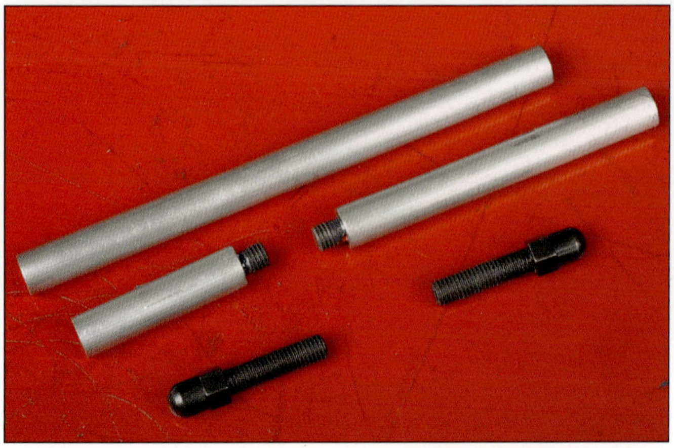

Checking pushrods are available in a variety of designs and lengths. This checker has a three-piece shaft and two thread-adjustable tips.

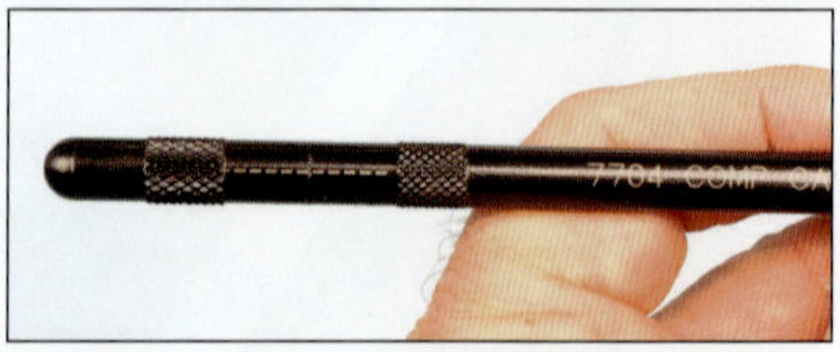

Some checking pushrod designs (such as this one from Comp cams) have reference marks laser-etched onto the body on each side of the adjustment parting line. When adjusted to flush with the lines matched, the checking pushrod is at the minimum length listed by the manufacturer.

When measuring for pushrod length, switching to light checking springs eases the task. Many adjustable checking pushrods don't withstand high valvespring pressures, especially when adjusting to lengthen. Using light checking springs makes your measurements more accurate.

If you temporarily install a light checking valvespring, be sure to retain a hardened spring seat (if applicable). If you intend to use lash caps on the valve stem tips, have these in place as well. Duplicate the same distance dimensions that will appear at final assembly.

Here a checking pushrod is installed between the lifter and rocker arm. Make sure that both ends of the pushrod are seated in their respective locations: in the center of the lifter and centered at the rocker cup or adjuster.

Before installing the rockers, clean the valve stem tips and paint them with a black marker. This provides a witness mark when the rocker sweeps across the valve tip.

If you're using guideplates, test fit the plates to the head and check pushrod clearance. Guide plates usually offer some adjustment via the stud holes to achieve optimum pushrod centering. Be sure to check and polish the pushrod slots.

Also inspect the rocker tip location's vertical center relative to the valve stem. The rocker tip should be as close to the center (left to right as seen here) as possible, to maximize the contact footprint.

With the target cam lobe at its base circle (with the lifter on the base circle), adjust the valve lash to spec.

Watch the rocker-to-valve contact path as you rotate the camshaft through its complete lift cycle.

adjustment. Some are laser etched with reference marks (instructions may say that one full turn is equal to .050 inch in length) and the minimum length.

With the checking pushrod fully collapsed (threaded to its shortest length), you can determine final length by counting the number of revolutions as you adjust the pushrod (increasing length).

In the midst of the lift cycle, the rocker tip should contact the center of the valve tip.

After rotating the camshaft through a full lift cycle, remove the rocker arm, and inspect the witness mark on the valve stem tip. This mark shows that the rocker tip concentrated its contact in the center area of the valve during the lift cycle. If the mark is too far inboard, the pushrod needs to be shorter. If the mark is outboard, the pushrod needs to be longer. If needed, re-adjust the checking pushrod and check again. It may take a few tries before finding a nice center.

If, for example, the short length is 7.000 inches and you rotated it three full turns to achieve proper length (assuming one full turn equals .050 inch), add .150 inch to the starting length of 7.000 inches. This indicates the need for a 7.150-inch-long pushrod.

Regardless of the style of checking pushrod, you can use a digital caliper to measure the adjusted length with the checking pushrod. This means that you need a long caliper (with a range of zero to 10 inches or so).

Perform pushrod length checking one pushrod location at a time. Ideally, for blueprinting, you perform pushrod length determination at each intake and exhaust location. By performing the measurement check at only one cylinder, you are assuming that the remaining locations require the same length. Using the same length for all intake positions and all exhaust positions is acceptable for most builds, but with blueprinting you are attempting to eliminate all variables. In the case of pushrod length, variables could include camshaft bore position and parallelism to the decks, valve length, rocker arm length and ratio variances, and camshaft lobe variances.

Mark the valve stem tip with a marker. With the pushrod and rocker arm in place, roll the cam through a full cycle. Remove the rocker arm and check the witness mark that indicates tip travel. The width of the mark should be .080 inch or less. (Photo Courtesy Trick Flow Specialties)

If your intent is to perform a blueprint build, each pushrod location should be checked for absolute verification. In most cases, the same-length intake pushrod is ideal for each intake location, and the same-length exhaust pushrod is ideal for all exhaust locations. But by checking each, you're working with a quantified approach instead of making an assumption.

Clean the valvestem tip to remove any oils. Using a black marker, paint the entire valvestem tip surface. This provides a witness mark during the check (when checking for rocker arm to valve centering).

Rotate the cam so that the lobe is at its base circle (the shortest lobe area). Adjust the checking pushrod to its shortest length, insert the checking pushrod into the lifter, and install the rocker arm. Adjust the checking pushrod (increasing length) until the rocker arm tip contacts the valvestem tip with zero clearance (the valve must remain closed, with no pressure pushing the valve down), then

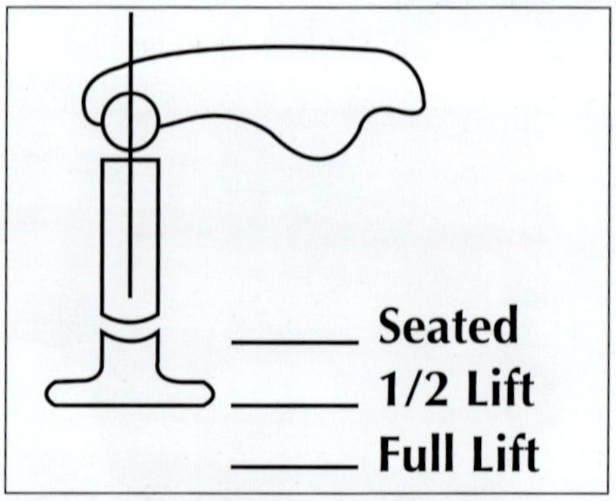

When using roller-tip rockers, note roller position on the valvestem tip at one half of net lift. The centerline of the roller tip should be located at (or very close to) the center of the valvestem tip. This verifies that the roller tip has an equal amount of travel on each half of the valvestem tip. (Illustration Courtesy Trick Flow Specialties)

adjust for the valve lash that you intend to run (check your cam card info for valve lash recommendation).

Verify that each end of the checking pushrod is properly seated in the lifter and in the rocker. If the checking pushrod is double-nutted, tighten to lock-in the current length. Double-check for lash and re-adjust if needed. If you plan to use hardened lash caps, be sure that the caps are in place during the checking phase.

Rocker Arm Sweep

Once you have this base measurement of pushrod length (zero lash with valve closed), check for rocker arm sweep across the valvestem tip. Rotate the camshaft as you watch the rocker arm tip contact at the valve, as the valve opens and closes. When the valve is closed, the rocker arm should contact the inboard side of the valvestem tip. As the valve opens at half of maximum lift, the rocker arm should sweep across the center of the valvestem tip, moving slightly outboard.

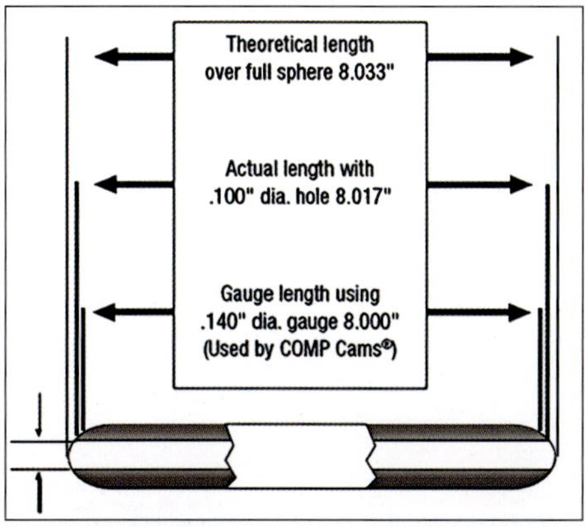

This illustration from Comp Cams shows the relative difference in length measurements. Note that the actual length, which is what you'd measure if using a long caliper, does not account for the full tip radius. In most cases, when you order custom pushrods simply tell the pushrod manufacturer that you're supplying the actual length. They can then adjust for the theoretical length. (Illustration Courtesy Comp Cams)

A jam nut is provided. Once the length is adjusted, the jam nut may be snugged to retain the adjusted length.

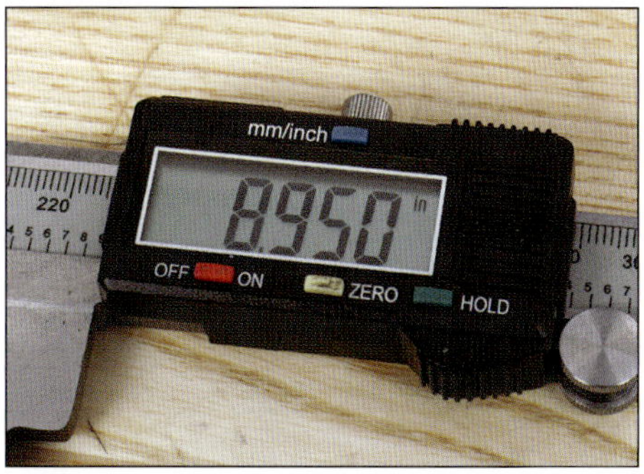

This particular checking pushrod was adjusted to 8.950 inches. After two additional rechecks, all results achieved the identical length, so I felt good about ordering that length. It may take a few more minutes, but rechecking any measurement is always a good idea.

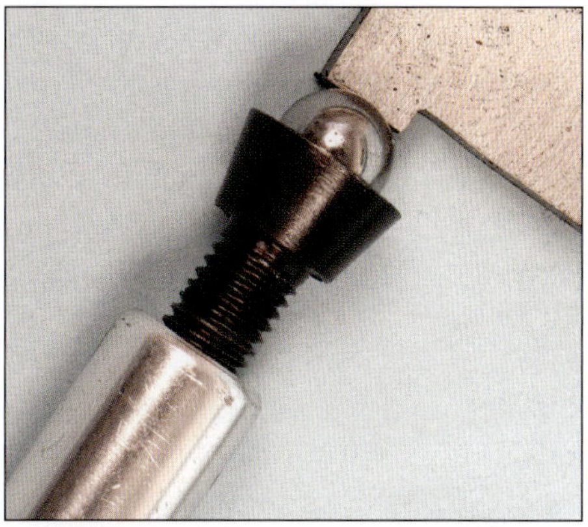

With the steel ball placed into the pushrod cup, measure the checking pushrod length with the caliper contacting the ball.

Here I'm using a 0- to 12-inch digital caliper from Goodson Tools. I rarely use this long caliper for anything except measuring pushrods, but when I need it, it's handy. Just in case you bumped the checking pushrod during handling, or just in case you contacted the pushrod with the caliper improperly, record the first length measurement and perform the length check once again.

Since it is difficult to measure to the bottom radius of the pushrod cup, insert a steel ball that is small enough to rest on the cup radius.

In order to obtain an accurate pushrod required length when a ball is inserted into the pushrod cup, measure the ball diameter, and subtract that diameter from the initial length (the length with the ball in the cup). For example, if the checking pushrod measured 7.312 inches and the ball diameter is .312 inch, the actual pushrod length (from lifter ball end to upper cup base) is 7.000 inches.

During the last half of the lift cycle, the rocker should again sweep across the valve center.

After rotating the cam through a couple of lift cycles (and carefully watching the rocker-to-valve contact sweep), bring the cam lobe to its base circle (zero lash) and remove the rocker arm. The witness mark on the valve tip should be concentrated at (or very close to) the center of the valve tip. The witness mark should measure about .080 inch or less in width (thickness of the mark/tip travel).

Regardless of which style checking pushrod you have, remove the rocker arm, and carefully remove the checking pushrod without disturbing the current length.

On a clean work surface, measure the pushrod length, using a caliper, and record the length. For safety, re-install and perform a second check to verify your measurement.

When using a checking pushrod, measure from the lifter contact to the rocker arm contact. If the style of pushrod for the engine has a ball end at each end, simply measure from ball tip to ball tip.

If the top end of the pushrod is cupped, measure from the lower (lifter) ball end to the base of the top (rocker) cup. Insert a precision steel ball into the pushrod cup, and measure from the lower ball tip of the pushrod to the top of the ball that you inserted into the top cup. To obtain the correct pushrod length, subtract the ball diameter. The ball must be small enough in diameter to properly seat into the base of the pushrod cup (depending on cup diameter, this might require a 1/4- or 5/16-inch ball).

Measuring Pushrods

Make flat contact with the caliper on each end of a pushrod that has oil holes. This allows you to measure the actual length. However, this does not include the theoretical length, since flat-anvil calipers don't capture the full radius of the tip due to the diameter and chamfer of the oil holes.

You measure at the small flat across the oil hole. Pushrod manufacturers use a special gauge to measure the exact length that includes the full radius. Because you probably don't own this special gauge (and different manufacturers use different gauges), simply tell the pushrod manufacturer that you measured actual length. They can then determine the finished-product length.

The theoretical length assumes that the pushrod has no oil hole and that the tip radius is complete. To achieve the theoretical length you'll have to add a little when you measure the actual length of an oiled pushrod. For example, in the case of a 5/16-inch-diameter pushrod, you add approximately .017 inch.

If you measure the overall length of a cup-type pushrod, you get a false reading, and the subsequent pushrod will be too long. Remember to keep a record of each pushrod length by location. Once you have obtained the measurements for all pushrod locations, chances are, depending on the style of engine, you need one length for all intake valve locations and a different length for all exhaust valve locations, or one common length for all. On rare occasions, you may need specific lengths for each valve location.

After you've determined the pushrod length for your engine, you can purchase off-the-shelf pushrods of the required length or order custom-length pushrods if your length is not available.

Several valvetrain parts manufacturers offer custom-order pushrods (Crane, Comp Cams, Crower, Trend, Manton, Scorpion, etc.). Simply tell them what diameter you need (5/16-, 3/8-, 7/16-inch, etc.), what wall thickness you want (choices are generally .080 or .100 inch but are available up to .188 inch depending on application), and what length. Off-the-shelf pushrods are typically available in length increments of .050 inch

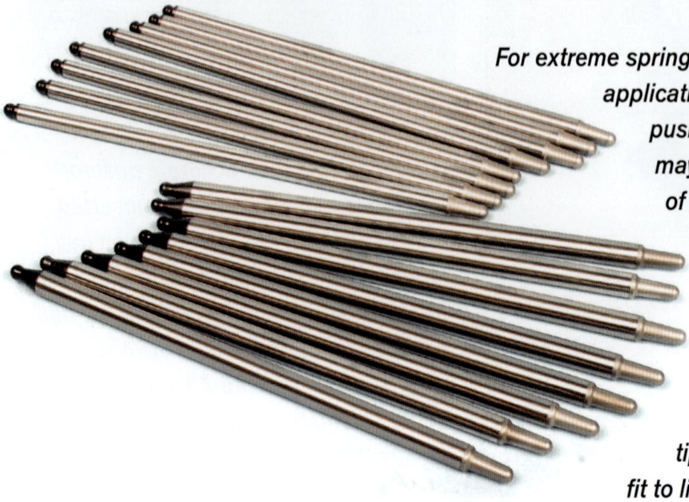

For extreme spring pressure, lift, and rpm applications, large-diameter pushrods with tapered tips may be required. A group of intakes and exhausts made by Manton are shown here. The larger diameter provides added resistance to deflection, while the tapered tips provide clearance fit to lifters and rockers.

This 3/8-inch-diameter pushrod from Comp Cams measures 7.400-inch length with .080-inch wall thickness. It is laser-etched for easy identification.

(.050 inch shorter than stock or longer than stock). Custom-order pushrods are commonly available in .050 inch increments, but some manufacturers offer even tighter increments. Also tell them what end style you need (ball or cup). Generally, custom pushrods can be delivered in as little as two days to two weeks.

Obtaining the optimum pushrod length helps to eliminate a few variables for valvetrain geometry, and aids in extending the life of the valvetrain and in optimizing valvetrain performance to maximize power output. Incorrect pushrod length can quickly cause a number of problems, including excessive and premature valveguide wear, reduced valve lift, coil spring bind, side loading on the valvestems, poor valve-to-piston clearance, and even rocker-arm-to-retainer clearance issues. Generally, longer pushrods decrease rocker arm tip travel. If you can't achieve tip travel of .080 inch or less after trying several pushrod lengths, you may need to switch to another brand of rocker arm and start over.

When checking pushrod length with roller-tip rocker arms, inspect roller tip position when the valve is at one half of its net lift. Measure with a dial indicator. Ideally, the centerline of the rocker arm roller tip should coincide with the centerline of the valve at one half of net lift (this provides an equal amount of rocker arm tip travel on each half of the valvestem tip.

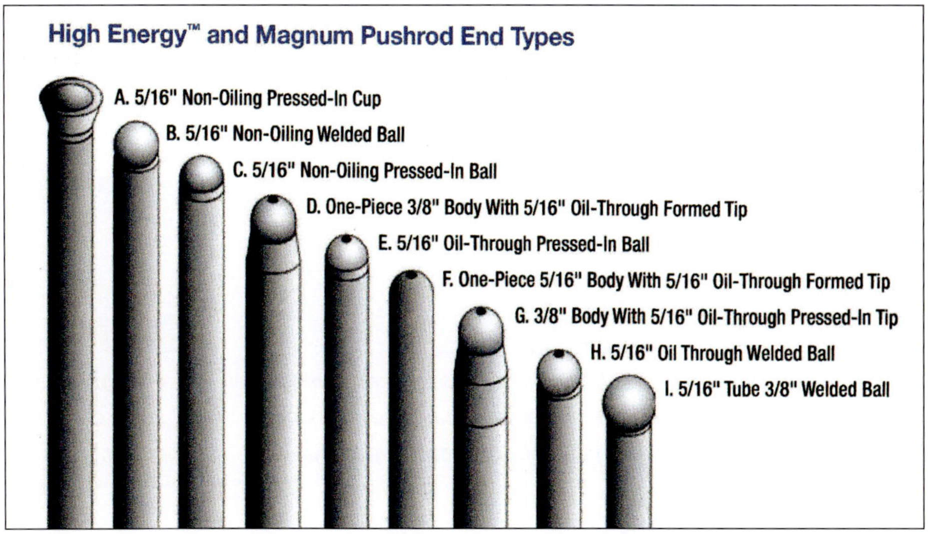

Pushrod tip styles differ depending on the application. Shown here are examples of Comp Cams magnum series. (Illustration Courtesy Comp Cams)

Guideplates

Pushrod guideplates limit the deflection of the pushrods during engine operation. If your heads require them, don't just take them out of the box and slap them onto the heads. Closely examine the edges of the pushrod slots.

Using an abrasive cylinder on a die grinder or similar tool, carefully deburr the edges of the slots to eliminate any sharp edges. Finish by polishing the slots with a fine emery cloth (you can wrap emery cloth onto an appropriate-diameter bolt and hand-lap the slots). When using guideplates, make sure that the pushrods have adequate surface hardening. Make sure that you order hardened pushrods. When checking for pushrod clearance (at the guideplates, through the head), don't get carried away by overclearancing the passage(s). Too much clearance allows more pushrod deflection. Generally speaking, a clearance of .010 inch is a safe minimum. When using guideplates, you must run hardened pushrods to avoid premature wear issues.

Pushrod Diameter

Run a pushrod with the largest diameter that the engine accommodates because it reduces deflection and increases valvetrain stability. However, that doesn't mean that you necessarily should move from a stock 5/16- to a 7/16-inch diameter. Today's pushrods are comprised of advanced materials and are available in wall thicknesses to enhance stability. I'm simply saying that if a 3/8-inch-diameter pushrod easily fits, it is preferable to a 5/16-inch, etc.

I recently built a 600-hp/604-ft-lb Pontiac 501 with 5/16-inch Trend pushrods and had no valvetrain issues at all. Of course, this engine was only being ramped up to about 5,600 rpm at peak. Don't be overly concerned about pushrod weight, since the pushrod is on the slower side of the valvetrain. When it comes to weight savings, it's more important to consider the faster side of the system (rockers and valves).

Pushrods are exposed to forces that try to make them deflect (bend) and that generates unwanted valvetrain harmonics. This is because of the eccentric loading that results from the angular load created as the rocker arm moves through its arc. Materials and design aside, the longer the pushrod, the more beneficial it is to use a larger pushrod diameter (theoretically, a large-diameter tube deflects less than a small-diameter tube). In a recent 632-ci build with extremely high lift and extreme spring pressure, I was forced to go to a combination of 7/16- and 1/2-inch pushrods, since lengths were in the high-10- to low-11-inch-length range.

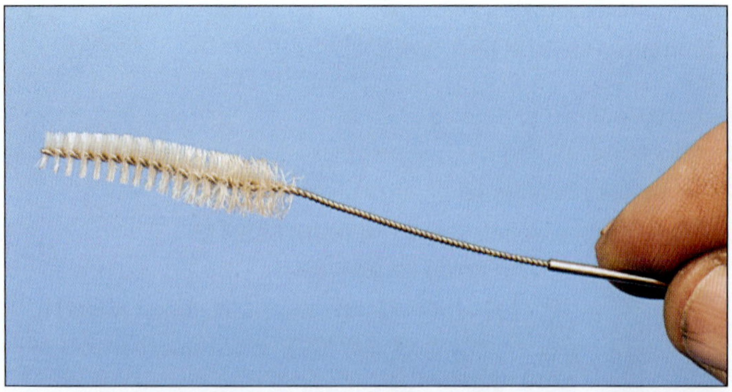

If your pushrods have through-body oiling passages, regardless of new or used pushrods, be sure to verify oil passage cleanliness. Small tapered pushrod cleaning brushes are readily available.

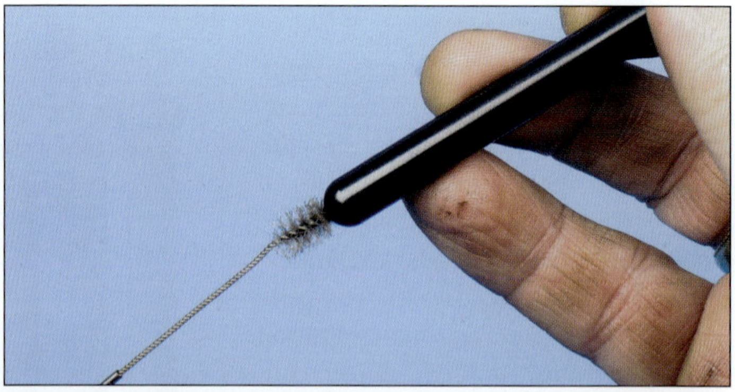

Use a bit of fast-evaporating solvent and a small pushrod brush and clean out the entire oil passage. Even new pushrods may contain oils or small contaminants from the manufacturing process.

Pushrods are available in straight (same diameter from end to end) or tapered designs. Tapered pushrods are available with a single taper (tapered at one end) or double taper (tapered at both ends). Tapered ends are often needed for clearance at lifters or rockers, and heavy-wall tapered designs also provide (theoretically) increased strength and decreased harmonics, especially in high-lift/ high-RPM applications.

Pushrod Wall Thickness

It's best to consult with the pushrod manufacturer when ordering custom pushrods. In very general terms, about .080-inch wall thickness should be adequate for most high-performance street engines and even many race applications. However, when building a race-only engine that experiences high lift and high-RPM use, moving up to .120 inch or thicker is prudent. Again, don't be concerned with pushrod weight; it's on the slower end of the valvetrain. Trying to save weight by going to a smaller diameter and/or thinner wall isn't going to gain anything, and may result in excessive pushrod deflection.

By moving to a single-taper pushrod (thicker area close to the lifter), you reduce deflection even more. This is of more benefit when using roller lifters, high-ratio rocker arms, and multiple valvesprings at high RPM. However, you can't run a tapered pushrod if you're using guideplates. With guideplates, you must run only straight pushrods.

Also, it's common for stock pushrods to be manufactured using 1018 steel tubing. For high-stress applications (high RPM, high spring pressures, etc.), you should always move up to chrome-moly tubing (4130 or 4140). 1018 steel pushrods should not be used with open spring pressures of about 400 psi or greater.

Pushrod End Types

Depending on the specific application (engine make, model, rocher style, etc.), pushrod ends have a ball or radius at the top (for rocker arm engagement). Radius ball ends (top and bottom) can also differ in diameter, so inform the manufacturer (when ordering customs) not only of length, diameter, and wall thickness, but of ball diameter as well. Also, depending on the application, the pushrod may or may not have an oil passage.

In some cases you may wish to restrict the oil being pushed through the pushrods (in applications where excessive oil is being delivered to the valvetrain and minimizing oil delivered to the bearings). One method of restricting oil is to use pushrods with a smaller oil passage (down to about .050 inch).

Clean the Pushrods

Before installing any pushrod (new or used), make sure that it's clean, and that means both inside and out. During manufacturing and packing, oils, machining contaminants, etc. may be present inside an oil-through pushrod.

Use a dedicated pushrod cleaning brush and solvent (followed by compressed air) to make sure that the oil passage inside the pushrod is clean and not obstructed. These brushes are tapered with a small diameter (similar to rifle-cleaning brushes). They are available from major cam and valvetrain manufacturers. They're inexpensive and somewhat delicate, so I suggest buying two or more.

After solvent-cleaning each pushrod, follow up with a hot soapy water cleaning, then rinse and dry with compressed air.

S-A DESIGN

Intake Manifolds

The intake manifold provides a pathway for the intake air charge. On an engine that is carbureted or that has throttle body injection, the manifold carries the air and fuel mixture to the cylinder head. On an engine with direct-multi-port fuel injection, the intake manifold's job is primarily responsible for ducting the intake air charge. Intake manifold design can have a huge impact on engine performance, affecting runner and port shape and volume.

Following a hydrographics treatment, a high-performance clearcoat finish is placed on this intake. Even after a full day on an engine dyno and after spilling fuel during carb changes, the urethane clearcoat held up without a single blemish.

Manifold Types

Intake manifold design affects peak power and the RPM band where the engine produces maximum torque and power.

Single-Planes

A single-plane manifold (depending on overall height) can be produced with longer runners and the ability to improve alignment of the cylinder head's intake port roofs. The larger plenum area with larger cross-sectional area provides a more direct shot at the cylinders, favoring the upper RPM band. The "straighter" runners tend to slow the mixture, which aids in reducing fuel separation (fuel droplets).

If the manifold has a divider plate, you can cut it down to increase plenum volume for better top-end performance. Rather than messing with the divided plate, try installing a carb spacer (experiment with thicknesses).

Single-plane intake manifolds are generally designed to allow optimum airflow at higher engine speeds. Single-plane manifolds have a larger plenum volume area, and the runners flow more directly at the cylinder head intake ports.

Dual-Planes

A dual-plane manifold has a split plenum with an upper plenum pocket directing the charge to four select cylinders. The lower plenum pocket directs the charge to the remaining four cylinders (delivering air/fuel to cylinders firing 180-degrees apart).

The runners are generally longer (better for low-end engine torque) and of a smaller cross-sectional area (which increases velocity).

Single-plane manifolds are offered in a variety of runner lengths and overall heights. Generally, the longer the runners, the more suited the engine becomes for top-end power.

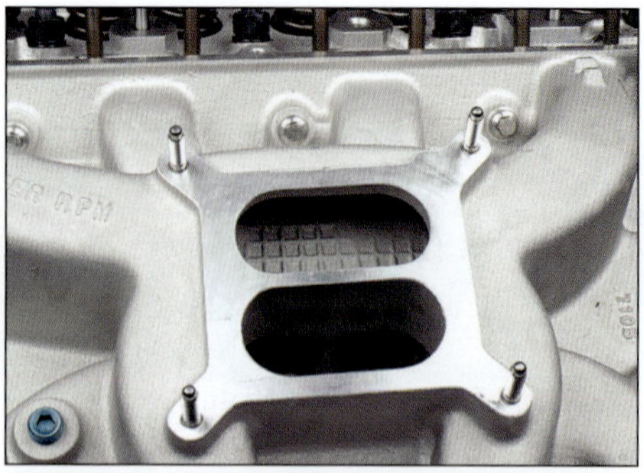

A dual-plane manifold with a divider wall. Each half of the plenum directs the fuel/air flow to half of the cylinders. Note the grid pattern in the cast floor of the plenum. This is intended to help bust up the mixture for better atomization.

A dual-plane intake manifold is usually the best choice for street driving because the design is targeted at idle, low-end torque, and low-end throttle response.

Intake manufacturers have devoted years of development and are always searching for ways to improve airflow and to maximize engine performance at specific ranges of engine speed. One example is Edelbrock's Performer RPM AirGap dual-plane manifold (developed from the Performer RPM). The runners are more isolated from engine heat and the height was increased for additional plenum volume. This improves the compromise between a dual-plane and single-plane, maintaining low-end performance of a dual-plane while increasing top-end power.

While the aftermarket offers dual-planes that focus on low-end idle quality, single-planes focus on the top end. Either design is slightly altered to broaden their ranges, and therefore multiple choices abound. I wish I had the space to devote to this subject in depth here, but suffice it to say that hood clearance issues aside, a dual-plane is likely the best choice for street cruising, and a single-plane is likely the best choice for high-RPM performance.

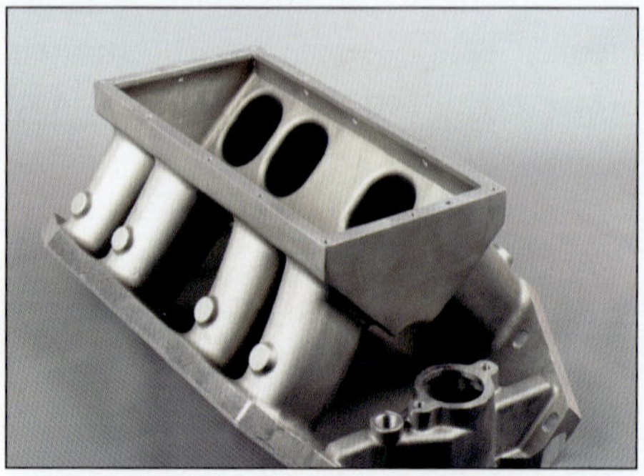

Some tunnel ram intake manifolds, such as this one from Pro-Filer, have a modular plenum top. This not only gives you a choice between using a single or dual carb setup, but the open access makes it very easy to inspect for port matching to the cylinder heads.

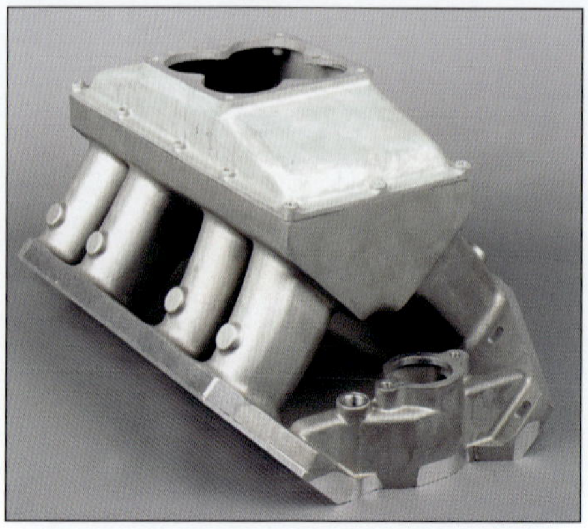

The plenum top attaches with a series of bolts and is sealed with a one-piece gasket.

Tunnel Rams

A tunnel ram intake manifold is simply a larger, longer, taller version of a single-plane manifold. Although a tunnel ram system really isn't a good choice for ordinary street driving, everybody thinks that a tunnel ram with dual carbs just looks cooler and they convince themselves that they can't live without this setup.

A tunnel ram can be suitable with careful tuning and component selection. A pair of 650-cfm carbs is a usual choice, but the tunnel ram manifold needs to

Increased flow was needed on this 632-ci Dart big-block. The Pro-Filer tunnel ram and 1,150-cfm Holley Pro Dominator carb helped this bad boy pull 1,115 hp on the dyno (all engine with no nitrous boost).

have fairly long runners in order to maintain acceptable low- and mid-range torque. These runners are typically 6 to 8 inches in length). This is actually good news for tunnel ram fans because a primary objective of running a tunnel ram intake on a street rod is to make it stick up toward the sky and thumb its nose to the world of conservatism.

A cross-ram intake manifold is another type of single-plane design, but with even longer runners than those on a tunnel ram.

Wet versus Dry Manifolds

One very basic difference between a carbureted intake manifold and a manifold used for fuel injection deals with the nature of flow. A carbureted intake manifold is referred to as a wet manifold because a mixture of air and fuel is fed through the manifold on its way to the cylinder head. A manifold for fuel injection is referred to as a dry manifold, since only air is flowed through the manifold (with injectors introducing fuel spray at the exit area of the manifold runners).

Because of this basic difference, the surface finish and shape requirements of the intake manifold runners can differ. Although many variables come into play in the specific build, the surface finish may be more important in a wet manifold. This is because you're dealing with flow, tumbling, and velocity of a wet (air and fuel) mixture (with the surface finish playing a role in how the fuel droplets can cling or atomize during flow). With an injected manifold, surface finish is likely less critical due to the absence of fuel handling.

If the manifold is designed for a tall-deck aftermarket block, there may be a gap at the front and rear manifold rails. In this case, a manifold rail spacer kit is available to fill this gap.

The rail spacers are predrilled for use with locating pins. Each rail is machined to conform to the front and rear block rails.

Port Matching

Port matching refers to achieving a proper alignment and shape between the intake manifold intake ports and the cylinder head intake ports. Depending on the cylinder head and manifold selection, a slight mismatch can occur, resulting in an airflow interruption or turbulence as the charge leaves the

Installing locating pins ensures that the spacer installs at the correct location, and that the spacer can't accidentally move during assembly or during engine operation.

Verify that the installed rail spacer matches the block deck angle.

A neat way to seal is with RTV. Before applying RTV to block/manifold end rails, carefully mask off the block face and manifold rail face (or in this example, the rail spacer). Apply the RTV, allow a few minutes for partial setup, and install the manifold or spacer. The excess RTV can be wiped off without smearing the block or manifold. Once the excess is removed, carefully remove the masking tape. The result is a sealed joint that looks neat and tidy.

manifold runner and enters the cylinder head port. Most commonly, the intake manifold port exit is then modified in order to match the exact location and shape of the head intake port. This usually involves grinding material from the manifold runner port.

The first step in port matching is to carefully measure the cylinder head intake ports and the intake manifold ports. If the intake manifold ports are wider (for example) than the intake ports on the head, first grind to widen the head ports to match the same width. Then grind the intake manifold roof and floor areas to match the head. The goal is to have the same port size at manifold and head, and to have them align with no steps or interruptions. However, it's common to size the intake manifold ports just a bit smaller (by about .015 inch or so) to accommodate any play or slop in the manifold bolt holes.

Before you start grinding, establish the fixed reference points; use the block that you intend to use with deck height finished. Install the heads with the exact type of head gaskets to be used during final assembly, or shim the heads to mimic the head gasket thickness. Also shim the space between the manifold and heads with the same thickness of the crushed intake gaskets (it's best not to use actual intake gaskets because they may interfere with precise outline scribing).

Place the manifold onto the block and heads and tap it down to make sure it's fully seated. Check to see if either the floor or roof of the ports aligns with the head ports. It's best to establish roof alignment. You can change shims to raise or lower the manifold to achieve roof alignment (just remember to use intake gaskets of the same thickness during final assembly). Slide the manifold fore and aft to check for common-wall alignment (the thin walls that separate ports).

Apply machinist's dye to the manifold deck around each port. Using calipers, measure the head ports (height and width). Using the previously established roof height as the index, use a precision straightedge and scribe a horizontal line across the entire manifold deck, in-line with the roofs. Using the height and width measurements taken from the cylinder head ports, use the straightedge and a scribe to mark the horizontal (floor) and vertical (wall) guides onto the manifold decks (you're simply transferring the head port locations to the manifold).

Select a cutter bit with the same radius as the port corner radii (if in doubt, apply machinist's dye to the head port corners and hand roll the cutter in the corner to see if the cutter makes full contact with the corner radius). Using a radius-nosed cutter bit on a controllable-speed electric die grinder (more controllable than pneumatic), begin to cut the ports edges exactly to the scribed outlines. Once each edge is cut open to the scribe line (and straight), blend the grinding into the port at a depth of about 1 inch (possibly shorter, depending on the manifold design). Do not grind beyond your scribe marks. Remember, for aluminum cutting, you must use a wide-flute bit designed for aluminum (otherwise you clog the flutes).

Once all ports have been cut, finish by smoothing and polishing with a 60-grit abrasive roll on the die-grinder tool. If you want a smoother finish, follow this with an 80-grit roll.

You can also take advantage of intake manifold gaskets in your attempt to port match. For instance, if the floor of the manifold ports are a bit lower than the floor of the head intake ports, you could move to a thicker intake gasket to slightly raise the manifold. For popular engines, manifold gaskets are available in a variety of thicknesses. Consider this before you start hacking away at the manifold or head.

If using different intake gasket thicknesses doesn't solve the initial alignment (for manifold installed height), it may be necessary to remove material from the intake manifold mounting flanges to achieve good port alignment and sealing surfaces.

Port Matching and Machining

The following are some general rules for decking and milling blocks, heads, and intakes for port alignment:

If a cylinder head is angle milled, the intake face on the cylinder head must be adjusted at the same angle, with minimal material removed. It then may be necessary to increase the diameter or create slotted bolt holes in the intake manifold for attachment to the cylinder heads.

For 90-degree V-8 engine blocks, for each .010 inch removed from the head or block deck, the intake port opening is expanded by .007 inch. In order to bring the port alignment back into position, the intake manifold needs .005 inch of material removed from each side to effectively make the manifold seat lower on the heads.

In most instances you can just divide the total amount of decking on one bank in half to calculate intake manifold adjustment.

Port matching really isn't necessary unless you're trying to optimize engine performance or encounter serious misalignment for tolerance stack-up. If you're simply toodling along on your way to a show or the local malt shop, don't fret over this.

Intake Plenum

For a carbureted manifold (or a carb-style manifold that uses a throttle body), take a look at the plenum divider walls. Remove any imperfections (casting bumps, flashings, etc.). In order to aid in airflow or fuel/air flow, address the

Before you start checking port alignment, apply machinists' dye to each port opening. This allows you to easily see any scribe lines.

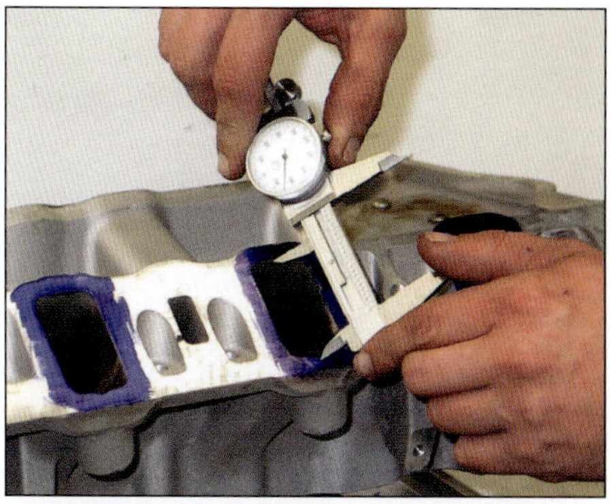

After using the port roofs as the reference points, and after measuring the cylinder head ports, the dimensions are transferred to the intake manifold.

In some cases, where the manifold ports are smaller than the head's ports, using a U-bent wire allows you to reach into the manifold, using the short vertical section to guide you around the head port, while scribing outlines around the manifold ports. Here Jim, of Fox Lake Racing Heads, prepares to scribe a lower intake manifold for a 5.0L injected Ford.

A machinists' precision straightedge is used to scribe common floor lines across the entire bank of ports.

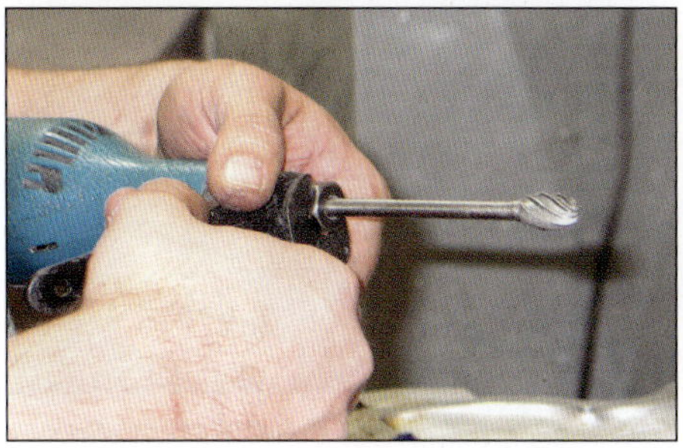

When cutting aluminum, be sure to use a cutter with large flutes designed for aluminum.

Proceed carefully. Make sure that you don't remove any material beyond your scribe lines.

After removing material around the port edges to your scribe lines, taper blend the cut inside the runner at a depth of about 1 inch.

Blend into the port to fully smooth out the surface, again, at about a 1-inch depth. If you prefer a smoother surface (more of a polish), then go over the surfaces with 80-grit.

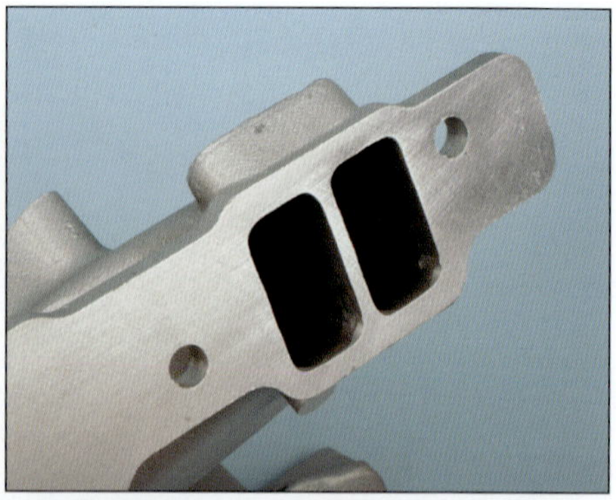

Depending on engine design/type, the manifold may have port pairs that are placed very close together. When port matching, align these divider walls to the heads first, and be careful not to eliminate the gasket sealing area.

The popular F.A.S.T. composite intake manifold for fuel-injected applications consists of a lower section (runners) and a top section that incorporates the throttle body inlet. Shown here is a unit for late-model Corvette.

Plastic intake manifolds for fuel-injection setups (OEM or aftermarket) are basically bolt-ons. Due to the weight-saving designs and preregistered elastomeric-seal grooves, you really can't modify the ports.

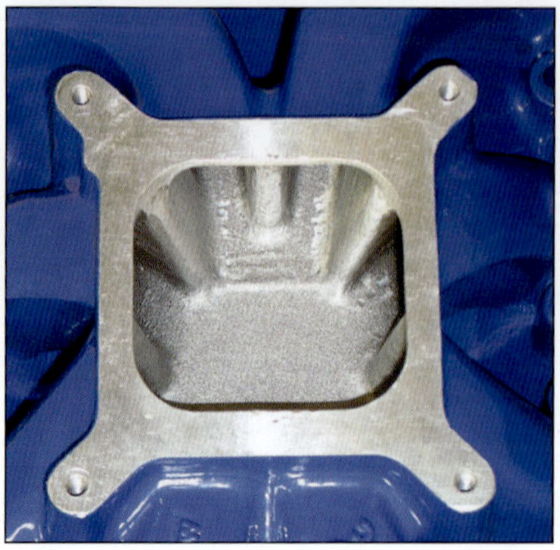

Round off the edges of the port dividers to achieve a radiused profile. There's no need to knife-edge these walls.

dividers. The dividers typically (not always) have rough or almost squared off edges. Using a grinder or abrasive roll, radius these edges to a bull-nose shape (not to a sharp knife edge). The goal is to remove sharp edges and/or abrupt surfaces that could create excess boundary layers (turbulence).

If you plan to use a carburetor spacer (to increase plenum volume for better top-end performance), apply machinist's dye to the manifold's carb mounting pad and install the spacer. If any manifold pad material extends beyond the inside walls of the spacer, scribe a line using the inside of the spacer as a template. Remove any exposed material from the carb pad to remove any obstructions to flow (match the manifold's plenum opening to the spacer). Gently blend this area of the carb into the plenum opening.

Intake Port Surface

Although a fully polished surface that looks chrome plated may look really cool, this is usually not necessary. Finishing with an 80-grit abrasive roll is adequate polishing. Polishing is more important for any sharp turns in the flow path, which is where flow speeds are the highest. The shorter the turn, the more need for polishing.

Also, the need (or benefit from) polishing can vary depending in part on the size of the manifold runners. Smaller runners can benefit more from surface polishing than larger port runners. Small-volume runners may be more sensitive to turbulence factors that can result from casting surface boundary layers.

Manifold Bolts

It's common for a novice to undertighten, overtighten, or unevenly tighten intake manifold bolts. Any of these can result in vacuum, oil and/or coolant leaks, and a warped or cracked manifold.

Always follow the bolt torque specs and the specific tightening sequence recommended by the intake manifold manufacturer. A fairly common challenge, simply due to the design of the manifold, is gaining adequate access to certain manifold bolt locations. In some cases it may be difficult to access the bolt heads with a straight socket wrench.

It is common for installers to use an open-end wrench and to *guess* at torque value. In locations were you cannot access a bolt directly with a socket wrench, the answer is to obtain an offset wrench extension. This places the wrench (at the bolt head) away from the centerline of the tool's drive head (making the effective overall length of the torque wrench tool longer or shorter).

When using an offset wrench, you must make a compensation adjustment in order to achieve the desired torque value. Otherwise, you unknowingly overtighten or undertighten the bolt. If the extension points out and away (but in-line with the torque wrench body) from the torque wrench drive, this obviously lengthens the overall tool length. If the extension is installed to the torque wrench 180-degrees (in-line with the torque wrench body but now underneath the tool), the effective length is shorter.

Use this formula so the adapter makes the wrench longer:

$$TW = (L \div L + E) \times \text{desired } TE$$

Use this formula so the adapter makes the wrench shorter:

$$TW = (L \div L - E) \times \text{desired } TE$$

Where:

TW = Torque setting on the torque wrench

L = Lever length of the torque wrench itself (from center of the wrench drive to the center of the torque wrench hand-grip area)

E = Effective length of extension, from the center of its square drive hole to the center of its wrench head

TE = Torque applied by the extension to the fastener

If you want to know where to set the torque wrench when using an adapter that alters the effective length of the wrench, you must calculate to compensate for the adapter. If the distance from the wrench drive to the center of the bolt makes the wrench longer, the final wrench setting must be adjusted to a lower value in order to compensate.

As an example, an intake manifold bolt torque value is listed as 30 ft-lbs. In order to access a hard-to-reach bolt, you may need to use a 2-inch offset extension wrench. In this case, the torque wrench measures 12 inches from the center of the drive to the center of the wrench handle. With the wrench extension pointing away from the wrench drive, this changes the distance from the center of the bolt to the center of the torque wrench grip to 14 inches (making the torque wrench 2 inches longer).

For this example, the formula works out like this:

$$TW = (L \div L + E) \times \text{desired } TE$$
$$12 \div (12 + 2) \times 30$$
$$12 \div 14 \times 30$$
$$.9 \times 30$$
$$27$$

In this example, the wrench is set at 27 ft-lbs, in order to actually achieve 30 ft-lbs.

If the wrench extension is aimed *toward* the handle (turned 180-degrees from the prior example), and you still want to achieve 30 ft-lbs of torque, you

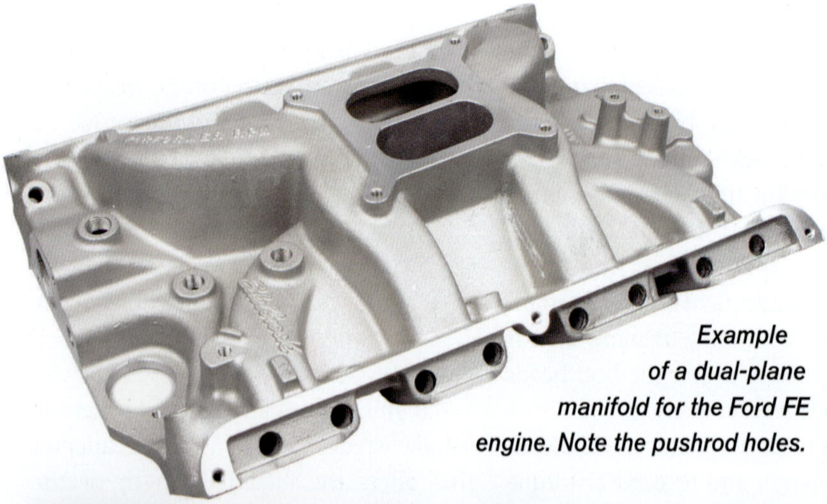

Example of a dual-plane manifold for the Ford FE engine. Note the pushrod holes.

Wrench extensions are readily available from quality tool sources. Just remember to adjust the torque wrench accordingly to compensate for the added leverage.

The vintage Ford FE big-block manifolds have pushrod holes. For high-performance applications where larger diameter pushrods and/or high-lift cams are used, these holes may require slight enlargement or elongation. This must be checked during test fitting, with an assembled short block, heads, and rockers installed.

For those situations where you simply don't have access with a socket wrench, a wrench extension makes it easy to apply the desired torque value to an intake manifold bolt, rather than just guessing at bolt tightness.

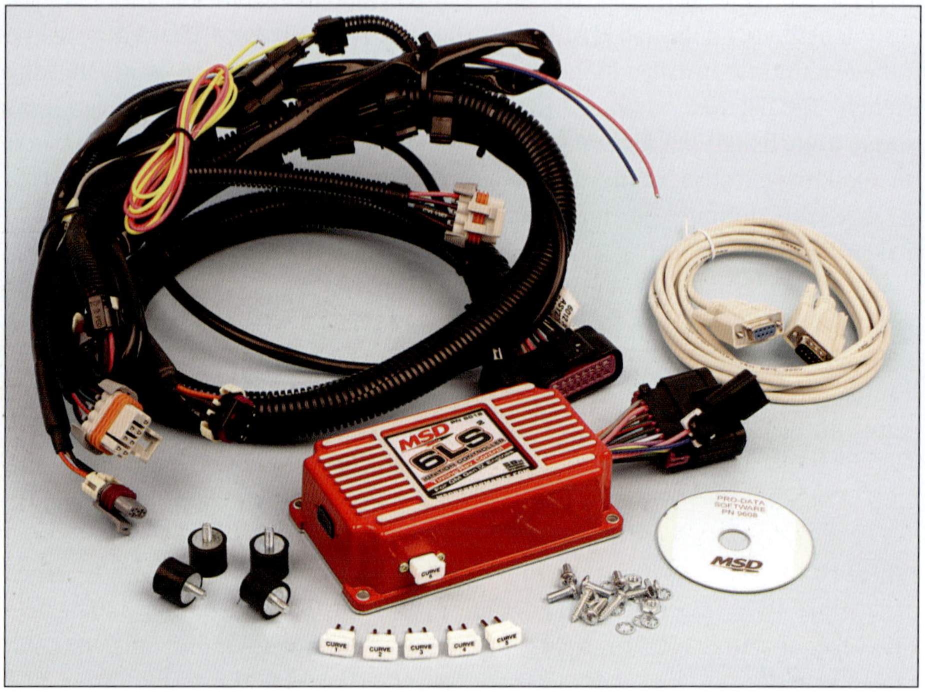

The MSD 6LS ignition module and harness makes converting an LS engine to carburetor extremely easy. There are two versions: the 6LS (PN 6010), designed for use with the 24-tooth crankshaft reluctor wheel (earlier LS1/LS6) and the 6LS-2 (PN 6012) for the later 58-tooth reluctor wheel (LS2 and others). Just make sure that you purchase the correct controller based on your reluctor wheel tooth count.

Converting a GM LS engine to old-school carburetion is embarrassingly simple, requiring only a manifold, carb, and an MSD ignition controller. No ECM required.

know that the adapter has now made the wrench shorter (because the center of the bolt is now closer to the center of the wrench handle).

For this example, the formula works out like this:

$$TW = (L \div L + E) \times \text{desired TE}$$
$$12 \div (12 - 2) \times 30$$
$$12 \div 10 \times 30$$
$$1.2 \times 30$$
$$36$$

In this example, the wrench is set at 36 ft-lbs, in order to actually achieve 30 ft-lbs.

If the adapter makes the torque wrench longer, you must *reduce* the setting on the torque wrench. If the adapter makes the torque wrench shorter, you must increase the setting on the torque wrench.

Converting LS to Carb

The Gen-3 and -4 GM LS engine was originally designed to feature electronic multi-port fuel injection. For those who prefer to run a carburetor, for street rod, custom car, or racing, the swap is relatively easy. The only components required include a carburetor, intake manifold, and ignition controller system. No on-board computer is needed. You retain the engine's ignition coils, crankshaft position sensor, camshaft position sensor, water temperature sensor, and manifold absolute pressure (MAP) sensor.

Because the carb and manifold handles fuel/air delivery, no electronics are involved. You simply need a way to time ignition. The commonly used controller for this application (and the only one I'm aware of) is the MSD 6LS controller, which includes a selection of six plug-in chips, each with its own ignition curve. Refer to the MSD instructions, pick the curve you want to try, plug it in, and go play. It's that simple. If you prefer to map your own curve, a CD is included that lets you program the curve on your PC.

Be aware that you need to buy the correct controller for the tooth count on your crank's reluctor wheel. The MSD 6LS (PN 6010) is compatible with a 24-tooth wheel (LS1/LS6 and early LS2), and the 6LS-2 (PN 6012) is for the later 58-tooth wheel (later LS2, LS7, LS3, and LS9).

Manifold Surface Protections

Okay, so you purchased a hot-dog aluminum intake that's gonna boost the pony count, and it looks way cool squatting between those trick aluminum heads. However, a few months down the road, the manifold starts looking funky. It's either scuzzing up with a white-hued film that rubs off like chalk or it starts looking brownish, as though it were rusting. How can this be? After all, it's aluminum.

If the manifold begins to turn brown (as though a light surface rust were occurring), you're not imagining this. If the aluminum manifold was *steel*-shot blasted during the manufacturing process, particles of steel may have been imbedded in the aluminum surface. Exposed to the elements, the steel begins to oxidize (if clean stainless steel shot is used, this does not happen).

If the aluminum surface begins to oxidize on its own (a result of moisture due to humidity or being exposed to water in some other manner), a white film begins to develop. Left unattended, this can lead to long-term pitting.

Penetrating Lube

In order to keep the new manifold looking, well ... new, you have a few pre-installation options for treating the surface to prevent oxidation.

The cheap (and somewhat effective) route is to carefully wipe the exterior surface with a soft steel wool pad (or red Scotchbrite pad) that has been soaked with a penetrating lubricant such as WD-40. Using a bit of effort, rub the lube into the entire outer surface, including every nook and cranny. This changes the appearance, providing a slight burnished look and ever-so-slightly darkening the aluminum, yet providing a still-attractive appearance.

Once you've applied the penetrating lube into the surface with the aid of a light abrasive pad, rinse the manifold thoroughly with hot water to remove all abrasive particles. Dry the surface and immediately re-apply the thin penetrating oil using a soft clean rag. This generally keeps the intake looking spiffy for a season or two, and helps to prevent future staining. Down the road, you can always clean the installed manifold surfaces and re-apply the lubricant with a clean rag. When the light oil is applied, don't leave it wet; dry and buff with a soft, clean rag (otherwise, the oil attracts dust/dirt particles). This may seem to be an archaic method, but it works as long as you maintain it via routine wipe-downs (detail cleaning).

Cleaning a bare (as-cast) intake manifold can be tricky if stained by fuel or road and weather elements because of the porous nature of an as-cast surface. Commercially available aluminum cleaners include OxiSolv and Evapo-Rust. It's important to follow the instructions included with these cleaning products.

Tumble Treatment

Another option is to have the manifold tumble-treated (where the manifold is gently tumbled in an immersion of small polishing stones). Depending on the size, shape, and composition of the media, this burnishing process smooths the surface to a satin, semi-polished, or full-polished finish, depending on what result you want. Although this doesn't provide a protective coating, it diminishes or eliminates casting surface texture, which makes it much easier to clean. Of course, this can always be followed up by the application of a protective film or coating.

Coatings

You can also have the manifold professionally treated or coated with a protective finish. This can be done by adding a Teflon coating (the surface darkens but prevents moisture and other deposits from sticking to the surface), a ceramic coating (which generally brightens the surface, depending on the formulation), or a powder coating.

A good powder-coating shop can provide virtually any finish you want, including clear, a color , a smooth finish, a wrinkly finish, pebble finish, etc. There are plenty of good powder coating shops around.

A race manifold (usually) shouldn't have a barrier coating (for heat dispersion), but for a street/show rod, appearance is paramount, so do what you need to do in order to maintain the beauty factor. Discoloration, stains, or oxidation can occur from weather conditions (humidity, airborne pollutants), fuel leaks, coolant leaks, etc. The proper protective coating eliminates surface oxidation and allows easy cleaning of other surface contaminants.

If a performance coating is desired, check with the major engine-component coating services. They offer a wide range of specialty coatings designed for whatever goal you have in mind, including anti-friction, heat barrier, heat dispersion, faster oil drainback, etc. These sources include (but are not limited to) Swain Tech Coatings, Polydyne, Calico Coatings, and TechLine Coatings

Concerning intake manifolds specifically, underside coatings (where the bottom of the manifold faces the lifter valley) are available that provide a heat barrier (keeping the manifold cooler) and coatings that prevent oil from clinging (for quicker oil drainback).

Some intake manifold manufacturers offer their manifolds already treated

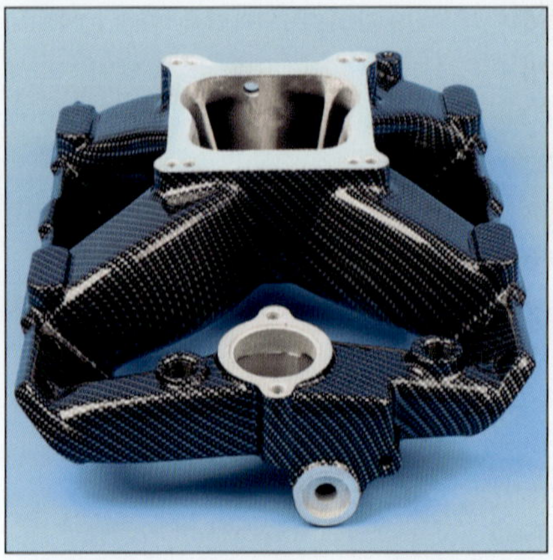

Hydrographics (wet ink film graphics) provides a simple and relatively inexpensive way to create a unique visual impact. This manifold was ground smooth, polished, then delivered to a dipping service. The techs at Dip 'N Designs (Wooster, Ohio) applied a black basecoat, carefully dipped the manifold through a water surface suspended with carbon-fiber ink film, rinsed, dried, and then clear coated the manifold. Since the graphic ink film is somewhat translucent, the black base coat was needed to achieve the desired hue.

Overhead view of the "carbon fiber" manifold on a Pontiac 455 that we bored/stroked to 501 ci.

A tunnel ram, or hi-ram, intake manifold such as the Holley Hi-Ram on this custom LS engine features long runners for added volume and air/fuel velocity, benefiting top-end power.

Using the Edelbrock dual-plane intake and a 650-cfm Holley (and a mild-cam change), we simply overbored this 5.3-liter LS iron-block to 327 ci and hit 450 hp. The manifold provided snappy throttle response and favored bottom-end and mid-range power.

Topped off with a dual-snorkel vintage-style air cleaner, the intake/carb swap from its previous fuel injected format resulted in taking advantage of current-day LS long-block technology while retaining a vintage performance look.

with some type of protective coating as standard or as an option. If yours is delivered bare, seriously consider applying some type of surface protection simply to maintain a like-new appearance.

Actual chrome plating, while attractive, probably isn't a wise choice, simply because the plating process (copper, nickel, chromium) can trap heat within the manifold, more so than other treatments. If you want a chrome-like finish, a good powdercoater can achieve this for you. The coating shop may also be able to apply a chrome-type finish in other colors than nickel-chrome.

Hydrographics

This is also called wet-dip. It involves a preprinted ink film (graphics of your choice, such as carbon fiber, camouflage, etc.). The film is laid onto the surface of water in a temperature-controlled tank. The component (in this case a manifold) is carefully lowered onto and through the ink film, much like dipping an Easter egg. The film adheres to the manifold surfaces, wrapping all contours. The manifold is then removed, rinsed, dried, and treated to a protective urethane clearcoat. For best results, the manifold's exterior surfaces need to be fully smoothed and polished prior to dipping.

Faux Fiber

I recently had a 4-barrel manifold (for a 501-ci Pontiac engine build) treated to a carbon-fiber graphic. I spent a few hours deburring the intake (removing casting flashings, casting-mold lines, etc.) and fully polishing the entire outer surface. I delivered the manifold to Dip 'N Designs. Because I wanted a black carbon-fiber appearance, they first applied a black basecoat (the film is somewhat translucent, so the undercoat influences the final hue), followed by dipping, rinsing, drying, and clearcoating. The result was spectacular. At a major performance trade show, everyone who inspected the Pontiac engine thought that the intake manifold was actually made of carbon fiber.

This graphic treatment does hold up to engine heat and contact with fuel. We ran this engine on a dyno for a full day, with no visible effects from the heat (no discoloration, no cracking, no lifting). Even with fuel spilled onto the surface during carb changes, the urethane clearcoat seemed impervious.

The only tip that I'd like to pass along deals with the intake manifold bolt-hole locations. The clamping force of the intake manifold bolt heads and washers tend to compress and raise the clearcoat around the edges of the washers. To avoid this, lightly spotface each bolt hole (flat-spot-face each bolt hole to slightly exceed the outer diameter of the washer). The spot-face doesn't need to be very deep, just enough to register the washer. The spotface should be a few thousandths of an inch larger in diameter than the washer (for example, if the washer OD is .450 inch, the spotfacing should be about .470 inch). After ink-dipping and drying, the graphics technician can then carefully mask each spotfaced recess before applying the clearcoat.

Carb Spacers

Although some builders take advantage of a carb spacer simply to provide needed clearance between the carb's fuel log and intake manifold (in cases where fit poses a problem), spacers are normally used to help tune the engine's performance band. Spacers don't add power; rather, a spacer can be used to tune the power band in the RPM range can only use generalities about each style of spacer, since each specific engine application has its own set of variables (a setup that works well on one engine might not act the same on a different engine).

Designs

Spacers are available in different designs, including four-hole, single-hole, open, combination, and with a plenum divider. There's more to choosing a spacer than simply basing the selection decision on thickness.

A four-hole spacer (four holes that align with the carburetor barrels) generally increases throttle response and acceleration, and generally increases torque by moving the power band to a lower RPM range. The four-hole design forces the column of air moving from the carb into the intake plenum to take a longer path (flowing longer), which increases air velocity.

A spacer with a single, large, opening tends to raise the power band to a higher RPM range (less bottom end but more top end). This type of spacer also has a center divider plate that splits the plenum path left-to-right. For applications where rich/lean conditions exist on the engine's left and right sides during turns, the divider plate helps to equalize fuel/air distribution for a more even feed to all cylinders. This is generally not required for street application, and is more targeted to oval track or some road-race applications. More advanced spacers are also available with tapered holes, where the taper angle has been developed for optimum performance on specific manifold/carb/cam setups.

An open spacer increases plenum volume by increasing the distance between the carb and plenum floor. The open type generally decreases throttle response.

A combination spacer has a four-hole design along with a relieved floor area (basically, a combination of a four-hole and an open style). The combination spacer design generally helps to increase throttle response, while also widening the torque and power band throughout the RPM range (sort of the best of both worlds). The top of the spacer (mated to the carb) has the four holes flush to the carb, while the underside of the spacer is relieved in a square pattern (encompassing the entire group of the four holes) to slightly increase plenum volume.

Materials

Spacers are offered in a variety of materials, including wood, plastic, phenolic, and aluminum. Stay away from wood and plastic (wood is a good heat insulator but can absorb fuel, and plastic isn't very strong and can crack).

Phenolic fiber is a good heat insulator and is a good choice. But if you plan to modify the spacer (custom porting), phenolic isn't very workable.

Aluminum (cast or billet) isn't the best heat insulator, but it's easily modified and, in the case of today's precision-machined billet and color-anodized choices, is a great choice, especially for a custom application where appearance is key (a sexy red or blue anodized spacer can add some real pizzazz).

Thickness

Aside from providing mounting clearance, spacer thickness is a tuning variable. The thicker the spacer, the more you increase plenum volume.

For the typical owner who doesn't have access to a dyno or flow bench, determining the optimum spacer thickness for a specific application involves some trial and error. Luckily, swapping spacers is easy. Remember, the shorter the spacer, the more low-end torque and power available. The thicker the spacer, the torque and power band moves to a higher RPM range.

If you do plan to play with spacer thicknesses, start with carburetor mounting studs that accommodate the thickest spacer that you have in mind. This eliminates the need to install specific-length studs during each spacer swap.

BALANCING

For any engine to operate at its peak, all vital components of the rotating assembly must be balanced. You need to perform the balancing procedure once all components have been verified for use and all primary machining and fitment has been accomplished. Don't waste time and money by balancing the crank until you're certain that *all* machining and test fitting has been performed.

Discussion about engine balancing is essentially referring to balancing the crankshaft. The factors that affect crankshaft balance include all of the rotating and reciprocating weight, so in order to balance the crank, you also need to balance the pistons and connecting rods. Rotating weight involves the crankshaft and the big ends of the rods as well as and the rod bearings. Reciprocating weight involves the small ends of the rods, the pistons, wrist pins, pin locks, and rings.

When balancing a crankshaft, the common approach is to consider 100 percent of the rotating weight and 50 percent of the reciprocating weight. Overbalancing is an approach that considers a higher percentage of reciprocating weight. Instead of weighing only a single piston, a single rod, etc., take the time to weigh *each* component in order to weight-match all pistons and all rods.

This Pro-Bal balancer features an overhead belt drive that contacts the crank's center main journal from above.

Follow basic weight-matching guidelines (use the lightest piston as the reference and remove weight from all remaining pistons to match, etc.). Even when performing an internal balance, don't assume that the zero-balance dampener or flywheel is in fact balanced to zero. Spin balance the dampener and flywheel independently as well, correcting it if needed. If, by chance, the dampener will be painted, spin it again after the paint job to verify that no severe paint-thickness inconsistencies are present that affect balance.

Engine balancing refers to balancing the crankshaft to accommodate the reciprocating and rotating forces that it encounters during operation. Reciprocating force is presented by the forces that go up and down and act upon the crankshaft centerline. This includes the weight provided by the pistons, piston pins, piston rings, and pin locks. The rotating weight attached to the crankshaft includes the rod big ends, rod bearings, and the amount of oil that clings to the rods.

Although the goal is to balance the crankshaft, you must first weight match the components that attach to the crankshaft. This means that each piston and pin must weigh the same, all rod small ends must weigh the same, and all rod big ends must weigh the same. Additional items such as rod bearings, pin locks, and rings are manufactured so consistently that by weighing a sample of each, you can assume that the rest weigh the same.

Once the pistons are weight matched and the rods are weight matched, a bobweight can be created, which duplicates the components that attach to the crankshaft. This bobweight is then secured to the crankshaft. The crank is spun on a balancing machine, and weight is then removed from or added to the crankshaft counterweights.

The goal is to eliminate unwanted forces at the crankshaft geometric centerline, creating a rotating and reciprocating package that allows the crankshaft to run as smoothly as possible.

Internal versus External Balance

Crankshafts may be balance corrected internally or externally. In either case the attached parts (pistons, rods, etc.) must first be weight matched. When a crankshaft has internal balance, balance weight correction is performed on the crankshaft itself, adding or removing weight from its counterweights. This requires a zero-balanced dampener and flywheel. The dampener and flywheel for this example are designed with a zero balance, so neither affects crankshaft balance. This is one advantage of internal balance. The dampener and flywheel may be changed in the future without affecting crank balance.

When a crankshaft is designed for external balance, the offset-weighted dampener (or balancer) and weighted flywheel must be attached to the crankshaft during balance spinning. Corrections are then made by adding or removing weight at the crankshaft's counterweights.

Two common types of crankshaft dampeners include the rubber/elastomer type and the viscous type. An OEM-style elastomer/rubber dampened balancer is designed to reduce crankshaft harmonics at a pre-engineered frequency range. A viscous dampener has a cavity partially filled with a viscous gel that is designed to maintain balance at all engine speeds.

If you are balancing an externally balanced crankshaft, where you need to include both the flywheel and dampener on the crankshaft during balancing, using a viscous dampener requires a different approach. Since the viscous dampener constantly works to maintain balance, this can mask slight crankshaft imbalance conditions. If the viscous dampener is designed for use on an externally balanced crankshaft, it was likely designed as a two-piece unit. Remove the viscous balancer ring from the hub and mount only the hub to the crankshaft during balancing.

Although some original equipment engines were designed for external balance, it is possible to internally balance any crank. The shop equipment required to perform crankshaft balancing includes: a professional-level digital scale, which is used to weigh individual parts including pistons, pins, rings, locks, rods, and rod bearings; a connecting-rod support stand used in conjunction with the digital scale; an electronic spin balancer designed for crankshaft balancing; an overhead drill press to drill holes into the counterweights; and a set of bobweights, which are used to simulate the weight of the pistons, rods, etc. Add a flywheel adapter hub to allow balance corrections to a flywheel that has been serviced or is in question.

While a new aftermarket crankshaft may be near-zero balanced at the factory, this provides only a starting point. Consider the mass of everything hung onto the crank: rods, rod bearings, pistons, wrist pins, pin locks, and piston ring package. Any crank must be balanced with rotating and reciprocating weight to create a balanced assembly package. Regardless of crank origin (OEM or aftermarket, new or used), the entire rotating/reciprocating package must be balanced. If at any point you change any of the mass-weight components from a previously running engine (such as pistons or rods), the crank must be rebalanced.

Weight Matching

The first order of business is to determine the weight of the pistons and connecting rods. You need to know how much each part weighs, in order to create a bobweight card, and to make sure that all piston and rod weights are identical.

Your digital scale must be absolutely clean. Astute engine builders always keep a dust cover on their scales when not in use. The scale should also be located in an area free of moving air. These scales are so sensitive that a breeze caused by a nearby door opening can easily create a false reading.

After pressing the ZERO button to calibrate the scale, weigh each piston and label with a marker pen. With the entire set of pistons weighed, you then examine the set to look for variations. If weights do not match, the lightest piston becomes the reference point, so you remove material from the remaining pistons in order to match the lightest.

If weight must be removed, do this carefully to avoid compromising the structural integrity of the piston. With the piston secured in a dedicated piston pin vise, material can be removed (by milling) at the underside of the pin boss. However, today's high-quality high-performance pistons are manufactured with such a high degree of precision that the need to perform weight correction is rare. When all pistons in the set weigh the same, record the weight of a single piston on the bobweight card.

Pin-to-Piston Weight

Next note each piston pin weight. Again, it is rare to find a set of quality piston pins that are not already weight matched. You should not consider modifying a piston pin, but if any slight variation in weight is found with the piston set and the pin set, before attempting to remove weight from any of the pistons, you can try to match pins to pistons in an effort to correct any weight deviation.

By mixing and matching combinations of heaviest and lightest, you may be able to create piston and pin sets that create a closely matched combined weight. If so, then each pin must remain with its piston all the way to final assembly.

Connecting Rod Weight

Next weigh each connecting rod. Place a specialty rod stand that accommodates the rod small end onto the digital scale and TARE (zero) it in order to ignore or negate the weight of this stand.

Place another rod stand next to (but not on) the scale. This stand supports the rod big end. Each stand is adjustable for height. Mount the rod to the stands in a horizontal manner and adjust so that the big-end center is level with the small-end center. Note and record the small-end weight.

Then reverse the rod in order to weigh the big end. Again, the rod must be level. Rod stand designs differ; if a different stand or stand component now rests on the scale, the scale must again be TAREd to negate the weight of the stand. Record the big-end weight.

After recording all rod weights, compare and determine if weight correction is needed. Today's high-quality performance aftermarket forged and billet rods are usually extremely well weight matched, so chances are no correction is needed. Older-design OEM rods may have weight bosses at the small end and at the cap, allowing material removal. Today's rods are typically free of these bosses, and you may not have any available excess material for removal.

If you have a set of aftermarket rods that are not weight matched, you're better off contacting the manufacturer to discuss the problem, which may mean exchanging for another set. Once all rods have been weighed and recorded, remove the stand from the scale and ZERO the scale.

Weigh and record one piston's set of pin locks (if your pistons have full-floating pins).

Weigh and record the weight of one piston's full set of rings (including the oil ring support rail, if so equipped).

Finally, weigh and record the weight of one rod's pair of bearing shells.

Bobweights

Bobweights have heavy-duty aluminum V-clamps that fasten together onto the crankshaft rod journals. Each side of the V-clamp has weights. One half of the total bobweight is placed on each side of the journal. Using the information recorded during component scale weighing, the bobweights are assembled on the digital scale to duplicate the determined weight. The total weight of the bobweight must match that

Each component (rods, rod bearings, pistons, pins, locks, and rings) is weighed on a precision digital scale. The first step is to place the weight of the on-scale stand on the scale and tare the scale. This will return the scale reading to zero with the bare stand to remove the weight of the stand from the equation. When weighing a rod big end, the rod must be mounted with the bore centers horizontal and parallel (note the bubble level on the rod big end). The small end is supported on a stand off the scale, while the big end is supported on a stand that rests on the scale.

The small end of the rod is weighed the same way as the big end. The on-scale support stand is first tared, and the rod is mounted horizontally.

of the bobweight card, which includes the V-clamps, adjusted weights, and nuts that secure the bobweight halves together.

Adjusting

The weight is adjusted by adding lead shot to capped barrels that attach to the V-blocks or by adding weighted discs onto threaded rods on the V-clamps (designs differ). To create a bobweight on a V-8 or 90-degree V-6, 100 percent of the rod throw's rotating weight (the big end of the rod and the rod bearing) and 50 percent of the reciprocating weight is factored in, which includes the pistons, small rod ends, rings, pins, and locks. That means both rod big ends and both sets of rod bearings, but only one piston/pin/ring set, is factored in for the reciprocating weight.

For other engine configurations, different percentages of the reciprocating weight may be required. The balancing-equipment manufacturer usually provides a reference chart, or has this

Each piston pin is weighed and recorded. Sets of pins generally weigh the same. However, if any small deviations are present, you may be able to mix or match lighter or heavier pins to pistons to create a piston/pin combination that provides a matching-weight set. This could reduce or eliminate any need to remove weight from pistons.

If weight must be removed from a piston, material can only be removed from an area where strength isn't compromised. The underside of the pin bosses can be shallow spot-faced, or a radial cutter can be used to remove an equal amount of weight from both bosses.

Considering the high-quality and uniform precision manufacturing of today's aftermarket performance pistons and pins, some builders opt to simply weigh the piston/pin and locks together as a package.

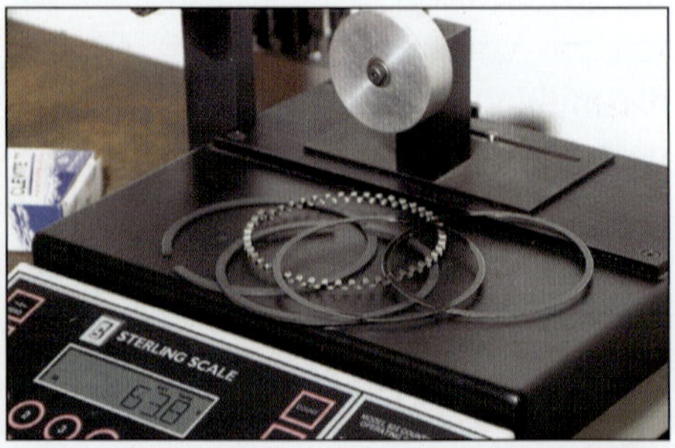

Ring packages from the same set weigh the same, so there's no need to weigh the complete set of rings. Weigh the package for one piston (top ring, second ring, and oil ring package). If your oil ring package requires a support rail, don't forget to include the support rail in the weighed package.

TECH TIP: Bobweight Card

This is a sample for a 90-degree V-8.

Component	*Weight (grams)*
Piston	390
Piston pin	106
Pin locks	3
Rings (1-piston set)	41
Rod small end	174
Rod big end	452
Rod big end	452
Rod bearing (44 x 2)	88
Total	1,706
50 percent of total	853
Oil	4
1/2 Bobweight	857

In order to complete the bobweight card, you add your measured component weights as follows:

- Weight of one piston
- Weight of one piston pin
- Weight of one piston pin's locks (depending on design, this may involve either two or four locks)
- Weight of one piston's set of rings
- Weight of one rod's small end
- Weight of rod big end x 2
- Weight of 2 pair of rod bearings

Once these weights are added, divide the total in half, and add 4 grams for "clinging" oil to determine the weight of each bobweight half (two bobweight halves are attached to each crank rod journal). The 4 grams of oil weight is the anticipated weight of parasitic oil that will cling to the rods and crank counterweights during engine operation.

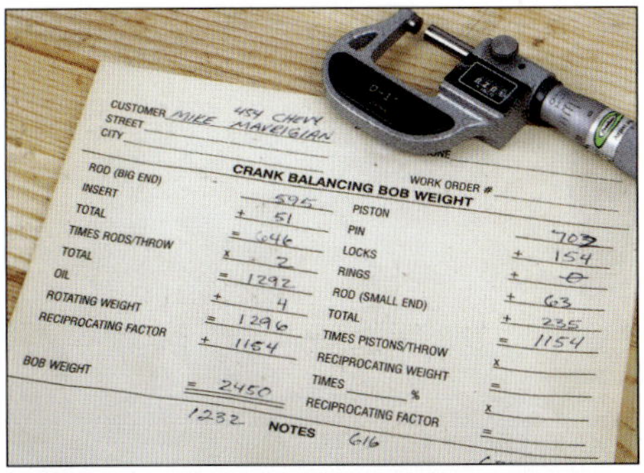

A sample bobweight card. This provides the technician with the necessary information to create the bobweights. This information should be kept on file in case this assembly requires service in the future.

During makeup of the bobweights, they're weighed on the digital scale and adjusted (with more or fewer weights to match the required weight).

Once the bobweight card has been completed, the individual bobweights are assembled.

information programmed into its computer software.

Once the total bobweight is determined, the weights are assembled to duplicate the real-life reciprocating mass.

Overbalancing

A 50-percent factor is normally used for most balancing jobs. But for certain racing applications with high compression ratios and/or high cylinder pressures, some builders prefer to slightly overbalance, using a factor of 51 percent. This adds a bit more weight to the crankshaft counterweights, which counteracts high-compression resistance. This theoretically reduces the negative torque created by the compression force, aiding in further compacting the air/fuel charge.

In certain instances, a high-revving race engine may experience a slight out-of-balance vibration at a specific RPM. In this situation, a slight overbalance may be beneficial by increasing the reciprocating weight factor to 51 percent or more. In doing so, the out-of-balance point may be moved to a lower engine speed range that isn't critical to the race engine's use. Overbalancing allows you to optimize crankshaft balance within the most important range of engine speed (the sweet spot) for the specific type of racing.

If you are balancing a four-cylinder in-line crankshaft, there is no need to create bobweights. Since the crankshaft has two opposing strokes, the dynamic forces tend to cancel each other. In order to balance this type of crankshaft, weight match all pistons and rods as you normally do for a V-type engine. Without bobweights attached to the crankshaft, spin and balance the crank.

Installing

The crankshaft is positioned on the spin balancer's V-blocks at the crankshaft's front and rear main journals. The V-blocks have nylon rubbing blocks to protect the main journal surface, which must be lightly oiled before positioning the crankshaft. The bobweights are installed to the crankshaft rod journals, with each journal's weights set 90 degrees from each other.

The bobweights must also be centered along the width of the journal. Most spin balancers have a drive belt that contacts the center main journal. With the crankshaft fitted with its bobweights, the spin balancer is turned on to spin the crank. The balancing machine displays the existing heavy or light areas of the crankshaft relative to zero, and indicates where weight must be removed or added relative to the front-to-rear counterweights as well as the radial point of the counterweights.

If weight must be removed, it is done by drilling to remove material from the counterweights. This can be accomplished with an overhead drill press, drilling into the outer edges of the counterweight, or, as determined by the balancing technician, by machining material from the outer faces of the counterweights on a precision lathe.

Bobweights should be centered at the width of each rod journal. To save measuring time, machinists often insert an aluminum spacer of a specific thickness to avoid the time required for measuring to center the bobweight.

Each bobweight is placed with the weight side 90 degrees to each rod journal.

Example of a bobweight centered on the journal width.

Bobweight bases have V-cuts that rest against the rod journal. Each end of the bobweight is tightened evenly.

Shown here is a spacer inserted on one side of a bobweight. With the bobweight pushed up against the spacer, the bobweight's halves are secured together, and then the spacer is removed.

Depending on the type of balancing machine, information is provided on the display screen regarding the drilled-hole diameter and depth required.

If weight is to be added, the crankshaft counterweight is first drilled to a specific diameter and depth, then filled with a heavy-metal tungsten slug. If the slug is installed to the radial edge of the counterweight, the technician then welds the slug in place to prevent it from slinging out of the counterweight. Most balancing technicians prefer to drill and install a tungsten slug horizontally through the face of the counterweight, eliminating the potential for accidental slug departure due to centrifugal force.

Each time weight is removed or added, the crankshaft is spin checked again to verify the change in balance because performing a weight change at one end of the crank can affect the opposite end. In some cases, as many as six (or even more) balance corrections may be needed as the technician chases the balance.

Balancing Procedures

Before attempting a crankshaft balance, always inspect the crankshaft for runout. If it is bent (excessive runout), the balancing machine displays a substantial difference in weight between the front and rear counterweights. Always check for runout before wasting your time trying to correct balance.

Inspect the crankshaft stroke and verify that each stroke distance (the centerline of the rod journal to the centerline of the main journal) is identical. This is especially important if the crankshaft has been reground. Slight differences in stroke affect balance.

Balancing is the final step in crankshaft preparation (aside from a final polish of the journals), so be sure that all machining that could affect balance has already been performed. This includes crankshaft machining, piston machining, rod bearing selection, etc.

It's important to note that any change in the weight of the reciprocating parts changes the overall balance. If, during a rebuild, you need to replace even one piston, do not assume that the replacement piston weighs the same as the original, even if it is the same size, style, and brand. Whenever any part of the reciprocating assembly is changed, the crankshaft must be rebalanced.

A flywheel or clutch pressure plate for an internally balanced crankshaft should have a zero balance right out of the box. But if the goal is to follow a blueprinting approach, don't make that assumption. Separately check the flywheel (or flexplate) and the clutch pressure plate and correct the balance if needed. Although an approximate flywheel and/or pressure-plate state of balance may be adequate for routine street driving, blueprinting requires trying to eliminate all variables and deviations.

However, is not critical, and often not practical, to achieve a perfect-zero crankshaft balance on the balancing machine. Even the movement and migration of slinging and clinging oil can slightly alter dynamic balance. Balancing the crankshaft to within 2 to 4 grams is acceptable. Just because you might achieve a zero balance on the machine, it doesn't mean that the crankshaft will experience zero balance during engine operation. Be realistic. You can spend hours in chasing a zero balance, often with no additional benefit.

When weight must be removed from a crank counterweight, it is done on an overhead drill press. Some balancing machines incorporate a drill while a separate drill press is required for other systems.

If it's necessary to add weight to a counterweight, it can be done at the edge or at the face. If weight is added to the outer edge, it's necessary to tackweld the weight in place to eliminate the possibility of it being slung outward. If weight is added to the face, a press-fit is adequate. Because of the required location of the weight shown here, the rear weight was drilled to access the target counterweight for drilling. A heavy-metal slug (tungsten, also known as Mallory metal) is press-fit into the target hole. Again, the technician refers to a chart specifying the diameter and length of tungsten required for a specific weight factor (in grams).

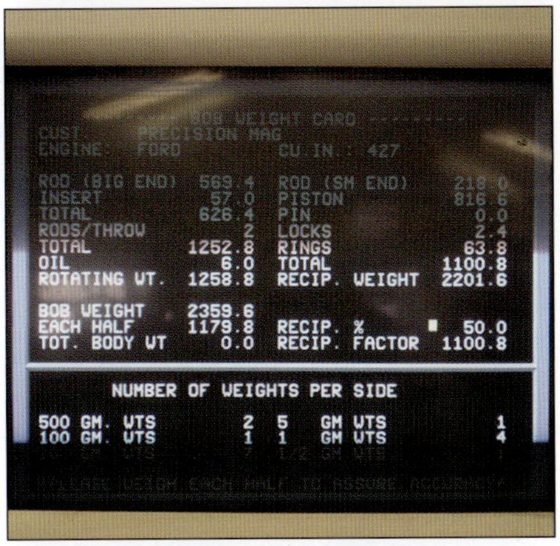

The monitor on this balancer's screen displays all of the weight data input by the technician.

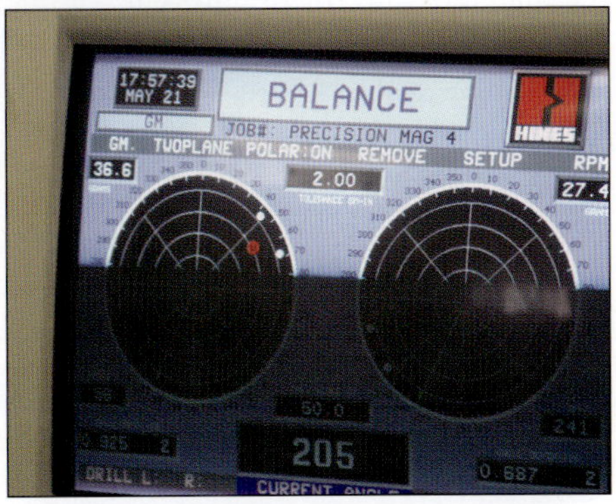

This Hines computer balancer provides a dynamic graph that shows the location required for weight removal or addition, in terms of clock position and depth at the counterweight.

CHAPTER 17

TIMING SYSTEMS

In this chapter I discuss the three most commonly used methods of camshaft drive (the manner in which the camshaft is driven by the crankshaft). These are chain drive, belt drive, and gear drive. Chain drives are the norm for most traditional OHV V-type engines in the performance arena. Many late-model OHC and DOHC OEM engines have belt drives. You commonly find gear drives in specialty applications or race engines. Each style offers distinctive design and benefits.

Chain Drive

Chain drives are offered in single- and double-tooth sprocket designs; double rollers are the preferred choice for any high-performance application. Relatively less expensive OEM systems often used aluminum or composite cam sprockets with nylon teeth to reduce noise. The nylon teeth, exposed to age and abuse (or both), can become very brittle and can fail, sending sharp bits of nylon into the oil pan sump. Basically, the nylon tooth setups are worthless for any engine that's spitting out some respectable power.

Timing chains are made in two basic formats: links and bushings. The bushings eventually wear and create sloppy cam timing. Roller chains with rollers provide superior contact with the sprocket.

Both single-roller and double-roller setups are available with crank and cam sprockets featuring single- or double-tooth rows. Double rollers offer more strength, less stretch, and more consistent timing. In some cases, you're limited to using a single-roller chain due to timing cover clearance.

If you're planning to use a chain drive to run your cam, go with a double roller. But be aware that it's the *quality* of the materials and the design that counts. Roller chains may be made with seamed or seamless rollers. Seamless offer greater concentricity and greater durability. Avoid bargain-priced overseas junk and stick with a major brand such as Melling, Cloyes, Comp, etc. Cam sprockets may be made from cast iron, powdered metal, or billet steel. For the street, and for engines making fewer than 300 hp, the higher cost of a billet gear simply isn't necessary. PM is stronger than you might think, and is used in many OEM high-horsepower engines. But if you're dealing with a reasonably powerful build

This is a double roller chain setup featuring a billet-steel cam gear. Billet cam sprockets offer increased strength, but considering the higher-than-cast cost, it really isn't needed for stock or mild engine builds. However, it's a sensible choice for a chain drive in high-horsepower applications. Note that this set has advance/retard changes of 2, 4, and 6 degrees. (Photo Courtesy Comp Cams)

High-quality chain-drive systems feature true rollers (not fixed bushings) that contact the sprocket grooves for reduced friction and longer life. Seamless rollers and precision-machined teeth/grooves are what sets high-quality high-performance chain drives apart from anonymous-brand kits. (Photo Courtesy Comp Cams)

An example of a belt drive; this one is a Jesel for big-block Chevy. The kit includes the backing plate, cam hub, bronze thrust washers, crank gear, and belt. The backing plate covers the front of the engine, sealing it off; the cam hub allows the cam gear to be mounted on the forward side of the cover. Cam timing adjustability via a two-piece cam gear offers +8/–8 degrees of change.

Comp Cams' Number-6100 Hi-Tech belt drive. It's a typical belt drive with a cover that seals off oil, achieving a dry belt design. (Photo Courtesy Comp Cams)

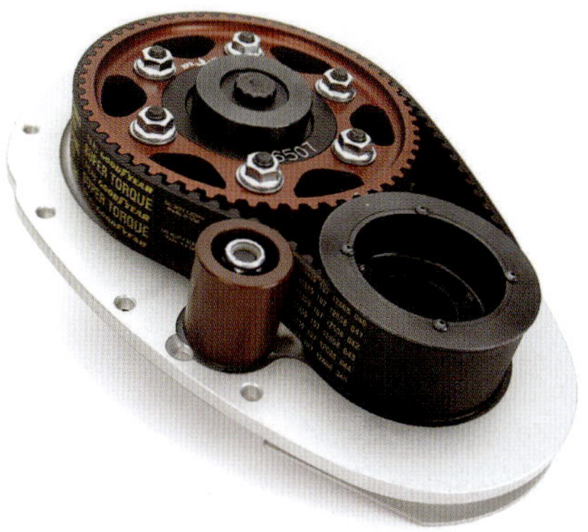

Comp Cam's Xtreme Duty Hi-Tech belt drive (PN 6507) has a 1.250-inch belt for extreme cylinder pressures encountered in nitrous and blower applications as well as a belt idler. (Photo Courtesy Comp cams)

(500-plus horsepower), paying a few extra bucks for a billet setup makes sense. If nothing else, it provides more peace of mind. Always buy the highest quality that you can afford.

Belt Drive

Swapping from a timing chain to a belt drive provides the possibility to increase power. A belt drive has reduced frictional loss, theoretically more precise timing, and smoother valvetrain motion. A dry belt system also eliminates windage variables (compared to a timing chain and gears running in oil). A belt drive does a better job of isolating crankshaft torsional vibrations. It can be viewed as a second harmonic dampener for the rotating assembly because it absorbs a degree of operating harmonics

You might assume that a belt stretches over time, but this isn't a concern. Performance drive belts have substantial reinforcement to prevent stretch. If a belt fails, it is likely due to age.

Belt drive systems offer rugged performance, accurate timing, quiet operation, and help to dampen harmonics. Aside from the relatively higher price tag (depending on brand and application, around $800 to $1,200), there's really no downside to going with a quality belt drive system. Even Pro Stock racers use belt drives, and *that* is an endorsement.

Installing a Belt Drive

Here I use a Jesel belt drive installation on a big-block Chevy application as an example. I installed this drive system onto a 632-ci build with a Dart Big M block.

The first step is to install a stock-type big-block Chevy timing cover gasket to the block face. I smeared a thin coat of RTV on both sides of a Victor-Reinz gasket for insurance (habit). Then I installed the Jesel aluminum cover housing. This cover is secured with six 1/4-inch x 20 socket-head cap screws, which I tightened to 50 inch-pounds.

Next I oiled a brass thrust washer (on both sides) and slipped it over the cam nose. This was followed by the cam nose adapter. For sealing purposes, a light coat of RTV was applied to the rear face of the adapter where it meets the cam nose face.

Jesel supplies a handy spanner wrench to hold the cam steady during bolt tightening. Three 5/16-inch x 18 torx-drive bolts were supplied, as well as a handy T-45 torx bit. Jesel recommends applying a light coat of RTV to the three bolt threads. These torx bolts were

With the cover mounted to the block, the cam hub is mounted to the cam. The kit includes a hub-securing tool to hold the cam steady while tightening the mounting bolts.

Bronze thrust washer(s) are installed at the hub to adjust cam thrust. The bolts' threads are coated with RTV and torqued to spec.

A retaining plate is mounted; it has a seal at the cam hub. The thrust washer(s) is/are located between the hub and retaining plate.

Alignment dots are found on both gears. The cam gear offers an adjustment range of as much as 8-degrees advance or 8-degrees retard.

Finish-installed Jesel belt drive. Since the belt runs dry and oil is sealed by the backing cover, the belt system can run exposed. This makes service for cam timing or cam swaps much easier.

tightened to 30 ft-lbs (Jesel's range is 28 to 30).

Next I installed another brass thrust washer to the outside perimeter face of the adapter (oiled on both sides).

The next step was to install the shim pack supplied in the kit. I installed all three shims (these are located over the six 1/4-inch x 20 studs on the face of the cover), followed by the seal-fitted shim cover flange. In order to avoid damage to the rubber seal, I first removed the small key from the cam nose adapter.

According to Jesel's instructions, camshaft endplay should be .010 to .015 inch. The supplied shims must be added or removed in order to achieve a cam endplay of .010 to .012 inch. When I set up my dial indicator, I found endplay to be .017 inch. I called Jesel to ask their advice, and one of the techs assured me that .017 inch is acceptable.

With cam endplay verified, I removed the thrust flange and the three shims. I lightly coated all shims and the rear face of the flange with RTV and reassembled, using the supplied 1/4-inch x 20 nyloc jam nuts, tightening them to 50 in-lbs. With the shim pack and flange installed, I re-installed the small key to the cam nose adapter.

Next I installed the crank gear drive (I installed a pair of keys measuring .185 inch thick by .735 inch long and .300 inch high into the snout's key slots). This is an interference fit, requiring an aluminum driver that slips over the crank snout driver, which was supplied in the kit. The crank snout had a slight step up in diameter. When the gear was slipped onto the crank and stopped with hand pressure it was then driven on another .384 inch (in this case) until it seated.

At this time, I rotated the crank to place the crank gear's timing dot at 12 o'clock.

The cam adjuster plate dropped over the four studs on the black cam gear and was assembled with the cam timing mark at the top, aligning with the center (zero) timing mark on the adjuster. I installed the cam gear and adjuster plate onto the keyed cam-nose adapter.

Next I installed the cam locking flange using the left-hand-thread bolt provided. It was snugged just enough to keep the cam gear keyed in place.

With the four adjuster plate to cam gear nuts installed with about 1½ threads, I tilted the top of the cam gear forward to slip the timing belt into place.

This took a few minutes because I needed to keep the timing dot on the bottom of the cam gear aligned with the top dot on the crank gear. The belt was a tight fit, so patience was definitely required. Once the belt was in place, I snugged the four 12-point nuts to secure the adjuster plate to the cam gear.

Once the timing belt was in place, with both dots aligned (cam gear dot at 6 o'clock and crank gear dot at 12 o'clock) and with the cam adjuster marks set at zero, I tightened the cam gear's center left-hand-thread bolt to 70 ft-lbs (I needed a 12-point, 5/8-inch socket for this bolt).

I was then ready to degree the cam.

Although the job seemed daunting at first glance, it actually wasn't that big of a deal. Because the oil was sealed by the rear cover, the belt drive was exposed, making cam adjustment easy.

Gear Drive

The theory behind a gear drive is simple. Crank motion is directly transferred to the cam via a set of meshed gears. With no chains to stretch or wear out, gear drives provide superior and consistent cam timing. Gear drives are also stronger and last longer than the average timing chain set. Those are the pros.

The cons include potentially greater harmonics transmitted to the cam and (in most cases) a notable whine that results from gear engagement. The whine is somewhat similar to the sound created by a supercharger blower, which can be really cool or really annoying, depending on your frame of mind.

If you want a quiet engine (aside from the exhaust note, of course), you probably won't like a gear drive. However, some gear drive manufacturers now offer a "quiet" version that produces less whine. Gear drives and related parts are available from Edelbrock, Milodon, BHJ, and others.

Gear drive with two idler gears. Gear drives offer accurate cam timing/drive, but can transfer more crank harmonics. Noisy and quiet versions are available. The noisy versions produce a gear-to-gear sound similar to a blower whine. This noise is matter of personal preference. (Photo Courtesy Comp Cams)

Edelbrock

You can replace a stock timing chain with Edelbrock's Accu-Drive camshaft gear drive kit. This gear drive system transmits power from the crankshaft gear to a full-floating main idler, which drives the camshaft gear. A unique feature of this system is the ability of the main idler gear to float to an optimum position between the crankshaft and camshaft gears, ensuring absolutely equal load sharing between them.

The gears are made from billet SAE-1144 steel with induction-hardened teeth. Gear teeth are shaved for precision operation. The set has hardened and ground idler pins from billet steel.

To install, just replace the stock crank and camshaft sprockets with Accu-Drive gears and slip in the idler assembly. Most Accu-Drives require no modifications to the engine block, although some fitting of the axles and front cover may be required.

Milodon

The gear drive system from Milodon has a three-gear setup with easy cam timing adjustment. All Milodon gear drives are three-gear, "fixed idler" models. The idler gear mounts solidly to the block (under cover) or to the cover (full cover).

This system does not rob any power from the engine and (more important) does not allow cam timing to vary. The full-cover drive uses an adjustable cam gear and hub assembly to set cam timing, accessible through the removable cam cover. The cam itself is bolted to the hub.

The Milodon gear drive uses the cam gear and hub to advance or retard the timing. You simply unbolt the hub and turn it until the indicator mark is lined up with one of the seven bolt positions on the cam gear, with no offset bushings or keyways to mess with. The cam must be installed straight up (no advance or retard) as follows:

With no cam gear in place, locate TDC for the number-one cylinder using a degree wheel. Rotate the crankshaft until the degree wheel indicates the intake valve opening per the cam manufacturer's specs.

Place a dial indicator on the intake valve and rotate the crank until the valve opens to the recommended checking clearance (usually .050 inch of lift).

At this point, the camshaft is now installed straight up with no advance or retard. Install the cam gear. Adjust for your desired advance or retard. Place a mark on the cam hub next to any cam gear bolt hole position and label that hole number-1. In a clockwise travel, label the remaining bolt holes two through seven.

You're now ready to set cam timing. Each bolt hole has gear-tooth positions for advance and retard. The chart in the provided instructions shows the settings for each hole. To advance the cam, turn it clockwise until the indicator mark on the hub is lined up with the proper bolt. To retard the cam, turn it

BHJ's timing set length gauge. This specialty gauge allows you to check and/or determine proper chain or belt length for your engine. Mounting posts are adjustable to set your exact center-of-crank-to-center-of-cam distance. (Photo Courtesy BHJ Products)

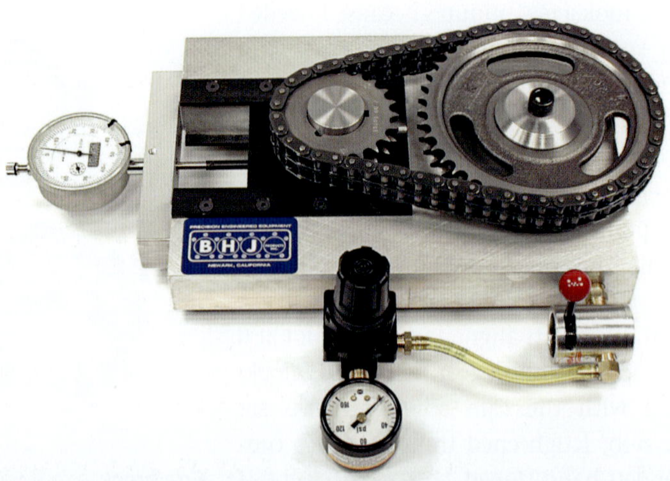

Once the engine's precise center-to-center distance is set, the timing system is placed in position with the dial indicator zeroed. Compressed air at 100 psi is applied, and the dial indicator is watched for +/–. Shown here is a chain drive check. (Photo Courtesy BHJ Products)

Belt drive check on the same length gauge. (Photo Courtesy BHJ Products)

counterclockwise until the mark lines up with the hole you want.

Tighten the hub bolts and your cam timing is set.

BHJ

BHJ offers a timing set length gauge (TSG-1) that's adjusted to the desired center-to-center distance using a micrometer and the setup pins or an optional setup standard. The dial indicator, included with the gauge, is then zeroed at the desired length. Timing sets can then be dropped onto the appropriate mounting plugs.

The TSG-1 requires 100 psi minimum air supply. Activating the air switch applies exactly 100 pounds of tension to the chain, as specified by leading chain manufacturers. The timing set length is read as a plus or minus dimension from the desired length on the dial indicator.

This precision measuring fixture eliminates the need to special order custom-length timing sets or waste time with tedious trial-and-error methods of obtaining proper timing chain fit.

TSG-1 accepts timing sets with center distances ranging from 4.125 to 6.125 inches, using the appropriate mounting plugs. The gauge includes a precision dial indicator, setup pins, and one pair of mounting plugs.

Also available is a timing set gauge made specifically for the Mopar R5 or the small-block Chevrolet using the Jesel belt drive. Additional mounting plugs and optional setup standards are available for all engines.

Ford small-block cam gears have one cam alignment dowel. Considering today's high-lift cams and high valvespring pressures used with roller cams, this timing gear design really needs an upgrade for better timing control. BHJ makes a timing-set drilling fixture (PN TSDF-FOS) that allows you to add a second alignment dowel placed 180 degrees from the existing dowel location. This requires a simple three-step process that can easily be handled with a bench vise and hand-held drill (or on a drill press or milling machine).

The kit includes a fixture plate, end mill, drill, finishing reamer, drill bushings guides, and mounting hardware. This is a worthwhile and easy-to-perform upgrade if you're running a nasty cam in your small-block Ford and plan to use a chain drive.

Engine Performance Coatings and Treatments

Professional racers have used internal engine component coatings for more than 20 years. These coatings offer increased horsepower, reliability, and engine longevity. They are not, however, intended for everyday commuter engines (since the cost likely isn't justified). Instead, these coatings should be considered for high-performance street and racing engines only, where power and engine life needs to be optimized.

Current coatings on the market deliver enhanced thermal control, improved lubrication, and oil-shedding. Depending on the type of coating and the area in which the coating is applied, the benefits can include increased power, component durability, or both.

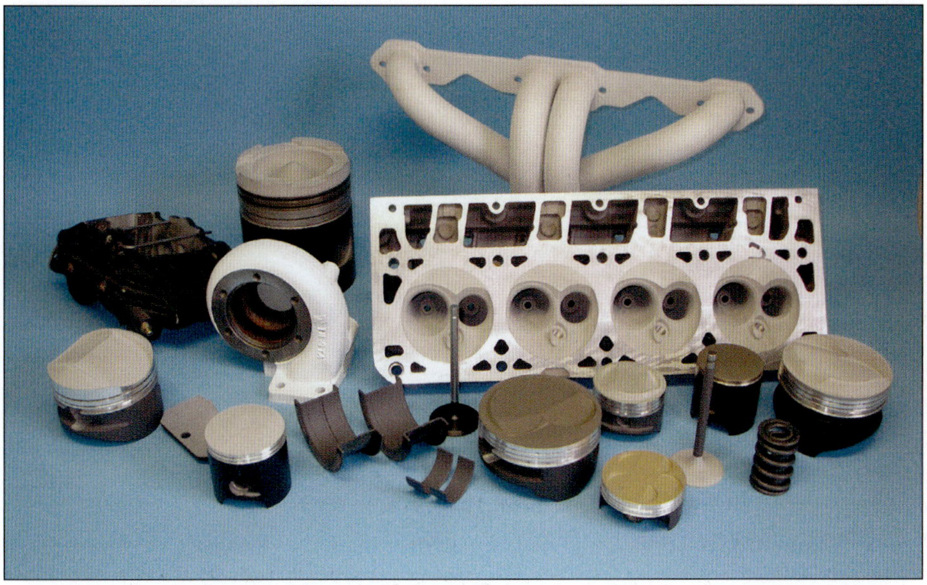

This group of parts helps to illustrate heat barrier and anti-friction coating applications. The examples shown here include high-temperature thermal-barrier coatings on head chambers, piston domes, turbo housing, and headers; anti-friction coatings on skirts, valvestems and bearings; and heat-release (or shedding) coatings on valvesprings. (Photo Courtesy Swain Tech Coatings)

Coating Types

Suppliers typically offer four types of coatings for high-performance engines: low-friction, oil-shedding, thermal-barrier, and heat-emitting. Each has its own set of parameters for use.

Thermal Barrier Coatings

Also called ceramic coatings, they act as a shield to prevent heat from passing into a part. For instance, a thermal barrier coating on a piston dome prevents combustion heat from being lost beyond the dome. This prevents heat from dissipating through the dome, causing the piston to expand. If there is a barrier coating, combustion heat is greatly reduced, and does not sink into the piston and cause expansion. Since the piston doesn't expand as much you can then run a tighter piston-to-wall clearance.

If the combustion chambers are also coated, combustion heat stays in the chamber and promotes more efficient use of thermal energy. To aid in this, the valve faces can also be coated to help seal the heat in as well as to prevent excess heat from traveling up the stems.

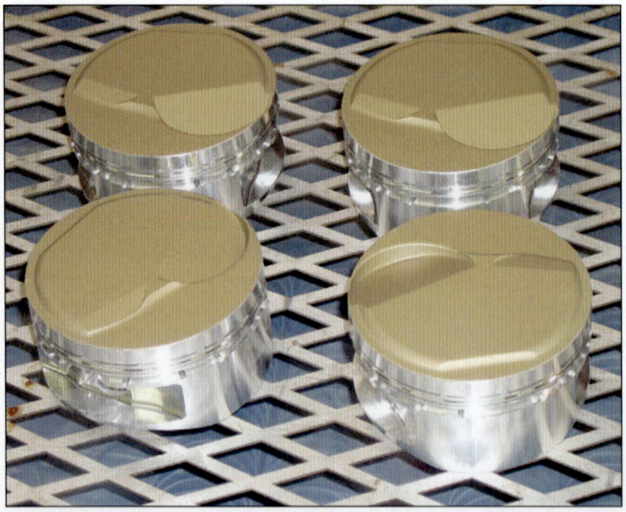

The set of pistons in this shot have had only the domes treated to a ceramic thermal barrier coating.

In addition to piston dome coating, the same high-temperature heat-barrier coating may be applied to the combustion chamber, all valve faces, exhaust valve throats, and exhaust valve ports. This completely encapsulates the combustion stream path. (This head was coated by Dart.)

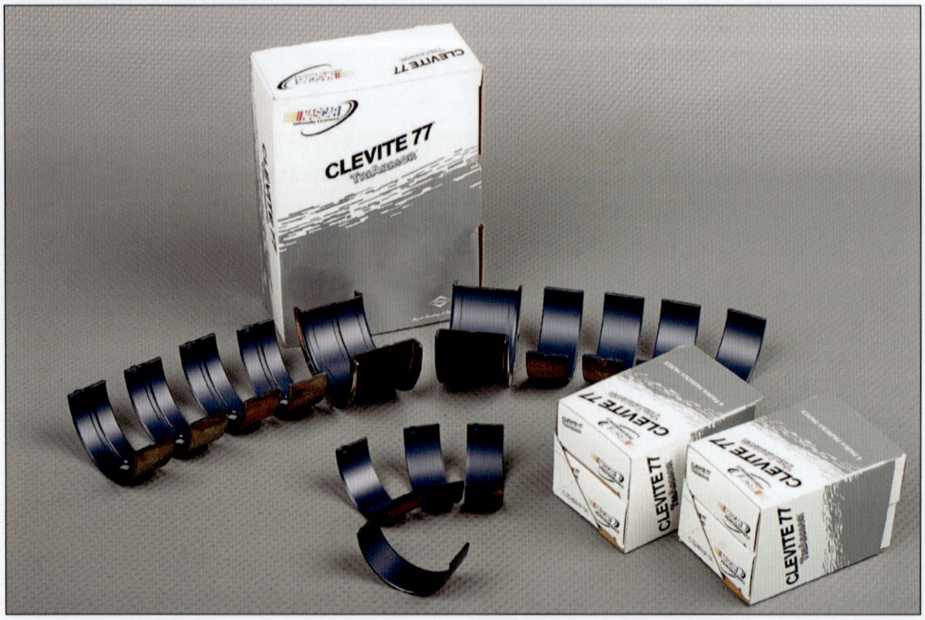

Precoated engine bearings, that is rod bearings, main bearings, and cam bearings, are now available already coated from various bearing manufacturers.

MAHLE Clevite, for example, uses a polymer-based moly-graphite coating on its performance bearings.

An oil-shedding coating is extremely useful for crankshaft counterweights, allowing oil to more quickly depart from the counterweight surfaces. This reduces the drag that exists because of clinging oil, which in turn reduces parasitic power loss.

If the exhaust path is coated, combustion heat is contained inside the exhaust path. This prevents loss of thermal energy and adjacent areas from being subjected to excess heat. Coating the exhaust ports and outside of the exhaust manifold also creates a scavenging effect, contributing to more-efficient exhaust flow. In essence, the combustion energy and heat stays in a more-contained path, doing its job, instead of being allowed to sink into surrounding areas, where it does nothing but waste energy.

Anti-Friction and Oil Management Coatings

Anti-friction coatings provide a low coefficient of friction between two moving surfaces and serve to retain surface oil between moving parts. This oil is applied to components, such as bearings and piston skirts. It's exactly what you want at bearing and skirt locations.

Oil-shedding coatings provide a slick and non-porous surface treatment to aid in slinging oil off a component. A shedding coating applied to connecting rods and crank counterweights

An oil-shedding coating applied to connecting rods serves the same purpose: to reduce parasitic power loss caused by unwanted oil clinging to the surfaces. After all, the engine doesn't need oil on the exterior surfaces of the crank and rods for lubrication, so by promoting oil departure from these surfaces, we reduce unnecessary drag, which helps to free-up horsepower from the rotating assembly.

prevents unwanted oil weight from clinging to the surfaces. As a result, parasitic drag is reduced as the rotating assembly whips through the air and a mechanical advantage is realized.

Oil-shedding coatings also prevent hot oil from clinging to the stationary wall surfaces, and this helps reduce radiated heat, promoting much faster drainback to the sump. Interior walls include roof of valve covers, oil pan walls, (some) dry sump pans, and the underside of V-type engine intake manifolds.

A thermal-barrier coating can also be applied to the cover exterior. This coating combination helps to reduce radiated cover temperature while preventing oil from clinging to the cover roof and walls, allowing oil to scoot back to the moving parts where it's needed.

The oil pump is an often overlooked but vital component. Applying an oil-shedding coating to the oil pump gears and passages allows the pump to more efficiently move oil, which reduces the amount of oil that clings to the passage surface areas. An oil-shedding coating lets oil move through the pump faster and reduces the amount of oil that hangs around by sticking to passage walls.

Heat-emitter coatings promote the release of heat. Common applications for this specially formulated coating include valvesprings, with the coating intended to prolong spring life.

Piston Skirt Coatings

The moly coating's true benefit is reduction of friction, which prolongs part life and reduces operating friction, and it naturally frees available horsepower. A moly skirt coating is not specifically designed to reduce heat, but because it reduces friction, heat is potentially reduced as a direct by-product. A moly skirt coating improves cold-engine startups; the moly coating prevents skirt scuffing that might otherwise repeatedly occur in that situation.

Many approaches can be used to reduce friction with the cylinder bores, including synthetic oils, plasma-moly ring faces, oil retention grooves in the skirt areas, plateau cylinder-wall finishes, etc. All of these methods have merit, but so does coating the piston skirt areas with a long-wearing high-lubricity material.

Commonly, this is a molybdenum-based application applied to the skirts in an average thickness of about .0005 inch per side, which might provide about a .001 inch increase in piston overall skirt diameter. Since the moly is applied in such a thin layer, no additional bore-dimension changes are required. In other words, you should hone the cylinder to the size dictated for proper piston-to-wall clearance for a non-coated piston. Whether you're installing non-coated or coated pistons, the finished bore size is identical. Do not compensate for the added moly coating when finishing your bores.

Forged piston applications typically run a .005- to .006-inch clearance, which does not change if you opt for moly-coated pistons. If you plan to use

All of the coating shops offer thermal barrier dome coating and anti-friction skirt coatings. Here's an example of a piston treated with a thermal barrier coating on the dome and moly-coating on the skirts.

hypereutectic pistons that require a .001- to .0015-inch clearance, for example, finished bore size is not enlarged to compensate for the coating layer. With hypereutectics, the skirt coating might actually create what first appears to be a near-interference fit of the piston to the bore. This is not a concern because any excess moly is sacrificed during initial runs, and mixes with the engine oil in a completely compatible fashion. So install and run them. If you overhone by .001 inch in an effort to compensate for the moly coating, you simply defeat the benefit of the tighter and more efficient clearance.

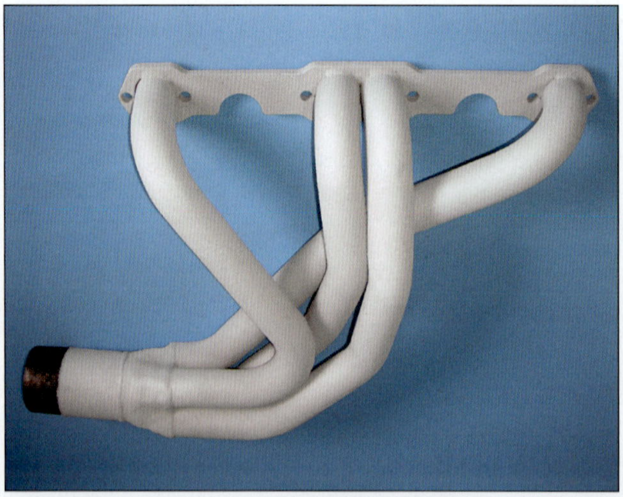

When you have exhaust ceramic-coated headers, pay attention to the collector area. If the header has a bolt-on flange, there's no need to mask off anything. However, if the collector has a slip-on end, ask the coating shop to mask off a sufficient area, since the thickness of the coating can result in a too-tight slip-on fit. (Photo Courtesy Swain Tech Coatings)

Coating Sources

Many engine bearing manufacturers offer anti-friction coatings as standard or as an option.

Mahle Clevite offers its TriArmor engine bearings already coated with a moly-graphite treatment in a low-friction PTFE polymer base.

Dart Machinery has a full array of coatings, including moly-based anti-friction coatings, synthetic ceramic and traditional aluminum-based thermal barrier coatings, oil-shedding polymers, moly-Teflon coatings for bearings, Teflon-based air-enhancement coatings for intake ports, etc.

Ceramic thermal barrier coating is ideal for turbocharger housings. This increases efficiency and helps to reduce underhood temperatures. (Photo Courtesy Swain Tech Coatings)

Dart's coating offerings include the following:

- DCI MOS2 Teflon Skirt Coating
- DC2 High Temperature Reflective Heat Barrier for thermal barrier applications on piston domes, combustion chambers, valve faces, exhaust ports, etc
- DCB-3 Engine Bearing Coating (a moly/Teflon based material with high-load/non-stick properties for bearings
- DC-4 Lubricating Pigments for wear and load capacity in applications such as valvesprings, oil pump gears, valvestems, timing gears, camshafts, and other friction related areas
- DC-5 Oil Shedding Coating for shedding parasitic oil from connecting rods, crank counterweights, windage trays, inside oil pans and valve covers, etc
- DC-7 Anti-Corrosive Protectant to protect surfaces from weather, gasoline, alcohol, nitro methane, brake fluid, etc.

In a turbo application, according to Swain Tech Coatings, a heat-emitter coating provides benefit to an oil-cooled, bearing turbo. A stellite coating on the turbo impeller shaft (which can experience 1,650 degrees F) reduces shaft wear and acts as a solid film lubricant while providing the needed hardness to protect the shaft.

Swain offers three different thermal barrier coatings, including the standard thermal barrier coating (TBC), Gold Coat, and White Lightening. TBC is used in most applications for piston domes, combustion chambers, valve faces, exhaust valve backs (radius), and exhaust ports. Where higher temperature extremes are encountered, such as in nitrous oxide or turbocharged applications, the Gold Coat version may be applied, which simply yields a higher thermal-barrier capability for these extreme situations.

Cryogenic Stress Relief

This process stress relieves a crankshaft, reduces the chances of warping an aluminum head, and prevents distortion in a block bores.

Cryogenic treatment involves freezing engine parts to make them stronger. The tool-and-die industry regularly uses this approach to temper and extend the life of steel tooling components.

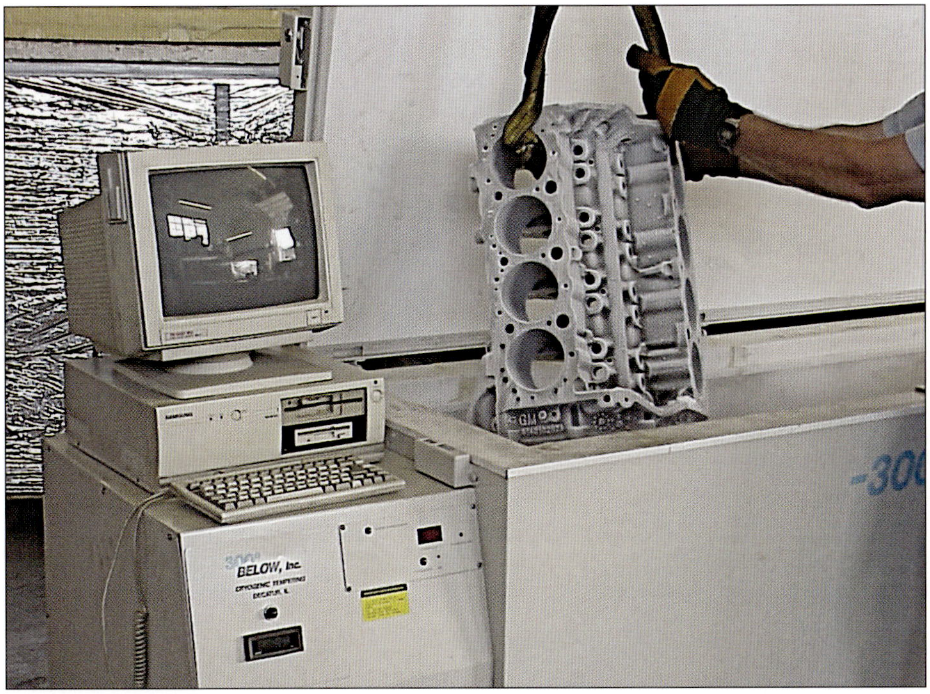

Cryogenic metal tempering requires a specialized, computer-controlled cryo tank that uses liquid nitrogen, in which components are exposed to temperatures at approximately –400 degree F. The pieces are cooled in a slow, controlled manner, then allowed to come back to ambient temperature in a controlled process. The engine block in this photo was removed from the tank only for the photo. Notice the white frosty surface, resulting from the frozen part exposed to room temperature.

What Does It Do?

In a nutshell, cryogenically tempering (deep-freezing) a metal part, makes the internal structure more uniform, more durable, and stronger.

The process of cryogenic freezing changes the structure of the metal being treated. Inside the metal, areas of weaker, potentially brittle deposits called austenites may exist. These are flaws that create the potential for cracking. Cryogenics changes these areas into harder, more uniform martensites. The process also creates a vast distribution of very fine carbide particles throughout the metal.

The Cryogenic Unit

Advanced Cryogenics uses the –300 Cryo-Processor, and has conducted an enormous amount of research in the field of cryogenic tempering of engine parts. For purposes of R&D, Advanced Cryo maintains its own test car, a modified D.I.R.T. race car with a Chevy big-block 467-ci bowtie engine. The block is .060-inch over, fitted with Chevy aluminum heads and produces a compression ratio of 13.5:1 and a dyno output of 650 hp. The engine runs on alcohol and lives at 7,200 rpm.

TECH TIP

Cryogenic Treatment Cost

The cost of cryogenic treatment is based on the amount of nitrogen used. The larger the part, the more nitrogen used. Treating aluminum sometimes uses less nitrogen (due to the higher conductivity of alloy).

In order to give you a rough idea of the pricing structure, I use Advanced Cryogenics as an example. Typical retail prices are as follows:

- Typical cast-iron V-8 block $150
- Aluminum block $125
- Crankshaft $50
- Set of eight connecting rods $45
- Set of eight pistons $35
- Camshaft $35
- Cast iron or aluminum head $50
- Set of 16 rockers $35

A typical complete engine assembly costs $350 to $400.

A wide range of components can benefit from cryogenic tempering, including blocks, heads, rods, cranks, cams, rockers, and brake rotors. (Photo Courtesy 300 Below)

"When we began the testing program with this motor," noted co-owner Joe Troya, "Constant attention to valve lash adjustment was needed. After the entire engine assembly was cryo'd, we've noticed that cylinder and ring wear have been dramatically reduced, and lash adjustments are almost a thing of the past; everything stays so much more stable." In addition to the engine block, heads, and all internal parts, the Advanced Cryo team has also started to treat and test the car's suspension and brake parts in the same fashion.

In all, the Advanced Cryogenics' race-testing program has involved the treatment of some 15 engines, in D.I.R.T., Busch Grand National, NASCAR, and a few SCCA road racing applications. They're also performing cryo testing for a few aftermarket engine parts manufacturers.

Benefits

Cryogenic treatment finds internal flaws missed by external Magnaflux bench detection. However, an internal pocket flaw may actually cause the part to snap open during freezing. Certainly, most engine builders agree that it's better to find a bad part during testing rather than discovering the problem in a running engine. In other words, if the part is badly flawed, this process may break it.

With this process, the tensile and ductile strength of the part is improved, an obvious benefit for items such as connecting rods and head bolts. Cryogenic treatment also improves the bond of a weld, by producing a more uniform metallurgy structure.

Since this treatment stabilizes the metal, it reportedly improves the machinability of the part. It is a one-time, permanent process that changes the metal structure throughout the part. Once a part has been cryogenically treated, there's no surface layer to break through during future machining. The only materials that don't benefit from Cryogenic treatment are plastics (fiberglass, ABS, various composites) and engine bearings. "The process doesn't hurt those parts, but it just doesn't seem to help either," says Troya. He's treated blocks, heads, pistons, rings, cams, lifters, valves, pushrods, connecting rods, valvesprings, head bolts, retainers, crankshafts, flywheels, and even frozen spark plugs.

The electrical conductivity of the plugs was increased measurably as a result of the process. "The go-kart guys really notice the difference," he noted. Although the process creates a slightly higher Rockwell hardness, that isn't the real directive. Cryogenic processing benefits a part by stabilizing the metal structure, creating a stronger molecular grain pattern throughout the metal.

Vibratory Stress Relief

Vibratory stress relief involves inducing a controlled series of vibrations to metal components (in this case, engine components). This process provides an effective and non-destructive method of achieving stress relief, which results in longer life and increased component stability. Bonal Technologies (to my knowledge, the only provider of this technology) designs and manufactures a complete system for subharmonic stress relief.

Two forms of stress can exist within a metal part: mechanical and thermal. Subharmonic vibratory stress relief (VSR) is a method that relaxes metal (commonly known as Meta-Lax), but only for the metal part's thermal stress, leaving the mechanical stresses unchanged.

Vibrating a part relieves any stresses that were created due to the heat involved in welding, casting, machining, or forging, but it does not alter the metal's strength. VSR does not generate heat and doesn't alter the part's hardness. As a result, VSR is safe to use on a repeated basis.

How It Works

Let's say that you have a part rated with a strength of 50,000 psi. Because of internal thermal stress that may be present, the actual strength of the part *might* be reduced by 20 percent (for example). If the part is heat relieved in an oven, the internal stress is relieved but the part might now only offer a strength of, say, 45,000 psi due to the softening created by the heat. (In contrast, VSR removes internal stress while maintaining the metal's full 50,000-psi strength.)

The vibration process searches for the harmonic peak of the workpiece by vibrating it. The peak is where the piece tends to create the maximum harmonic disturbance, just as a tuning fork vibrates when subjected to a force or a fishing rod whips and vibrates when dynamic force is present.

As a harmonic disturbance is sent through the workpiece, those signals are sent back to the system's controller (via the transducer). The force inducer is then automatically adjusted to vibrate the workpiece in a frequency range that begins just before the peak. This energy area at the base of the peak is where maximum energy dampening occurs. Sending these adjusted (or tuned) vibrations through the workpiece removes stress.

According to Bonal Technologies' president, Tom Hebel, "We first find the harmonic peak to establish where the subharmonic area is. We then induce vibration at that frequency until the part stabilizes. It's as though the part tells you that, up to a point, it can dampen the vibrations on its own; but beyond a certain point, it can't. This process shifts the harmonic peak to its natural location, from an artificial frequency to a stress-free frequency."

Conventionally, if you find the harmonic peak, you might add weight to dampen that vibration. With VSR, you relax the metal in order to alter the harmonics that can take place. Think of a stressed engine component as a musical instrument that's out of tune. VSR brings the part back into tune.

Instead of waiting for an engine block, for example, to settle and remain more consistent in its molecular structure through exposure to time, heat, and harmonics, subharmonic vibratory stress relief essentially seasons a part on an accelerated basis. A seasoned engine block is favored by many builders because it's been subjected to harmonic vibration and has settled and relaxed to a degree, which makes it a good candidate for precision machining for a more stable platform.

Benefits

Hebel cited a simple experiment that helps to illustrate the benefit of VSR: "A local industrial machine shop was curious about the concept, so we performed a demonstration using a solid aluminum bar, 2 inches in diameter and 36 inches long, supplied by the shop. They cut the bar in half (two 18-inch sections). We Meta-Lax treated one section only. The shop then machined each bar section on a lathe, by exactly the same amount.

"Each section was then measured for distortion. The untreated section distorted (warped) by .020 inch along its length, while the treated section moved a mere .002. The dramatic increase in metal stability was obvious."

Thermal stresses can also be induced as a result of distortion immediately after grinding, welding, or machining. This creates a tendency to prematurely crack. Hebel advises that the optimum time to vibrate the part is immediately before the final machining phase (just before the last task that could cause distortion). All four types of stress relief (heat treating, cryogenic freezing, vibratory relief, and natural seasoning) address the issue of thermal stress. VSR offers an advantage because it's faster and doesn't affect mechanical stresses. You already know that a stress-relieved part distorts less during use. When a relieved part is trued during machining, it tends to retain this trueness. For machining advantages, the process reportedly reduces machining time and reduces distortional effects during machining.

If the part distorts less during use, wear is reduced and engine efficiency is enhanced. A quality stress relieving helps to keep bores round and decks flat, something that every performance-engine builder strives for, and is central to the blueprinting goal.

How to Use It

A VSR system includes a force inducer (this unit attaches to the workpiece and applies the vibrations), a transducer (this sends a feedback signal from the workpiece to the control computer), and a control console unit (computer). System prices start at around $7,000 and can climb as high as $35,000, depending on size and application. Some controller units offer an analog readout, while others provide digital information and a hardcopy plotting printout. The object is to transmit the programmed vibrational forces to the engine part(s).

This stress relief system can be used on-site as a portable setup, or in the shop as a permanent fixture. The force inducer can be attached directly to the piece being treated. For example, the inducer can be clamped to a bare engine block, or even a complete engine assembly, while on a stand or even in the vehicle. As a non-portable shop system, the workpieces are clamped to a heavy steel-platform table.

If a number of parts are to be stress relieved at the same time, all of the parts can be mounted to the precision steel table. A block and cylinder heads can be bolted to the table, and smaller items such as a crankshaft, connecting rods, valvesprings, etc. can be secured to the tabletop by means of accessory brackets and clamping fixtures. As long as the vibrations can be transmitted to the engine parts (by direct attachment to the force inducer or by being rigidly mounted to the tabletop with the force inducer secured to the table), the vibrational force is able to affect the engine parts' metallurgy.

Portable systems have been successfully used by race chassis builders. While a race car tubular chassis is being welded, or even after welding has taken place, the portable force inducer is bolted to the chassis. The resulting vibrations stress relieve the tubular network, reducing the chance for frame fatigue and failure. Portable setups have even been used at race sites on completely assembled engines.

Hard Cylinder Coating

This is an application of an optimized deposit of dispersed nickel and a uniform inclusion of hard silicon carbide particles, known today as Mahle's patented Nikasil.

According to Mahle, with a roughness average of 15 Ra, typical Nikasil-coated cylinders can operate at higher temperatures with less wear. This allows clearance from cylinder wall to piston to be optimized to better control oil consumption with lower ring tension, for a reduction in friction. This coating also provides superior heat conductivity and allows increased oil retention during cold starts.

Benefits

The performance benefits of the Nikasil coating include low wear rate, low operating noise, thermal consistency, and faster ring break-in. When selecting honing stones, the rule is: the harder the cylinder surface, the softer the bond of the stones.

Nickel is oilafilic, which means that oil is attracted to it, resulting in a constant oil layer. Another benefit is Nikasil's uniformity and repeatability. For teams with multiple engines, this greatly reduces any variances between cylinders and between engines.

Nikasil and similar coatings are electro-deposited. Ceramic coatings are similar. For example, Perfect Bore uses a 16-percent silica-carbide particle in the 2- to 3-micron range, with a balance of nickel. A number of engine builders are seeing their work run two or three races without the need for honing during a freshening rebuild, with only the need to merely run a silica carbide brush for minor touch-up. Some engines have run several thousand race miles without being rehoned. The current trick is to hot hone to minimize bore distortion to eliminate potential blow-by.

For racing applications, this hard coating provides the opportunity to run tighter wall clearances. Wear resistance is higher, cylinder wall lubricity is better, and wear is minimal. In a Winston Cup engine application, a cleanup of only .0002 inch might be required following a grueling 600-mile race, allowing even the original rings to be reused.

These hard coatings really don't offer a performance benefit in terms of gaining increased power. The benefit lies in being able to run tighter clearances and a substantial gain in engine longevity.

Honing the Coating

According to extensive research conducted by Sunnen Products, Nikasil-coated cylinders should be honed with diamond stones. Hard cylinder coatings can be applied to the bore (in iron and alloy blocks), but coated cylinder liners are more popular. If the bores have been coated and now must be recoated as a result of wall damage, the block must be shipped away, which eats time. Coated liners allow the engine builder to much more quickly refresh a block in his shop.

With chrome coating, it is more likely to hone through the chrome (potential of fracturing of the plating) due to any distortion of the cylinder in the areas adjacent to the head bolt areas, even with a dry torque plate. Nikasil and similar coatings don't share this potential problem since they're not as brittle as chrome.

TECH TIP

VSR Cost

In today's world, prices are always subject to change, but here is an estimate:

Component	*Cost*
Engine block	$75
Cylinder head	$25
Crankshaft	$4
Camshaft	$25
Connecting rods (ea)	$8
Pistons (ea)	$8
Valves (ea)	$5
Valvesprings (ea)	$5
8-cylinder engine package	$295

The force inducer generates the vibrations that are transmitted to the work piece. This inducer is clamped to a steel worktable upon which the parts are secured.

Nikasil, on the other hand, is very consistent for thickness and hardness uniformity. Because Nikasil allows tighter clearances and its oil retention is much higher, you can get away with creating a smoother finish. Typical Nikasil cylinder finishes are in the Ra range of 6 to 10.

Rings

It's absolutely critical to use the correct type of rings in conjunction with Nikasil-coated cylinders. The top ring is typically moly or PVD; the second ring is ductile iron. Chrome rings cannot be used (including oil control rails) because chrome and Nikasil react to each other and result in scuffing. So just run plasma-moly top rings, ductile-iron second rings, and unplated oil rails. Ring seating occurs very quickly. Nikasil and other cylinder coatings are applied to aluminum or iron walls (on parent bores or on liners).

Typically, the coating is about .008 inch and can be sprayed on between .012 and .015 inch thick. This type of coating visually appears to be chrome, very shiny and difficult to scratch. In order to touch up Nikasil in order to re-ring, a 150-grit aluminum-oxide stone usually suffices, just to clean and remove any high points. However, if you find excessive wear or if new sleeves are to be installed, finishing requires a 500-grit diamond stone.

Cost

For a four-cylinder block, cost could typically be around $750 for coating the metal bores, or $200 per liner if you need sleeves. Coated liners for an eight-cylinder block may run about $1,600.

Nikasil and similar coatings are high-end processes for builders with the budgets to handle the expense. However, when you consider the reduced inventory required for blocks, cranks, rings, etc., and the extremely minimized need to rehone, the initial investment quickly pays for itself.

SOURCE GUIDE

300 Below
2999 E. Parkway Dr.
Decatur, IL 62526
800-550-2796
www.300below.com

A1 Technologies
7022 Alondra Blvd.
Paramount, CA 90723-3926
562-408-1808
www.a1technologies.com

Advanced Cryogenics
1250 Piper Rd.
Leechburg, PA 15656
412-845-8708

Aeromotive
7805 Barton St.
Lenexa, KS 66214
913-647-7300
www.aeromotiveinc.com

Aeroquip Performance Products
14615 Lone Oak Rd.
Eden Prairie, MN 55344
888-258-0222
www.aeroquip.com

Airflow Research
28611 W. Industry Dr.
Valencia, CA 91355
877-892-8844
www.airflowresearch.com

Arias Pistons
13420 S. Normandie Ave.
Gardena, CA 90249-2212
310-532-9737
www.ariaspistons.com

ARP
1863 Eastman Ave.
Ventura, CA 93003
800-826-3045
www.arp-bolts.com

ATI Performance Products
6747 Whitestone Rd.
Gwynn Oak, MD 21207
877-298-4817
www.atiracing.com

Aviaid Oil Systems
10041 Canoga Ave.
Chatsworth, CA 91311
818-998-8991
www.aviaid.com

Barnes Systems
3162 Kashiwa St.
Torrance, CA 90505
310-534-3844
www.barnessystems.com

BBK Performance
27440 Bostik Ct.
Temecula, CA 92590
951-296-1771
www.bbkperformance.com

BHJ Products
37530 Enterprise Ct.
Newark, CA 94560
510-797-6780
www.bhjproducts.com

Bill Miller Engineering
4895 Convair Dr.
Carson City, NV 89706
775-887-1299
www.bmeltd.com

Bonal Technologies
1300 N. Campbell Rd.
Royal Oak, MI 48607
800-638-2529
www.bonal.com

Brad Penn Lubricants
77 N. Kendall Ave.
Bradford, PA 16701
814-368-1340
www.bradpennracing.com

Braswell Carburetors
7671 N. Business Park Dr.
Tucson, AZ 85743
520-579-9177
www.braswell.com

Brodix Blocks
301 Maple Ave.
Mena, AR 71953
479-394-1075
www.brodix.com

Bryant Racing, Inc.
1600 E. Winston Rd.
Anaheim, CA 92805
714-535-2695
www.bryantracing.com

Bullet Racing Cams
8785 Old Craft Rd.
Olive Branch, MS 38654
662-893-5670
www.bulletcams.com

Calico Coatings
P.O. Box 901
Denver, NC 28037
704-483-2202
www.calicocoatings.com

Callies Performance Products
P.O. Box 926
Fostoria, OH 44830
419-435-2711
www.callies.com

Canton Racing Products
232 Branford Rd.
North Branford, CT 06471
203-481-9460
www.cantonracingproducts.com

Carrillo Industries
1902 McGaw Ave.
Irvine, CA 96214
949-567-9000
www.carrilloind.com

Cartel
1632 E Northfield Dr., Ste. 200
Brownsburg, IN 46112
317-645-8786
www.theindycartel.com

Central Tools
456 Wellington Ave.
Cranston, RI 02910
1-800-TOOLCTR
www.centraltools.com

Centroid Performance Racing
159 Gates Rd.
Howard, PA 16841
814-353-9290
www.centroidperformanceracing.com

Challenger Engine Software
115 Jeanette Dr.
Granite City, IL 62040
618-797-1770
www.virtualengine2000.com

Clausing Industrial
1819 N. Pitcher St.
Kalamazoo, MI 49007
800-323-0972
www.clausing-industrial.com

Cloyes Gear & Products
6101 Phoenix Ave., Ste. 2
Ft. Smith, AR 72903
248-365-0363
www.cloyes.com

Cometic Gaskets
8090 Auburn Rd.
Concord, OH 44077
800-752-9850
www.cometic.com

Comp Cams
COMP Performance Group
3400 Democrat Rd., Ste. 207
Memphis, TN 38118
800-999-0853
www.compcams.com
www.compperformancegroup.com

Controlled Thermal Processing
P.O. Box 4005
Antioch, IL 60002
847-651-5511
www.metal-wear.com

CP Pistons
1902 McGaw Ave.
Irvine, CA 92614
949-567-9000
www.cppistons.com

Crane Cams
1830 Holsonback Dr.
Daytona Beach, FL 32117
866-388-5120
www.cranecams.com

Crower Cams & Equipment
6180 Business Center Ct.
San Diego, CA 92154-5604
619-661-6477
www.crower.com

Cryogenic Institute of New England
78 Chilmark St.
Worcester, MA 01604
800-739-7949
www.nitrofreeze.com

Cryogenics International
14715 N. 78th Way
Suite 200
Scottsdale, AZ 85260
480-991-0299
www.cryogenicsinternational.com

Cryotron
P.O. Box 69060
13040 137 Ave. NW
Edmonton, Alberta, Canada T5L 5E0
888-879-2796
www.cryotron.com

CV Products
42 High Tech Blvd.
Thomasville, NC 27360-5560
800-448-1223
www.cvproducts.com

CWT Industries
4708 Old Peachtree Rd.
Building #300
Norcross, GA 30071
800-449-1849
www.cwtindustries.com

Dart Machinery
353 Oliver St.
Troy, MI 48084
248-362-1188
www.dartheads.com

Del West Engineering
28128 W. Livingston Ave.
Valencia, CA 91355
800-990-2779
www.delwestusa.com

Depac Dyno Systems
201 Mill St.
Rome, NY 13440
315-339-1265
www.depac.com

Detroit Racing Components
250 N. Hadley Rd.
Ortonville, MI 48462
248-627-5320

Diamond Racing Products
23003 Diamond Dr.
Clinton Twp., MI 48035
877-552-2112
www.diamondracing.net

DTS
26400 Groesbeck Hwy.
Warren, MI 48089
800-521-0628
www.drivetrainspecialists.com

Dyer's Top Rods
P.O. Box 596
619 N. Center St.
Forrest, IL 61741
800-867-7637
www.dyersrods.com

DYNOmite Dynamometer
25 Henniker St.
Concord, NH 03301-8528
603-226-3966
866-dynomite
www.land-and-sea.com

Eagle Specialty Products
8530 Aaron Ln.
Southaven, MS 38671
662-796-7373
www.eaglerod.com

Earl's Indy
302 Gasoline Alley
Indianapolis, IN 46222-3967
800-331-4639
www.earlsindy.com

Earl's Plumbing
19302 S. Laurel Park Rd.
Rancho Dominguez, CA 90220
310-609-1602
www.earlsplumbing.com

Edelbrock
2700 California St.
Torrance, CA 90503
310-781-2222
www.edelbrock.com

Electronic Chrome & Grinding
9128-32 Dice Rd.
Santa Fe Springs, CA 90670
562-946-6671
www.ecgrinding.com

Elgin Industries
1100 Jansen Farm Dr.
Elgin, IL 60123-2555
800-323-6764
www.elginind.com

Erson Cams
16 Kit Kat Dr.
Carson City, NV 89706
800-641-7920
www.pbm-erson.com

FAST
COMP Performance Group
3406 Democrat Rd.
Memphis, TN 38118
800-999-0853
www.fuelairspark.com
www.compperformancegroup.com

Federal-Mogul
26555 Northwestern Hwy.
Southfield, MI 48033
800-325-8886
www.federal-mogul.com

Ferrea Racing Components
2600 NW 55th Ct., Ste. 234
Ft. Lauderdale, FL 33309
888-733-2505
www.ferrea.com

Fluidampr-Horschel Motorsports
180 Zoar Valley Rd.
Springville, NY 14141
716-592-1000
www.fluidampr.com

Forster Tool & Manufacturing
1135 Industrial Dr.
Bensenville, IL 60106
800-369-8177
www.forstertool.com

Fragola Performance Systems
888 West Queen St.
Southington, CT 06487
866-337-2739
www.fragolaperformancesystems.com

Gear Head Tools
P.O. Box 21887
Carson City, NV 89721
800-733-4463
www.gearheadtools.com

GM Performance Parts
6200 Grand Pointe Dr.
Grand Blanc, MI 48439-3314
877+245-9981
www.gmperformanceparts.com

Goodson Tools & Supplies
156 Galewski Dr.
P.O. Box 847
Winona, MN 55987
800-533-8010
www.goodson.com

Granatelli Motorsports
1000 Yarnell Pl.
Oxnard, CA 93033-2454
805-486-6644
www.granatellimotorsports.com

GRP Rods
333 W. 48th St.
Denver, CO 80216
303-935-7565
www.grpconrods.com

Haas Automation
2800 Sturgis Rd.
Oxnard, CA 93030-8901
800-331-6746
www.HaasCNC.com

Harland Sharp
19769 Progress Dr.
Strongsville, OH 44149
440-238-3260
www.harlandsharp.com

Hastings Manufacturing
325 N. Hanover St.
Hastings, MI 49058
800-776-1088
www.hastingsmfg.com

Hines Industries
240 Metty Dr.
Ann Arbor, MI 48103-9498
734-769-2300
www.hinesindustries.com

Holley Performance Products
1801 Russellville Rd.
Bowling Green, KY 42102
270-782-2900
www.holley.com

Howards Cams
280 W. 35th Ave.
Oshkosh, WI 54902
920-233-5228
www.howardscams.com

Isky Racing Cams
16020 S. Broadway
Gardena, CA 90248
323-770-0930
www.iskycams.com

JE Pistons
15312 Connector Ln.
Huntington Beach, CA 92649
714-898-9763
www.jepistons.com

Jesel
1985 Cedar Bridge Ave., Ste. 2
Lakewood, NJ 08701
732-901-1800
www.jeselonline.com

Jet
427 Sanford Rd.
La Vergne, TN 37086
800-274-6548
www.jettools.com

Joe Gibbs Driven
13201 Reese Blvd. West, Ste. 200
Huntersville, NC 28078
866-611-1820
www.joegibbsdriven.com

Jones Racing Products
72 Annawanda Rd.
Ottsville, PA 18942-9702
610-847-2028
www.jonesracingproducts.com

K1 Technologies
889 76th St., Unit 6
Byron Center, MI 49315
616-583-9700
www.k1technologies.com

Katech Performance
24324 Sorrentino Ct.
Clinton Twp., MI 48035
586-KATECH1
www.katechengines.com

Kaufman Racing Enterprises
22280 Temple Rd.
Glenmont, OH 44628
740-599-5000
www.krepower.com

KB Performance Pistons
1040 Corbett St.
P.O. Box 2680
Carson City, NV 89706
800-648-7970
www.kb-silvolite.com

Kellogg Crankshaft
3524 Wayland Dr.
Jackson, MI 49202
517-778-9200
www.kelloggcrankshaft.com

Kibblewhite Precision Machining
580H Crespi Dr., Ste. 1
Pacifica, CA 94044-3426
650-557-2046
www.blackdiamondvalves.com

King's Crankshafts
P.O. Box 662
4106 Sinclair St.
Denver, NC 28037
704-483-1005
www.kingscrankshaft.com

L.S. Starrett
121 Crescent St.
Athol, MA 01331-1915
978-249-3551
www.starrett.com

Langcourt
2080 McMillan St.
Auburn, AL 36832
334-887-9633
www.langcourt.com

Levil Technology
1704 Kennedy Pt., Ste. 1124
Oviedo, FL 32765
407-542-3971
www.levil.com

Lista International
106 Lowland St.
Holliston, MA 01746
877-465-4782
www.listaintl.com

Lock-In-Tool
2511 Providence Rd. S.
Waxhaw, NC 28173
704-843-5477
www.lockintool.com

Lunati
11126 Willow Ridge Dr.
Olive Branch, MS 38654
662-892-1500
www.lunaticamshafts.com

MAHLE Clevite
1350 Eisenhower Pl.
Ann Arbor, MI 48108
800-338-8786
www.mahle.com

Mallory Ignition
10601 Memphis Ave., Ste. 12
Cleveland, OH 44144
216-658-6413
www.mallory-ignition.com

Manley Performance Products
1960 Swarthmore Ave.
Lakewood, NJ 08701
732-905-3366
www.manleyperformance.com

Manton Pushrods
601 Crane St., Bldg. C
Lake Elsinore, CA 92530
951-245-6565
www.mantonpushrods.com

March Performance
16160 Performance Way
Naples, FL 34110
239-593-4074
www.marchperf.com

Mazak
P.O. Box 970
Florence, KY 41042
859-342-1700
www.mazakusa.com

Melling Select Performance
P.O. Box 1188
Jackson, MI 49204
517-787-8172
www.melling.com

Meziere Enterprises
220 S. Hale Ave.
Escondido, CA 92029
800-208-1755
www.meziere.com

Midwest Cryogenics
842 Industrial Park Dr. SE, Ste. A
Lonsdale, MN 55046-4015
507-744-3990
www.midwestcryo.com

Mile High Crankshafts
888 South Lipan St.
Denver, CO 80223
303-781-6764
www.milehighcranks.com

Milodon
2250 Agate Ct.
Simi Valley, CA 93065
805-577-5950
www.milodon.com

Mittler Brothers Machine & Tool
P.O. Box 110
Foristell, MO 63348-0110
800-467-2464
www.mittlerbros.com

Mitutoyo America
965 Corporate Blvd.
Aurora, IL 60502
888-648-8869
www.mitutoyo.com

Moroso Performance Products
80 Carter Dr.
Guilford, CT 06437-2116
203-453-6571
www.moroso.com

Mr. Gasket
10601 Memphis Ave., Ste. 12
Cleveland, OH 44144
216-658-6413
www.mrgasket.com

MSD Ignition
1490 Henry Brennan Dr.
El Paso, TX 79936
915-857-5200
www.msdignition.com

Newen
298-D Enterprise St.
Escondido, CA 92029
800-639-3693
www.newen.com

Oliver Racing Parts
1025 Clancy Ave. NE
Grand Rapids, MI 49503
800-253-8108
www.oliverconnectingrods.com

PAC Racing Springs
21200 Telegraph Rd.
Southfield, MI 48033
866-799-9417
www.racingsprings.com

PBM Performance Products
7301 Global Dr.
Louisville, KY 40258
800-588-9608
www.pbm-erson.com

Performance Distributors
2699 Barris Dr.
Memphis, TN 38132
901-396-5782
www.performancedistributors.com

PerTronix
440 East Arrow Hwy.
San Dimas, CA 91773
909-599-5955
www.pertronix.com

Peterson Fluid Systems
9801 Havana St.
Henderson, CO 80640
800-926-7867
www.petersonfluidsys.com

Peterson Machine Tool
Peterson Wash & Blast Systems
1100 N. Union St.
Council Grove, KS 66846
800-255-6308
www.petersonwashandblast.com

Pioneer
5184 Pioneer Rd.
Meridian, MS 39301
800-647-6272
www.pioneerautoinc.com

Polymer Dynamics
11211 Neeshaw Dr.
Houston, TX 77065
888-765-9396
www.polydyn.com

Precision Metalcraft
30575 Rockstone Rd.
Valley Center, CA 92082
760-751-0378
www.Precision-Metalcraft.com

Pro-Filer Performance Products
P.O. Box 217
New Carlisle, OH 45344
937-846-1333
www.profilerperformance.com

Pro-Gram Engineering
475 5th St. NE
P.O. Box 472
Barberton, OH 44203
330-745-1004
www.pro-gram.com

Probe Industries
2555 West 237th St.
Torrance, CA 90505
310-784-2977
www.probeindustries.com

Professional Products
2205 El Segundo Blvd.
Hawthorne, CA 90250
323-779-2020
www.professional-products.com

PRW Industries
193 West Orangethorpe Ave.
Placentia, CA 92870
714-792-1000
www.prw-usa.com

Quick Fuel Technology
129 Dishman Ln.
Bowling Green, KY 42101
270-793-0900
www.quickfueltechnology.com

Racing Parts Maximum
RPM International
16313-16319 Arthur St.
Cerritos, CA 90703
800-981-0776
www.racingpartsmaximum.com

RHS
COMP Performance Group
3406 Democrat Rd.
Memphis, TN 38118
800-999-0853
www.racingheadservice.com
www.compperformancegroup.com

RMC Engine Rebuilding Equipment
5775 Bridgeview Ctr.
Saginaw, MI 48604
800-248-5062
www.rmcengine.com

Ross Racing Pistons
625 S. Douglas St.
El Segundo, CA 90245-4812
310-536-0100
www.rosspistons.com

Rottler
8029 S. 200th St.
Kent, WA 98032-1114
800-452-0534
www.rottlermfg.com

Russell Performance
2700 California St.
Torrance, CA 90503
310-781-2222
www.russellperformance.com

Scat Enterprises
1400 Kingsdale Ave.
Redondo Beach, CA 90278
310-370-5501
www.scatcrankshafts.com

Scorpion Performance
5817 NW 44th Ave.
Ocala, FL 34482
352-512-0800
www.scorpionperformance.com

Serdi
1526 Litton Dr.
Stone Mountain, GA 30083
800-447-3790
www.serdi.com

SI Valves
4477 Shopping Ln.
Simi Valley, CA 93063
800-564-8258
www.sivalves.com

Stuska Dynamometer
N60 W22700 Silver Spring Dr.
Sussex, WI 53089
262-252-4091
www.stuskadyno.com

Summit Racing
P.O. Box 909
Akron, OH 44309-0909
800-230-3030
www.summitracing.com

Sunnen Products
7910 Manchester Rd.
St. Louis, MO 63143
800-325-3670
www.sunnen.com

Superflow Technologies Group
4060 Dixon St.
Des Moines, IA 50313
888-442-5546
www.superflow.com

Supertech Performance
3550 Charter Park Dr.
San Jose, CA 95136
408-448-2001
www.supertechperformance.com

Swain Tech Coatings
963 North Rd.
Scottsville, NY 14546
585-889-2786
www.swaintech.com

T&D Machine Products
4859 Convair Dr.
Carson City, NV 89706-0492
775-884-2292
www.tdmach.com

Tech Line Coatings
26844 Adams Ave.
Murrieta, CA 92562
972-775-6130
www.techlinecoatings.com

Total Engine Airflow
285 West Ave.
Tallmadge, OH 44278
330-634-2155
www.totalengineairflow.com

Total Seal
22642 N. 15th Ave.
Phoenix, AZ 85027
800-874-2753
www.totalseal.com

Totally Stainless
P.O. Box 3249
Gettysburg, PA 17325
800-767-4781
www.totallystainless.com

Trend Performance
23444 Schoenherr Rd.
Warren, MI 48089
800-326-8368
www.trendperform.com

Trick Flow Specialties
285 West Ave.
Tallmadge, OH 44278
330-630-1555
www.trickflow.com

Trick Titanium
2046 Depot St.
Building B
Holt, MI 48842
517-694-7449
www.tricktitanium.com

U.S. Chrome
175 Garfield Ave.
Stratford, CT 06615
203-378-9622
www.uschrome.com

Weinle Motorsports
4817 E. Miami River Rd.
Cleves, OH 45002
513-607-2592
www.weinlemotorsports.com

Wilson Manifolds
4700 NE 11th Ave.
Ft. Lauderdale, FL 33334
954-771-6216
www.wilsonmanifolds.com

Winberg Crankshafts
333 West 48th Ave.
Denver, CO 80216
303-783-2234
www.winbergcrankshafts.com

Wiseco Piston
7201 Industrial Park Blvd.
Mentor, OH 44060
800-321-1364
www.wiseco.com

World Products
51 Trade Zone Ct.
Ronkonkoma, NY 11779
877-475-6651
www.worldcastings.com

Xceldyne Technologies
37 High Tech Blvd.
Thomasville, NC 27360
336-475-0201
www.xceldyne.com

XRP
5630 Imperial Hwy.
South Gate, CA 90280
562-861-4765
www.xrp.com

Zanzi Spa
Corso Vercelli 159
10015 Ivrea (TO), Italy
+39 0125251540
www.zanzi.com